The Advent of Unmanned Electric Vehicles

S. Van Themsche

The Advent of Unmanned Electric Vehicles

The Choices between E-mobility and Immobility

Springer

S. Van Themsche
Berlin
Germany

ISBN 978-3-319-20665-3 ISBN 978-3-319-20666-0 (eBook)
DOI 10.1007/978-3-319-20666-0

Library of Congress Control Number: 2015947106

Springer Cham Heidelberg New York Dordrecht London

Printed on acid-free paper

Springer International Publishing AG Switzerland is part of Springer Science+Business Media
(www.springer.com)

Preface

For probably the first time in history, human beings are confronted with the fact that physical mobility is regressing. During peak hours, residents of big cities can lose 4 h in traffic jams and things will only get worse. Due to massive urbanization, cities with 15–30 million inhabitants are popping up all over developing countries. These cities are imploding and under this densification process, transport problems are increasing exponentially. Not only are there more people to transport, but with greater wealth and easier access to financing, the number of cars is exploding and will reach 1.7 billion 20 years from now. To add to this bleak vision, under galloping demography and the surge of millions of new middle-class citizens, the past 2 % average worldwide energy consumption—mainly CO_2 driven—is unlikely to change. Everyone realizes that temperature will rise by 2 °C and with it drastic changes will happen affecting the environment and the economy, but no one wants to change his or her lifestyle. Transportation is one of the main drivers of this energy consumption, accounting for 27 %. This is why any transport improvement will help to stabilize temperature rise and limit pollution emission.

If people won't change their lifestyle, relief will need to come from new transportation technologies, changes in work patterns and from personal choices on transportation means that the e-mobility revolution can bring. This revolution is happening under the convergence of IT and wireless telecommunication, and in association with improvements to power electronics and battery performance improvements.

People can already see this revolution in action. Electric and hybrid cars are starting to hit the roads in greater numbers. New plug-in technologies should reduce charging times to 5 mins and battery improvements should increase autonomy and reduce costs, making electric cars more attractive to consumers. For Governments, supporting transport electrification by supplying the necessary infrastructure is mandatory if they are to meet their CO_2 emission commitments.

Each year nearly 1.3 million people die as a result of road traffic accidents, a number likely to double following the same worldwide car growth trend. Under constant government scrutiny, the railways have been able to improve their safety track records. This was achieved by applying three safety principles—block interlocking, block signalling and system integrity—and by eliminating the human factor. By applying the same principles and with driverless technology leapfrogging from other industries, manufacturers and Governments cannot ignore the fact that going driverless will save millions of lives every year. Although most experts expect driverless cars to be sold around 2025, unmanned technology will not happen out of the blue. Instead, a continuous flow of new safer technologies will be presented to the market, ranging from driver based, semi-autonomous to unmanned classification.

This book is about explaining how the e-mobility revolution can help save millions of passengers from injuries, reduce millions of deaths from pollution-related diseases, bring better quality of life, reduce global warming and decrease transportation costs. It is also about showing with simple concepts how new business models, innovative financing solutions and technologies can increase infrastructure capacity while reducing transportation demand. By showing the challenges facing the transportation industries and demonstrating the solutions e-mobility can bring, it gives every reader insights into how to make the right choices to avoid immobility.

The first chapter will give an overview of the main principles and trends that affect transportation. It will provide the readers unfamiliar with transportation challenges, a vision on how e-mobility technologies are being developed to help society apprehend such fundamental changes. It will show which features of society will be impacted by these new technologies.

The following figure summarizes which element will be profoundly changed.

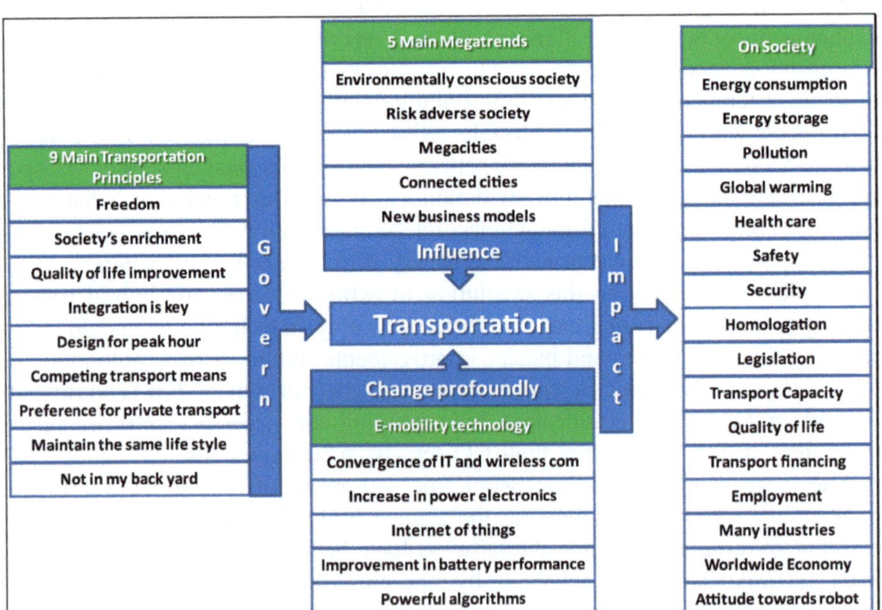

In the following chapters, we will explain in details how e-mobility technologies will directly address issues raised by the five megatrends and as a consequence affect the social features described above.

Contents

Chapter 1
Principles and Megatrends Affecting Transportation

Acronyms

BRT	Bus Rapid Transit
CAPEX	Capital Expenditure
CCTV	Closed-Circuit Television
EDA	Event-Driven Architecture
GPS	Global Positioning System
PPHPD	Passenger Per Hour and Per Direction
PPP	Public–Private Partnership
PRT	Private Rapid Transit
M2M	Machine to Machine
SIM	Subscriber Identity Module
SOA	Service-Oriented Architecture
TCP/IP	Transmission Control Protocol/Internet Protocol
V2I	Vehicle to Infrastructure
V2V	Vehicle to Vehicle

Society is confronted with a paradox. With the advent of Internet, everything gives the impression of moving extremely fast. Today's technology will be next year's passé. Today's news will be tomorrow's history. With the Internet revolution and the creation of new social media, such as Twitter and Facebook, every one is connected and sharing personal information in real time. Society and especially its newer generation live in an era of instantaneity where space and distances are irrelevant.

At the same time, for probably the first time in history, human beings are confronted with the fact that physical mobility is regressing. The number of cars is likely to double in the next 20 years to reach more than 1.7 billion cars and with it the doubling of fatalities unless new safety technologies and standards are implemented. Fifty years from now, about one hundred megacities will have around 30–35 million inhabitants. Just imagine what it is to organize the transport of Canada's population in an area of just 2500 km^2, equivalent to a country like Luxemburg. The issue is not just the share increase in demography but the fact that

S. Van Themsche, *The Advent of Unmanned Electric Vehicles*,
DOI 10.1007/978-3-319-20666-0_1

in many huge cities, there is just no place to add new roads. Transport infrastructures in developing countries are cracking from everywhere under the emergence of new middle class citizens wishing to get access to the benefits of the consumer society. In fact, there is a real possibility that megacities will come to a halt. Already, citizens of Mexico, Sao Paulo, or New Delhi are experiencing more than 200 km of daily congestion. It is just a matter of time before traffic jams double in size, increasing the traveling time to unbearable levels (Fig. 1.1).

Salvation will need to come from new transportation technologies, changes in work patterns, and, most importantly, in our personal choices on transportation means that the e-mobility revolution can bring.

Is e-mobility just another buzzword used by big corporations to sell a green image or is there behind the marketing appeal of this word, real fundamental technological breakthroughs? We believe that e-mobility will help save millions of passengers from injuries, reduce millions of death from pollution-related diseases, bring a higher level and better quality of life, reduce global warming trend, and decrease transportation costs. This might seem too good to be true, but the reality is that under the convergence of IT and wireless telecommunication, and associating the increase in power electronics together with the improvement in battery performance, fantastic new opportunities are being unleashed to reduce energy consumption and increase transport capacity, while improving the passengers' journey experience.

Fig. 1.1 Traffic jams in New Delhi. *Source* Nomad [*Picture* Copyright of NOMAD (CC BY 2.0 via Wikimedia Commons)]; http://creativecommons.org/licenses/by/2.0)

There are many articles on private and public urban transport but few documents in our view exist to help normal citizens understand the challenges facing transportation and potential solutions that can help make the right personal choices. Furthermore, few analyses of transportation means have been made in a holistic approach where transportation modes have been compared taking into consideration power consumption, environment impact, transport capacities, social and economical benefits and costs.

Many reasons could explain this. Firstly, the car industry has one of the strongest worldwide lobby groups and nothing to gain in promoting comparisons between various modes. Secondly, the simple fact that cars and trains use different combustion fuels made it very complex to even do this comparison. Obviously the fact that electrical cars are now being sold in serious quantities enables a more direct comparison. Thirdly, the concepts used to make such comparison rely on many engineering formulas and terminologies which are complex and hypothesis which may in some cases even seemed biased. Finally and more importantly, individuals believe their life is unique and thus their transportation needs are as well. In this context how can we be sure that we make the right decisions?

Environmentally conscious citizens might believe for instance that by switching to electrical cars they are making a small contribution, but are they really? Wouldn't society as a whole be better off should they buy a monthly metro pass? If transportation needs are so specific to each citizen, why bother trying to get a comprehensive view about such issues and solutions for a city, region or even country? Surely there can't be a single recipe to help understand the urban transportation challenges and apply solutions to improve current and future world transportation woes? And even if there was, shouldn't we just trust the authorities to make the right investment choices in transportation policies and infrastructure? Though there is no single worldwide recipe to solve all the transportation problems, we believe that by educating each and everyone we can help people make knowledgeable decisions each time they are confronted with the choice of buying or renting a new car, selecting the right technology, choosing between private and public transport or even voting for the right candidates.

Though there is no silver bullet to solve all transportation woes, we believe that a good way to start is to identify a few general principles applied to transportation that we can all agree upon. Surely if we can find the right connections between these principles and the relevant global trends we should be able to get a reasonably clear picture of what the transportation challenges are. This should help understand where companies are earmarking billions of R&D money or why cities must spend billions in new transportation infrastructure.

1.1 General Transportation Principles

What are these universal principles linked to transportation that everyone could agree upon? We've identified nine that have a significant impact on society and how people view transportation. Many of them are more philosophical or socio-behavioral than technological. The first and probably most important principle fulfills one of the basic human needs: freedom.

1.1.1 Transportation Brings Freedom to Citizens

If one defines freedom as the condition of being free of restraints, it is clear that transportation frees human beings from geographical constraints. Freedom and even emancipation is often associated with public transport in many minds.

In the USA, Mrs. Rosa Park an African American lady became famous when she refused to give up her seat to a Caucasian person. As a result of this fight and many others, in most countries there is today no more discrimination based on race or skin color in public transport. In most societies, anyone can enjoy the freedom of traveling wherever he wishes to go for a small fare. However, explicit or implicit gender discrimination still exists. For instance a 2015 French survey indicated that 100 % of women declared having been already harassed in public transport. In some countries such as Mexico, women's right to safe public transport has been taken into consideration with specific train compartments being made available to protect them from men harassment. On the contrary, in a few existing patriarchal societies such, as the one of a few Arabic Golf countries, women are still implicitly forbidden to drive and have no way of commuting alone. They are stranded at home and cannot travel without the presence of their husband, father or male family members. Things will change, as Governments are investing in Public Transport allowing women to use separate gender compartments. These new investments when they will be finished around 2016 will bring liberty of movement to many women and will change profoundly the life of millions of women of that region. Not surprisingly, some courageous Saudi women have openly manifested their right to drive in 2013, wishing to anticipate this freedom.

However, the freedom provided by transport that many Middle East women can still only dream of, is compromised elsewhere. Though citizens have taken for granted that they should be able to hop on a bus or drive their car to the grocery store and be there in a few minutes, in many big cities this freedom of travel is more and more being victim of its own success. Daily exodus of suburban residents to the inner city is there to remind us that though cars can reach a speed of 250 km/h, for at least 4 h per day one can travel quicker by bicycle than by a Ferrari. This is now representing a threat to the second principle, as cities' wealth can be now hampered by traffic jams.

1.1.2 Transportation Enriches Society

Our ancestors were nomads moving from one hunting ground to another. Traveling by foot was the only way to survive, to find food and avoid over hunting. Around 9000 years ago societies which relied on sedentary agriculture appeared. With agriculture came the social stratification development and economic specialization, as well as population densities sufficient to support urban life. As population in the cities expanded, transportation of food became more and more complicated. Luckily for human beings two fantastic developments happened more or less at the same time, creating the foundation of transportation: the domestication of wild horses, which most likely started in the Eurasian Steppes (centered probably in Ukraine) approximately 4000–3500 BC, together with the creation of the wheel around the 4th millennium BC (which appeared almost simultaneously in several regions: Mesopotamia, the Indus Valley, the Northern Caucasus and Central Europe). As city population density started growing to a point where citizens relied on supply from other regions further and further away, road access became increasingly important to the vitality and the wealth of the cities. More recently, with the apparition of trains, tramways and later on cars, citizens were for the first time able to work and live in different areas. Workers were suddenly able to work in plants far away from their home, fuelling even more the industrial revolution.

In modern society, working and living in two different regions of the cities or suburbs is more the rule than the exception. The possibility to meet co-workers, clients, and suppliers within a few kilometers of distance is an important factor explaining the increasing importance of cities in the countries' wealth generation process. However, massive city traffic jams are now seriously sapping the benefits that people and society as a whole are enjoying from urban life. Will there be a point in time where the big cities will come to a standstill or, on the contrary, will the modern communication tools generated by the Internet society reduce the needs for commuting? In many big cities, transportation which was for many years perceived as a quality-of-life enhancer, is rapidly becoming a burden on the citizen.

Can the third principle, which is that transportation improves the quality of life, still be true when poor people from the outskirts of the megacities lose 6 h per day in buses? Surely something needs to be done, as even in a rich country like the

USA, drivers of any of the 15 main US cities are now losing as much as 52 h in traffic jams (2011 figures) every year.

1.1.3 Transportation Improves Quality of Life

Transportation is meant to improve quality of life in two ways. First, it allows people to go rapidly where they want to in order for them to do whatever they wish to do or buy whatever goods they want. Private and public transport favored the advent of the leisure society enabling people to easily get access to libraries to borrow books, to theater to watch their favorite movie, get to a football stadium to play or cheer for their local team.

Secondly, it also favors high population density on smaller areas, which allows culture, knowledge, and leisure to prosper, and makes cities like New York, London, or Paris thrive.

> Journey time and quality of ride is fast becoming an important factor in the success of transportation policies. As society grows richer, people expect to be able to go to work quickly, comfortably and safely. Furthermore, cities internationally competing against each other must provide decent quality of ride from home to work to attract talent.

If quality and journey time are so important, why are so many cities failing to improve such factors? The reality is that there are three other principles in designing the transportation networks that are not always well apprehended and which can generate opposite results if badly applied: Integration, design for peak hour, and competing technologies.

1.1.4 Integration Is Key to Successful Transportation

When city planners design their transportation network they often forget about a simple factor: people don't travel from one bus or metro station to another but from one activity to another (for instance, from resting at home to going to work or dinning). Furthermore, it is unrealistic to think that most people will walk a long distance to hop on a bus or a metro every day. It goes back to the principle of journey time and quality of ride. To improve quality and reduce journey time, city planners must integrate the different transportation means.

This means buses feeding metro systems or even providing parking space to enable people to get easily from their home to the metro station. The image of a fishbone is often invoked to explain the best integration systems. In such well designed transportation systems, small bus lines come from the low density residential areas to feed the main metro lines and small residential roads connect to national roads or highways.

1.1.5 Networks Must be Designed for Peak Hour Capacity

Unfortunately and as for many systems, transportation networks are as good as their weakest part.

This means that you can build a 7-lane road but if there is a bridge reducing the available flow to only two lanes, the maximum capacity will be limited to these 2 lanes. Actually it even could be worst: stop and go movements could make the car outflow slower than if you had maintained a consistent 2-lane road.

Similarly, the capacity needs to be planned in function of its limit, which happens during peak hours.

A metro system and its fleet will be designed to meet the peak hour ridership requirements. This is evidently a waste of resources as the fleet might be working at full capacity during two hours in the morning and evening and at only one third of this capacity outside rush hours (commuting time occurs mainly during the two-hour morning and evening rush, creating over capacity during most of the day).

The system's over-dimensioning is an issue any city planner faces and this is why it is so important to understand the investment costs required by the various technologies and the fixed and variable costs incurred during operations. In terms of transport capacity, building a five-lane road or a metro can probably achieve the same result. If the end result of transporting people can be achieved either way, what are the factors that need to be taken into consideration?

1.1.6 *Transit Technologies are Competing with Each Other*

In an ideal world, there would be no space or cost restriction. Unfortunately, investments and space dedicated to private transport is competing against investment in public transport. Similarly, metros can replace bus lanes and thus mass transit technologies are also competing with each other. If this is the case what technical information could be used to help select the best transportation modes?

1.1.7 *People Prefer Private Over Public Transport*

For several reasons and all conditions being equal, citizens prefer to ride their car than use public transport.

One reason that might explain such a principle is that people, especially men, view their cars as extensions of themselves (2007 study conducted for BMW[1]). This isn't obviously the case for public transport. Furthermore, the car industry spends billion on advertising associating cars with status symbols, such as power, youth, sexiness, etc. This understandably reinforces the cars' appeal.

Private transport is also per nature more comfortable and flexible as drivers will always have a seat, can avoid walking or waiting to get onboard under cold, hot or rainy conditions. It is also often more practical as it allows citizens to avoid interconnections.

Thus, in order to encourage people to use public transport, authorities need to give incentives—gratuity, lower fare through subsidies—or exercise coercive measures, such as reducing parking places and imposing city tolls. Public transport has some cards to play. Journey time is obviously a key potential differentiator in big cities. Cost when parking and tolls are factored in, can be another key element.

1.1.8 *People Don't Want to Reduce Their Lifestyle*

The nature of humans is to react only when facing drastic consequences, and if given the choice, people won't reduce their lifestyle even if this would help avoid global warming.

[1] The secret life of cars and what they reveal about us; Authors Richard Benson, Dr Iain MacRury and Dr Peter Marsh; (2007).

Every human being from Canada to Senegal is today aware of the consequences of car pollution on global warming. However, given a choice and the means, most residents of poor countries would still buy a car, regardless of the impact. Furthermore, citizens of rich and hot countries will not switch their air conditioning off, even if they know it consumes more energy and thus is bad for the environment.

This is to say that only new technologies, better public transport services, higher costs, or coercive measures will make people switch to more environmentally or socially friendly transportation means.

1.1.9 Not in My Backyard Syndrome

People want the benefits of new transportation infrastructure but understandingly don't want to suffer the consequences of it.

In authoritarian States, transportation planners usually disregard this issue. After all, the good of the overall society is more important than the well being of a few. However, in democracy any citizen has the possibility to defend his rights and use the courts to block any new initiative.

With Internet and the possibility in many countries to review and block initiatives through a petition—even those generated by social networks or a referendum—big infrastructure projects can be delayed for years or killed all together.

These nine principles we've just indicated should in our view be considered by all the players involved in the transportation field, be it public authorities, city planners, consultants, public and private operators, technology partners and transport manufacturers.

In the following chapters, we will bring a vision of what could be the solutions to the new challenges facing society, taking these nine principles into account. As it is difficult to make predictions, especially about the future (a quote, for which there are several potential owners), we will select a few issues the world is now facing and new trends within society, which most futurologists agree upon are likely to have huge impact on mobility.

1.2 An Increasingly Environmentally Conscious Society

The first megatrend, which is affecting transportation, is the impact of human activities on the environment and the constantly more stringent solutions society is applying to reduce such impact. In fact, society as a whole is becoming more environmentally conscious.

Obviously some countries are so much more than others, but everywhere under the pressure of green activists or forced by the level of pollution and its health impact, authorities and simple citizens are making it known through their acts or their vote that society needs to stop degrading further the planet and hopefully, with clever insight, citizens can start making the right decisions and even start repairing it.

1.2.1 Greenhouse Gas Emission

One of the most obvious issues that impact decisions on transport is global warming. Massive consumption of fossil fuel and the greenhouse gas emission resulting from it is increasing the planet's average warmth. Though some scientists still doubt the impact's significance in terms of temperature increase, it is now widely accepted by the scientific community that there is an increase in greenhouse gas—especially carbon atom—due to human activities.

In the latest forums on global warming, specialists have now suggested measures, which aims' aren't anymore to reduce global warming but rather to maintain the temperature rise to only 2 °C. Though better than nothing, this would still have a dramatic effect on sea level, wild life and vegetation and generate more extreme climate with increased flooding and drought. But even maintaining such rise will require drastic efforts from most countries, bringing back their CO_2 emission level to levels known 15–20 years ago. These measures will need to touch most areas of human activity, and in particular transportation.

Cars and buses using fossil fuel combustion are often criticized as one of the main sources of such gas emission, while metros and other electrical vehicles are incensed by public authorities as a clean transport mean. However, this is a simplified view. In order to compare apples with apples and verify which transportation means has the lesser impact on global warming, one must compare vehicles' energy consumption not only at the point of consumption but also integrating all the production chain inefficiencies.

For diesel and gasoline vehicles, this requires not only to understand their engine efficiency, but to factor in the losses resulting from refining petrol and transporting it to the petrol stations. For electrical vehicles it requires calculating energy loss by power plants, including electrical losses in the electrical grid (but partially compensated by kinetic energy recuperated during braking). We know that this is a hot debate between transportation specialists, but our view is that electrical vehicles can only be as energy efficient as their regional energy network is, once again factoring in the recuperated braking energy.

Therefore, comparing vehicle energy efficiency means assessing not only the vehicle's own energy inefficiencies but also for electric cars calculating the network mix of clean generation technologies (hydro-electric, solar, wind, or nuclear) and dirty combustion fuel technology (thermal plants using coal, gas, or petrol) inefficiencies.

1.2.2 Air Pollution

Human activities also have a significant impact on quality of air. In many cities, pollution has reached levels that aren't just affecting their residents' health, but is starting to reduce their economical output. Obviously industries play a key role in air pollution, but fossil fuel combustion by cars or buses is also one of the main pollution villains.

Once again, is it fair to criticize conventional cars and buses while incensing metros, tramways or commuter trains? It is true that the place where the combustion occurs is important as there are more people in cities to be contaminated than on the countryside. Obviously the impact on people's health is also bigger when smog is concentrated in big cities with fewer possibilities for pollutants to be evacuated. However, can we consider more environmentally friendly measures to evacuate particles through a coal power plant chimney than through the exhausts of a car?

For pollution also, one must compare the direct pollutant emission of oil-based cars and buses with the indirect emission of electrical vehicles. By integrating the regional energy mix, we can really get an accurate picture of the more environmentally friendly transportation means.

1.2.3 Investment in Clean Generation Technologies

In order to reduce pollution and greenhouse gas effects, many governments have decided to promote clean or carbon-free generation technologies. Most governments are making political choices to quickly change their energy matrix mix, by investing massively in wind farms or solar plants, by encouraging the replacement of old coal and petrol thermal plants, or by making them more efficient.

In the USA and Canada, non-conventional gas and petrol drilling has had an impact on the power generation network, replacing coal thermal plant by more environmentally friendly gas thermal plants. Some countries such as China are investing in new nuclear power plants that don't affect global warming, though they are considered by many as pollutant. Germany has gone in the opposite direction, deciding to close as soon as possible its nuclear power plants and replacing them by renewable energy, but as a consequence also using cheap coal from the USA.

All these measures are changing the generation energy mix with three potential impacts for public transportation:

- The electricity consumed is likely to be more environmentally friendly;
- As solar and wind generated power increases, stability of electrical grids is likely to be problematical during certain period of the day or year (i.e., night for solar and periods with less wind) while transport power consumptions is consistent (mostly during peak hours); and
- There is still no easy and cheap way of storing energy and thus electrical car or bus fleets could be used as temporary storage facilities.

1.2.4 Investment in Clean Combustion Technologies

The automotive and bus industries must invest massively in cleaner technologies. Under the impulse of government initiatives such as the Californian free air act, automotive companies must improve tremendously their vehicles' oil consumption.

Some companies (i.e.: Nissan/Renault) decided to invest significantly in electrical cars while others are using hybrid solutions (i.e.: Toyota) to improve their overall carbon footprint. As progress in performance of energy consumption is steady, will there be a time when the energy consumed through a standard motor is better or similar to energy produced by power plants? Is it fair to assume that the environmental footprint of an electric Tesla car consumes as little as a metro or passenger vehicle?

Though there is no definite answer to the maximum efficiency that a conventional car will be able to achieve, we will prove that electrical vehicles are more energy efficient and cleaner than conventional vehicles.

Furthermore, we will show that the railway industry isn't caught flat footed and is also investing in cleaner and more efficient technologies.

These new technologies, such as silicon carbide inverters, permanent magnet motors, direct drive technology, as well as developing on-board and way-side technologies for recuperating the energy produced while braking, called kinetic energy have or will permeate the automotive industry, improving even more the efficiency of electrical systems.

1.2.5 Transport Modes Energy Comparison

Electric cars are more efficient than gasoline or diesel motorized vehicles, especially when taking into consideration kinetic energy recuperation, while driving in urban environments.

Car efficiency doesn't always translate into environmentally friendly transport mode. To use a similar comparison basis, we need to integrate the notion of capacity and occupancy. Because of mass production and much more relaxed safety standards, cars have a much lighter weight per surface than buses or trains. In fact the ratio is almost 50 and 33 % respectively. However, intuitively we feel that we can't compare in terms of capacity a car that can carry 5 passengers on an area of 8–10 m^2 with a train or a bus that can withstand at least 4 passengers per m^2 in smaller cities and up to 8 passengers per m^2 in the megacities of the developing world. Moreover, we are all too aware of the fact that during peak hours, most cars run almost empty. In fact, in North American and European cities, the average occupancy rate of a car is barely more than 1 passenger.

However, when capacity and occupancy are factored in, buses and especially trains have a track record of being much greener than cars, even electrical ones, especially in countries where there is a strong culture of public transport. What is true for energy consumption is obviously applicable as well to pollution emission, as there is always a strong correlation between energy consumption and pollution.

Is this to say that governments should systematically invest in public transport rather than in new roads? The simplistic answer would be to say yes, but most governments are broke and cannot afford new metro or tramway lines.

What is also new, and at the heart of this book, is that there is today a clear separation between Public and Private Transport. This concept will soon be challenged by the implication of the new IP technologies on the automotive industry. The day cars are electric, driverless, and belonging to an operator's fleet, there will be no more fundamental difference between these different transport modes, only semantic. This e-mobility revolution will obviously affect the transport world but will also have a significant impact on all sectors of the energy market, be it the generation, transmission, distribution, and storage facilities.

1.2.6 Economic Impact of Going Electric for Transportation

What would be the impact on the economy of major non-OPEP countries if their Transport Minister had a magic stick and could suddenly transform their current petrol-based fleet into an all electric one. Obviously, this would create huge opportunities for infrastructure and energy companies to build the new power plants required to add capacity.

For instance, France would need to add about 50 % of its current installed capacity to address the additional electrical demand. This would create thousands of jobs and depending on the energy source, foster growth of companies from the renewable energy or nuclear markets.

However, creating new power plants wouldn't make economical sense, if in the end the efficiency of these new power plants is lower than the efficiency of gasoline or diesel car engines. The fact is that new generation of power plants are more efficient than cars in terms of combustion and can reach as much as 60 % efficiency in co-generation mode.

In reality when we calculate the impact of such shift for three big countries—USA, France and Brazil, which have very different generation mix—we can calculate huge savings. What is true for these three countries would also be for any other country relying on reasonably recent power plants. The methodology applied in this book could easily be transposed for comparison sake to any other country the reader would be interested in.

How much money could this shift generate[2]? A lot, as for instance, the savings France would get from such change would be in the range of 13.6 billion euro (US$ 17.5 billion) per year. In the USA, yearly reduction in gasoline would represent around 141 billion liters of gasoline, which would result in annual savings of around $100 billion.

1.3 Risk Adverse Society

Society is increasingly risk adverse, especially when it comes to environments where fatalities can happen. Some people might say, rightfully so, that one person killed on the road or track is one person too many. However, there is a cost to this choice made by society.

For instance, the railway industry under the pressure of public authorities has put in place measures and safety principles that have reduced fatalities to a very few exceptional events. That wasn't always the case as history of signaling and safety principles have shown us. The railway industry paved the way for zero fatalities long ago. Efforts to reduce the number of people killed on roads have also taken place, but with much less drastic intensity. New laws limiting speed, regulating drinking habits and imposing points on licenses, as well as enhancements in safety car equipment have improved the safety record of the automotive industry. However, accidents are still tolerated because there is a general agreement within society that accidents can be caused by hazardous situations. The "no one's fault" excuse is still widely accepted.

Why did society impose level of investments in safety requirements on the train industry that have no common measure in the automotive industry? For instance, why do trains need intrinsically safe equipment or redundancy in software and hardware but cars can still avoid such costly solutions?

This situation will in our view change, as cars will become more and more autonomous. This will be especially true when cars will be completely unmanned. The same safety rules and principles will then need to be applied, to avoid trial lawyers and class actions.

Though some people might believe that the era of self-driving vehicles will never happen, we believe that it will come sooner than later. In fact, we don't believe that with technology available, car manufacturers will be legally allowed to continue with more than 1.3 million yearly fatalities worldwide indefinitely, without

[2]Note: these savings were calculated with 2014 oil prices and would be reduced in 2015 but would likely recuperate in the mid future.

being sued or having class action against them. Worst with the doubling of the fleet size within the next 20 years, this number of fatalities is likely to more than double.

The reality is that many technology and safety principles already exist in the railway industry, which could be adapted to the car market. Indeed for more than 10 years now, unmanned trains have been transporting safely billions of passengers. Furthermore, Private Rapid Transit (PRT) systems already exist and have proven that cars running on segregated lanes can be unmanned.

The point is that by eliminating altogether the human factor, the metro and PRT systems have shown that they are safer now than when drivers were at the command. What is true for trains and PRT will also show for cars running under general road conditions. Furthermore, other benefits coming from unmanned transport will permeate throughout society, making the shift from driven cars to unmanned vehicles even more appealing to society. Moreover, besides these additional benefits, at least three megatrends will push the advent of unmanned vehicles: The graying of society, an increasingly pro-litigation society with its impact on safety requirements and a world where mathematics play an increasingly fundamental role.

1.3.1 Graying of Society

The word has never seen so many people over 70. As people grow older they tend to use more often public transport instead of driving their own cars, though not necessarily by choice. This is possible for senior citizens who live in urban areas. But what happens to people with driving or health problems living in low density areas? They surely also would prefer to live at home as long as possible and still be able to travel wherever they need or wish, without depending on others. This is why in the future senior citizens will benefit tremendously from these self-driving cars and plebiscite their sale. As the number of senior citizens will grow exponentially, what could still be perceived as a niche market, will become a pace setter for the industry.

1.3.2 Litigation Society

The reader could probably expect such a megatrend to have a negative impact on the development of such unmanned technology. Our view is that provided there is a clear legal framework, unmanned technology is likely to prevail. Such a highly regulated market as the railway industry has started going that way and won't go back. Though in this book we will mainly use the railway industry as a way to show the impact of new technologies and safety concepts on the automotive market, let us focus for a few minutes on an entirely different industry to understand the ineluctable one-way road toward unmanned systems.

1.3.3 Safety Systems

In order to understand better the future of transportation, it is good to analyze the past.

> In 1852, Elisha Graves Otis, a mechanic from New York, invented the first safety hoist. By the end of that century the newly constituted Otis elevators company had brought many improvements to elevators, fuelling the "verticalization" process that turned big cities like Chicago and New York in what they are. Few people realize today the impact elevators had on modern architecture. By enabling residents or workers to avoid exhaustion when climbing stairways of sky-scrappers, they allowed the densification of people in down-town areas. Today's elevator design is more or less the same as the one introduced in 1903 by Otis, with its gearless-traction electric elevator, especially if you include the 1915 company improvement of a self-leveling device that allowed the elevator platform to stop exactly at floor level. With the introduction of the push-button in 1924, which made possible stops and automatic speed, a fundamental change in the industry occurred: elevator attendants became optional and were progressively made redundant. Today, only when traveling in developing countries can we still see occasionally people seating all day long and pushing the button for the clients. In developing countries one needs to be over 50 to remember seeing such attendants.

If we want to make analogies to other transportation means (an elevator is after all, a vertical transportation means), we need to ask ourselves why have elevators become unmanned? There are four factors at play:

- *Cost*: The obvious answer would be that it costs too much to put someone in an elevator just to push buttons. As a matter of fact, in countries were hourly rates are extremely low, we can still see a few attendants here and there.
- *Technological*: Development of a relay-based logic, which through push buttons and contactors allowed the passenger to select the right floor. Programmable Logic Controllers (a kind of industrial PC)-based elevators have now replaced the sequencing of floor selection, optimizing this selection in functions of other criteria such as speed of access, energy efficiency, etc.
- *Legal*: A transfer of operational responsibility from the company to the users.
- *Procedural*: Emergency procedures taking into consideration the absence of any trained attendants: fire evacuation, immobilization procedure, etc.

Why has the evolution from attended to semi-autonomous (people selecting the floor) and then to unmanned elevators (people getting access only to the floors they are cleared for) happened first within the elevator business and not in the automotive or mass transit markets? The probable answer is that it was far less risky and technically complicated to do this with elevators than with other transportation means.

Nature of the environment The evolution from an attended transport mean to a semi-autonomous system requires reducing uncertainty to the minimum acceptable, for a specific operational level (Fig. 1.2).

An elevator is after all a cabin clamped to a wire, which cannot collide with any other obstacle. This reduces uncertainty to its bare minimum. On the contrary, lanes or tracks are per nature open environments, where uncontrollable events (i.e.: pedestrian, next train's driver braking suddenly, snow or rain, or even an object parked in the middle of the lane) can happened. For instance in many metro environments, the level of automation is such that drivers are there only to manage the unexpected, such as someone jumping in front of a metro. In order to avoid such uncertainty, unmanned systems even introduced safety doors that can only open when trains have stopped. In such monotonous environment, the drivers are kept alert by having to close doors, a feature that could be done safely automatically.

The human risk factor Most accidents are due to drivers taking on a curve too quickly, driving too fast for operating conditions, or simply being in a state of

Fig. 1.2 Implementation of platform doors controlled by the signaling system removes operational uncertainty and allows for unmanned metro operation. *Source* Author

fatigue, stress, or under unauthorized substances. When they aren't due to drivers, accidents are caused by people in the control center making the wrong assessment. In other words, the human factor brings in risks that automated systems can reduce or eliminate.

Firstly, safety systems can be tested under all conditions, especially under emergency situations. Once tested and validated, the answer is systematically adopted. Secondly, through software upgrades any new finding or best practice can be incremented to the system, thus making the system even safer. On the contrary, any change to a process usually meets resistance from drivers. Thirdly at high speed, the drivers aren't able to assimilate quickly enough all the data and react in time for any emergency situation. Automating and getting rid altogether of the human factor is the best way to reduce accidents and fatalities.

Going from manned to driverless systems Under the convergence of several technological trends that we can regroup under the word of e-mobility—Information technology, Internet access, wireless telecommunication, and power electronics—Public and Private transport are going the same way as elevators: unmanned.

The railway industry has already implemented for the last century, technologies and concepts that have limited the harm caused by errоneous human decisions. By doing this, the safety track record of this industry has been improved tremendously and now completely outperforms the automotive industry. However, this is going to change not only for technological reasons but also under the scrutiny of the legal society.

Whenever cars will become autonomous, zero accident will be the norm. The question is can the concepts used by the railway industry be applied to the automotive environment? Will the automotive industry adopt a mixture of smart infrastructure and cars as it is in the railway industry supervised by an operational control center or will it be uniquely governed by smart cars and dumb infrastructure?

1.3.4 Mathematics and Algorithms

A third important trend that enables such e-mobility revolution is the increasing importance of mathematics in our daily life, resulting from the creation of more and more powerful algorithms. Though many people might not perceive it, because it is hidden behind software or incorporated into the equipment's structure or functionality, we live in a mathematical world governed by algorithms.

Most people are familiar with search engines from companies such as Google or Yahoo. The software behind these engines is composed of thousands of mathematical formulas linked together by a set of rules applying to the calculation formulas, such as optimization models. These set of rules are making the systems each day more intelligent. So much so, that the notion of artificial intelligence (AI) is more and more invoked to describe these powerful algorithms. In fact, among the traits that we associate with AI there are reasoning, knowledge, planning, learning, communication, perception and the ability to move and manipulate objects in the case of robots. Furthermore, what defines intelligence is not just one of these elements but the combination of several of these features working together.

Why is Artificial Intelligence important for transportation? In our view, today's unmanned systems from the railway industry and tomorrow's unmanned cars can be assimilated to systems relying on AI. Though this characteristic isn't frequently used to describe these systems, we believe it is very appropriate. In fact, whenever a technique reaches mainstream use, it is usually no longer considered as AI. However, there is a good reason why unmanned cars are also called robot-cars. They have to rely on several quantitative models for detecting the environment, understanding it, and making the right decisions accordingly.

Obviously AI raises ethical issues. In the case of unmanned vehicles, the fact that they don't look like human beings have so far not really antagonized the media and the ordinary citizen. This might change when these self-driving cars will start putting taxi drivers out of business. Will the bankruptcy of an entire industry raise ethical questions? The fear of robots creating a workless society will most certainly be used by the taxi industry to try to shield their protected market. Furthermore, it will also raise philosophical issues such as who will be responsible for paying the parking ticket of an unmanned car? From a legal perspective, robot-cars cannot be prosecuted so in the case of an accident will the owner, the manufacturer or the infrastructure provider be liable?

Governments will need to address the legal issues robot cars will raise, to unleash the full potential of these new IP driven technologies in the automotive industry.

1.3.5 Security

We make a difference between safety and security. A safety-related accident will be caused by a failure in the system coming from equipment or software or by the wrong decision of a human being. Security-related events will actually not be called an accident as it originates from the action of a human being wishing to harm someone intentionally.

In the last years, IP technologies have changed the entire security market. Most readers are familiar with IP cameras and network video recorders. For those working in high security areas, access codes with biometric detection are well known. These technologies have been introduced to the train environment. On the way-side, mainly the same technologies have been introduced to control key and sensitive areas of the networks. However, on the onboard environment new technologies allowing for real time view within the compartments have been implemented. New compression technologies have allowed for increasing recording capacity on hard disk caddies or even remotely on network servers.

The IP revolution, which hit like a storm the security industry, has been adopted in most railway and bus environments. The next wave of new technologies will be software driven.

Video or audio analytics, such as empty train detection, crowd detection or profiling are already under test in many operations. Problems due to poor lighting or light reverberation are been tackled with two different approaches. Simpler video analytics are being integrated at the level of the camera, for fast analysis and immediate response. More sophisticated video analytics are being integrated at the level of the Video Control Unit or Network Video Recording servers. The next phase will be to use such events to automate train functionalities.

The car industry has also started using CCTV system to monitor car accidents. In some countries like Russia, onboard CCTV is now the norm and used to protect citizens in court. In a not too distant future, our legal society will impose CCTV in all cars, as a way to identify car crash responsibility. With the introduction of onboard IP technologies, new security risks will be unleashed, such as hackers getting into the car system and tampering equipment to generate accidents. It won't be long before movies or TV series picture car fatality due to software manipulation.

1.4 Mega Cities

Another megatrend, which is affecting transportation, is the urbanization of society and the surge of megacities in developing and poor countries.

As countries industrialize, people living in rural areas are leaving in great numbers their farms to find better opportunities and living conditions in the cities. This creates huge issues. Commuting facilities become a vital necessity, as peasants who tend to be poorer, can only find housing on the cities' outskirts but need to travel to the center to find job opportunities.

After a few years of massive internal migration, transportation problems will increase exponentially within these urban areas.

1.4.1 Massive Urbanization of Developing and Poor Countries

The world is going through massive urbanization. Today more than 50 % of the world population is living in cities and this number is likely to hit the 70 % threshold. As people live in cities, they tend to work far from where they live and as a consequence, transport which was at the origin of urbanization, perceived as an occasional need, becomes an absolute necessity in a city where work and living area are decoupled.

In some rich countries such as the USA, people tend to live in the suburbs and work in the city by choice. In other countries, it can be based on necessity, because of the cost of housing in inner cities (i.e.: European cities).

The selection of such development model can have a significant social and economic impact on any city adopting this typical American lifestyle model. In cities that already have the commuting infrastructure, down-town area can become in the best case a ghost area during the evening. In the worst case, the downtown starts becoming a haven for criminality. As things get out of control, citizens and jobs move out to the suburbs, amplifying this trend. This phenomenon is often described as the "donut" effect.

The urban sprawl impact can be pretty negative for rich countries but is terrible for megacities of the developing world. In fact it is far from certain that these cities can add several peripheral roads or build new metro lines that would be necessary to cope with such development model.

Megacities in China are probably an exception. Indeed, China with its new wealth can tackle the problem by spending lavishly on new infrastructure. For instance, Beijing will be adding its seventh peripheral road in 2015, which totals 940 km and will be constructing its 15th metro line. Unfortunately, poor countries of Africa, Asia or South America can't even afford contemplating public transport investment at all.

The second problem is that megacities of developing or poor countries don't already have a mass transit or significant road and highway networks that can cope with the increased demand. Everything must almost be done from scratch, requiring tremendous efforts.

Third, the megacity expansion has usually been done without a master plan and can be described at best as disorganized. This means that usually there is no more physical space to build new roads, Bus Rapid Transit, or tramway lines from scratch.

Underground systems can be added but are extremely slow and expensive to build. The only solution is to expropriate existing houses or building, which is also extremely costly and socially difficult to do for city mayors. They can also confiscate existing roads to build a system with higher capacity, a concept for which we have introduced the notion of "appropriation". This concept is very often forgotten by city planners who propose BRTs or tramways. The reality is that these new segregated lanes in the middle of the city can have a negative impact on traffic.

Last but not the least, in democracies activists have several laws that they can use to kill projects.

As seen, any neighbor of a new BRT or a metro line can become a potential opponent, under the "not in my backyard" principle. Environmental laws that are there to protect nature are often used as a tool to kill projects even though the overall impact of a new metro line should be extremely beneficial. In megacity slums, the fact that land ownership often isn't clear makes it also a weapon to postpone indefinitely land release.

This makes it extremely difficult and painful to build new infrastructure within a reasonable time frame. Construction is stopped for long periods, often resulting in additional costs to civil construction companies that need to demobilize and remobilize often. These extra costs are then transferred back to society through claims or absorbed by the civil companies that then need to increase their risk provision on their future project costs.

1.4.2 Mega Transportation Problems of Megacities

In the last 10 years, the world has seen the emergence in developing countries of megacities, which can have between 10 and 30 million inhabitants. This is significant because transportation problems in megacities increase exponentially. Indeed not only does the city expand horizontally by urban sprawl, it also expands vertically by higher housing densification. In other words megacities don't explode, they implode. As they implode through a "verticalization" process, there is no possibility to build new roads to cope with additional inhabitants.

The consequence of this is that each day during peak hours, many megacities are at a standstill with 200–300 km of traffic jams and residents loose 2–6 h commuting. As seen, the only solutions are either to:

- Build underground infrastructure, which is extremely expensive;
- Expropriate houses, which is socially explosive and almost impossible in densely populated areas; and
- Appropriate existing roads to build new public transport, which creates other problems, such as new traffic jams, separation of cities along these corridors, increased noise and pollution.

As any potential solution would take time, many city politicians prefer reverting to palliative solutions, such as creating restriction on car use.

For instance in Mexico City or Sao Paulo, cars cannot be used one day out of the week, depending on their plate number. The reality is that such measures penalize the poor citizens, as the richer ones use taxi or buy a second hand car to avoid the restriction. As older cars pollute more than newer ones, the end result of such measure is often worst.

Politicians could use an easy fix for lack of transportation capacity. They could do more with the existing transportation infrastructure. There are basically two ways of achieving it:

- Increase the number of passengers transported per each vehicle; and
- Put more trains, buses, or cars on an existing line or lane.

1.4.3 Increased Vehicle Capacity

Vehicle capacity is a function of the number of passengers that can be carried by the vehicle. This is defined by four elements: physical characteristics, safety standard, comfort level, and occupancy rate.

To illustrate how these elements are intertwined, we will use an example that everyone will understand easily. A soccer mum's car could transport most of the team members if the children would accept to sit on each others' lap. However, the law which defines the safety criterion would impose that every child buckles up. Thus legally, the soccer mum's family car couldn't carry more than a maximum of 6 children. For comfort reason though, she could decide to reduce this number to 5 children. However, whenever she goes to work like millions of commuters, she would most likely be alone in her car.

Transport authorities have a few cards up their sleeve to improve vehicle capacity. They can buy larger, longer, or two-storey vehicles. Train or bus manufacturers can also improve the design by, for instance, putting equipment on the roof. Transport authorities could decide to change the train or bus comfort level by reducing the number of seats or inner space dedicated to seating area. They can also take a more drastic attitude, which is to buy trains or buses that can withstand higher volume of standees per area, affecting directly the level of passenger comfort.

In most countries, this level is 4 passengers/m^2 but in some megacities of South America and India, they apply an 8 passengers/m^2 ratio. This measure almost doubles capacity instantly, but at the detriment of the quality of ride. Some potential clients might then decide to revert to their car, defeating the purpose of such measure. They will then join the cohort of drivers with no other passengers.

In modern cities of Europe and North America, the occupancy rate is only 1.1 passengers per car during peak hours. Investing in car pooling is one of the cheapest and best ways to improve capacity. Another way of improving Public Transport occupancy rate is to increase subsidies.

1.4.4 Increased Network Capacity

Public authorities can also improve the network throughput. Larger and longer vehicles, running faster and closer to each other can make better use of existing infrastructure. Thus, they can play with four variables: the capacity of the vehicles, the number of vehicles running on the system, their average speed, and the headway for trains or interspacing between buses and cars. Though train, bus, and car obey to these same four variables, the way they are impacted can differ.

Vehicle capacity: Besides the own limitation of the trains that we've seen above, the capacity is also limited by the physical network characteristics. There are many physical limitations such as station clearance and length, tunnel envelop calculated with jerking, electric grid connection type, etc. This explains why railway manufacturers must mostly tailor-make their train to the network and don't really benefit from economy of scale. The environment of cars and buses is much more standardized. Roads always have a minimum width that doesn't really impact manufacturers. Buses maximum length, which is mainly function of the number of sections, can only be restricted by tight curves. For cities that envision using double-deck buses such as in London city, overpass height may also create issues.

Number of vehicles: There is a maximum number of vehicles that can run on a track or lane at any given time. However, till this number is reached there is always a possibility to add extra capacity. Once this number is reached, there is no point in adding more vehicles, as the extra vehicles would only stay idle or would need to slow down to maintain the same safety inter spacing.

Average speed: The speed of a line or lane is function of the inherent characteristic of the vehicle, the safety and comfort of passengers, as well as the limiting factors of the network operation. For instance, vehicle's acceleration, maximum speed and deceleration are key factors influencing the overall system speed. In most metro applications, because distances between two stations are short, trains are almost always accelerating and decelerating. Metro car motorization is usually adapted for those specific conditions. However, a metro operator must also take into consideration, the passenger's safety and comfort when accelerating and decelerating. After all they don't want passengers to fly through the windshield, when they apply the brakes. Another limiting factor to speed is the network operation. Average speed clearly changes according to railway application type. For instance, metro average speed will be in the range of 30–35 km/h, while high speed trains will hit the 380 km/h mark in operational mode.

Though car acceleration or speed possibilities are usually more a function of the driver's wealth and his willingness to show it off, in most countries even the cheapest automobile isn't limited by its inherent motorization characteristics. Infrastructure specificities, weather conditions or legal speed limitation are the main restricting factors. Traffic light and potentially any people, cars or objects blocking road access will also influence speed.

Bus speed is influenced by a mixture of all the above, which explains why generally buses running in urban areas have a problem going over 20 km/h.

Headway and interspacing: The distance between two vehicles moving in the same direction is important for network capacity because it limits the number of vehicles you can add to this system without reducing the overall speed. The terms interspacing and headway are usually used for cars and trains respectively. In both cases, distances are transformed into time. For cars, people maintain naturally the equivalent of a 2-s minimum distance, regardless of their speed. In the railway industry the minimum headway is an area of great investments. With new electronics and powerful software, signaling companies have been able to reduce significantly the time without increasing safety risks. This evolution of safety standards that we will describe in detail in Chap. 3, have enabled through moving block technology, headways in the range of 60 s. Going from 120 s, which was already difficult to achieve with fixed block technology, to 60 s can almost double the network throughput capacity.

1.4.5 Road Capacity

Maximum highway capacity seems quite straightforward to calculate. For conventional cars running on one lane, flow rate is limited to around 1800 vehicles/h.

However, this number isn't proportional to the number of lanes, as the possibility of people changing lane will affect throughput but is nevertheless a good indicator. The number of heavy vehicles, the width of the lanes, driving habits will also influence the throughput. For roads, it is a little more complicated to calculate as street lights, stops or possibilities to turn at junctions, as well as potential interference such as pedestrian crossings and bicycle lanes will also greatly impact throughput. Intelligent lighting systems, integrating video cameras on an IP network can improve capacity by reducing waiting time in function of the counted number of cars.

One interesting aspect of the e-mobility revolution that we will tackle in the book is the possibility to do platooning/convoying of cars. Under the leadership of a main vehicle, cars will be able to follow each other a few meters apart. In other words, thanks to Vehicle to Vehicle communication (V2V) technology, cars will also be adopting the moving block principles, where speed is a function of the front vehicle's own speed and the minimum distance required to brake safely.

Adopting such driving scenarios, could more than double highway capacity. We will argue that cars will need to adopt one of the safety principles called "brick wall", which could limit such huge capacity benefits. However, even if the brick wall principle were to be applied to unmanned cars, they would still increase throughput by around 30 %.

Once the road flow rate is calculated, we still need to factor in the occupancy rate. Thus in most European and North American countries, we need to add a 1.1 and 1.5 ratio, which leaves us with a real occupancy rate of around 2000 passengers per lane per hour at peak hour and 2700 passengers on average. For politicians, the easiest capacity win is thus to promote car pooling campaigns. Adding an extra passenger to all running cars would almost double the capacity instantaneously.

1.4.6 Bus Rapid Transit (BRT) Capacity

Since its first introduction in Brazil 40 years ago, operational concepts have been improved and have transformed BRTs into a direct competitor of metro systems.

One concept, which is difficult for railway people to understand, is that throughput of the BRT line isn't limited by the number of vehicles nor dwell time. This misunderstanding has often made discussions between bus and train specialists seem like people living on two different planets. In railway systems, trains stop at station and usually upcoming metros must wait for the train ahead to depart the station to be allowed into the corresponding block. Buses in BRT systems don't behave that way. They depart the lane and go to curbside spaces where they can stop to unload and load passengers, also called berths. There can be up to 5 berths per bus stop. Dwell time is thus not on the critical pass for overall BRT capacity, as load can be done in parallel. In other words and if we didn't take into consideration the departing and reinjection into the lane, all 5 buses could take only 20 s dwell time, if all these buses were synchronized perfectly. However, synchronization is impossible and the efficiency of the 5th berth can be as low as 10 % for random arrivals.

1.4.7 Mass Transit Capacity

The capacity of metro or commuter trains can vary substantially from one network to another.

As the globe-trotters of this world have probably noticed, trains and metros come in various shapes, size, height, length and width. For those globe-trotters who have a keen interest in train infrastructure, they probably also noticed that station length, platform size and height, tunnel diameter and other network characteristics were also very different from one city to another. Furthermore, they were probably astounded to observe that in some countries during peak hour, it seems that a metro departs from the station when a new metro arrives, with almost no time between trains. In a form or another, these are all variables that affect the mass transit's capacity. Although some changes can eventually be made to these network variables, it's not easy to adapt the network to the modifications in demography or in the neighborhood profile, expected over time.

Knowing that changing existing capacity cannot be done easily, what are the parameters that city planners should look at in priority when designing their mass transit network? Should they fix the infrastructure dimensions and let the train manufacturers come up with the best solution for train capacity? On the contrary, should they define the train characteristics and calculate the tunnel and station size? We are strong believers in turn-key approach and in our view they should avoid both tendering strategies. They should focus on defining what is the capacity required at the start of the operations, 15 and 30 years from now and let the system providers and civil construction companies come up together with their best solution. By focusing on immediate and future capacity requirements, the tendering outcome will most likely result in the most optimized solution for the tax payers and the users.

In fact, compiling origin-destination data allows city planners to identify the likely users, station per station, from one side of the network to the other and in both directions. By knowing where commuters will embark and get off the metro, they can calculate the maximum capacity required in terms of PPHPD. This PPHPD will then be analyzed hour per hour and define what is the operational fleet required during the entire day. Two other variables are used to size a network's capacity.

- Ridership expressed in daily number of passengers, will calculate how many passengers have indeed been transported during a given day.
- Passenger × kilometer travelled will also be used to calculate the number of trips performed and passengers transported by a given network.

We will explain in the section dedicated to mass transit capacity the main train, station, network, and headway features that have an impact on these indices. More importantly, we will describe how standards can impact capacity and how new technology can increase it.

1.4.8 System Price Comparison

Transportation networks can stretch from a few to thousands of kilometers. Of all the big infrastructure work, it is probably the most influenced by the geographical environment on which tracks or asphalt are laid on.

Even though each transport infrastructure is unique and can be influenced by thousands of criteria, there are some commonalities that allow comparison between the various transportation means. In order to use comparable data, it makes sense to avoid considering specificities, such as building a bridge, digging a tunnel running under a river, doing deep tunneling, building rails over a 6 % ramp, etc. To get an average worldwide cost per transport means, one must also factor in cost of land and property, regional salary level and other characteristics specific to a country, without forgetting costs that are usually not taken into consideration, such as health costs due to pollution, lost productivity due to traffic jams and appropriation of land. Even then, risks, project and network specificities still create huge differences, which means that any comparison would still suffer 10–20 % variability.

In Chap. 6, we will try to give budgetary information about the following inner city land transport means: road, highway, bus, tramway, Light Rail Vehicle, and metro.

1.5 Connected Cities

Many newspaper articles emphasize the passage from an industrial era to a civilization of Knowledge. In this new civilization, cities are playing a fundamental role, both as an environment fostering talent and creating new growth opportunities.

Many Marketing people, always happy to surf on a wave, have coined the term "intelligent cities" or "smart cities" to describe this new environment. However, we believe that the term "connected cities" is more appropriate, as there isn't yet a mega computer supervising all aspects of our urban daily lives. That said, Authorities are trying to associate their town with this "intelligent city" concept. In fact, many international cities have realized that they are competing with each other to attract talent. Educated people tend to look for quality of life when selecting their place to live in. Good transportation facilities, as well as good social infrastructures such as universities and hospital, score high on the agenda of such mobile crowds. Furthermore, young highly mobile people want to connect everywhere and at any time. They must be able to on the go—be it walking, standing in a bus or in a metro, or driving—to access all information that they deem necessary.

In order to attract the brightest and youngest, cities and, also companies, will want to invest in technology enhancing quality of life and its pendent for transportation—quality of the journey. Being able to know when the next train or bus will arrive and where the passengers must change line is one of these key attractive new technologies perceived as improving the quality of ride. This is why public authorities are favoring the implementation of new Internet applications in their city, aiming at fulfilling this need for real-time traveling information.

1.5.1 Constant Network Connection

For the newer generations, being informed everywhere and at any time is indispensable but also seems technically easy to achieve. However, it is only with the advent of Internet and newer mobile technologies that people can get access to information on the go.

Instant and constant connectivity changes not only our daily habits, it also impacts dramatically our social life, with the possibility to maintain contact with family, friends and co-workers regardless of the place on earth we are in. Mobile technologies are permeating throughout all spheres of society—business, politics, entertainment and education—changing our relationship to information. These technologies aren't just redefining how we access Knowledge but even what Knowledge is. Before the advent of internet, a cultivated man or woman was someone who could articulate his/her thoughts around information accumulated over many years of reading and studying. For instance, providing the information on how transport first started, as we just did, would have required many hours of sweating at public libraries going through several books or microfilms. Even then, we would have needed to be lucky enough to have a librarian with sufficient transportation background to guide us, or a library classification scheme. With internet, such knowledge is now available at the click of a mouse. With the early days of internet, Knowledge stopped being recalling the information and became knowing how to retrieve it. However, with new powerful search engines, even such searches are becoming easier. Even more so, in a mobile environment, where anyone can instantaneously and 24/7 search whatever subject they are interested in. As search engines become more and more powerful and information growing exponentially on the net, Knowledge will in our view become more and more a question of connecting information, events, or even bits of information in a way that makes sense. To illustrate our point of view, let's take an example. Suggesting that transport started with the advent of domestication of horses and the wheel creation as we just did, is more relevant than figuring when these two independent events occurred, as both dates are easily available on internet. However, it doesn't mean that we didn't miss other important factors, without which transportation wouldn't have happened.

This raises also the issue about the validation of information, especially in a web environment. The Italian semiotician and novelist Umberto Eco, in an interview[3] on the impact of Wikipedia (Wikinews 24-04-2010), raised the issue about validation of information in such a collaborative environment. Quality of reviewers and correctness of information are at the heart of the criticism of Knowledge acquired through internet.

Why is the validation of information in collaborative environment relevant to transportation? In fact, the "Internet of things" is build around these same collaboration principles. We could even say that the newer Internet technologies are inherently built around a collaborative model. Service Oriented Architecture, Service Delivery Platform, Event Driven Architecture (EDA), Plug and Play designs, are a few of these technologies. As they are an intrinsic part of this e-mobility revolution, explaining how they work and are changing the mobile world, makes it in our view mandatory, in order to apprehend the issues facing the "Internet of things" implementation in transportation.

Furthermore, as we will see, connectivity doesn't only apply anymore to human beings. In fact, Machine to Machine (M2M) communication or its specific transportation variant V2V, V2I and moving block will change profoundly transportation. What is true for an environment where people need to filter and assess information becomes even more crucial in a world where no human beings intervene.

[3]The Italian semiotician and novelist Umberto Eco, in an interview on the impact of Wikipedia (Wikinews 24-04-2010).

1.5.2 The "Internet of Things"

The concept of "internet of things" was introduced to society before the new millennium. The main idea behind this concept was that everything would be one day connected through Internet. At the time, the limitation of the IP technologies made that prediction bold. Today's Internet revolution is still based around the same old technology: TCP/IP suite of protocols.

Several times, IT experts have said that the internet was too limited for it to prosper. The reality is that the IT industry has been able to provide "patches" and new concepts that allows internet to thrive. With all their limitations, TCP/IP protocols can still provide end-to-end connectivity specifying successfully how data should be formatted, addressed, shipped, routed and delivered to the right destination.

However, this suite of protocols is being enhanced by new mobile, network, and software technologies, which will be explained in simple ways for non IT specialists in Chap. 5. What is important to understand is that these technologies are being developed conceptually using the "collaborative" principles we referred to.

Even centralized network design (i.e.: bus or tree network topologies) are being replaced by non hierarchical topologies (i.e.: ring, or mesh), where fall-back options are possible in case of problems. Service Oriented Architecture is one of these collaborative tools. In this new world, this architectural software separates functions into distinct units (services), which can be distributed over a network and can be combined and reused to create business applications. Any element connected to a network (i.e.: PC, mobile, any type of machine) can use the service and provide itself services for other the use of other network elements. EDA technology complements SOA architecture because it introduces the notion of events, which can trigger a service. This software architecture promotes production, detection, consumption of, and reaction to events. Plug and Play technology associated with SOA simplifies connectivity, as any element (i.e.: CCTV, camera, door, etc.) can automatically configure itself and decide which are the services they will use and provide to other elements. This allows for immediate equipment connectivity without the need for hard core programming, reducing cost and possibility of interface errors.

When all the "collaborative" technologies are combined, any equipment can seamlessly and automatically connect and know which of the million bits of information sent to the network is actually meant for it. The issue in this technological world is similar to that described by Umberto Eco, where validating information received is extremely critical. For instance, in the security world, receiving too much false information (also called false positive) is as bad as not detecting a dangerous situation. Furthermore and as we already indicated, in this Internet world, knowledge is not only validating the information but deciding which services can be associated together to create more added value. In a world where human beings are absent, data mining software will become crucial. Specialists with real knowledge in specific areas will develop algorithms associating events or bits of information that at first glance don't seem to have any link, unleashing the power of trillions of interconnected objects.

1.5.3 M2M Communication

One of the limiting factors of old Internet technology was linked to the addressing of devices or computers through a Unique Resource Identifier. Originally, this identifier was a 32-bit number (IPv4) but to allow for trillions of devices to be interconnected, a 128-bit identifier was launched in 2013 under the IPv6 initiative.

With IPv6, the "internet of things" exponential growth isn't limited anymore by software restrictions. Every device will have its own identification number and, when associated with collaborative technologies, will be able to receive, consume and transmit seamlessly information to other devices on the network. In the next ten years, M2M communication will grow exponentially. Though there is still no defined standard for M2M, most likely the technologies will evolve around the collaborative software and architecture we just mentioned.

M2M is changing profoundly railway operation. Automation of equipment, which requires M2M connectivity, is increasingly becoming a common denominator in railway yards and garages. Moreover, predictive maintenance which relies mainly on M2M information is an area where railway manufacturers and operators are investing a significant part of their R&D budget.

Through predictive maintenance, rolling stock reliability and availability can be greatly enhanced, allowing for a significant reduction in railway total cost of ownership and also reduced penalties for operators and manufacturers. This trend being pushed by Governments enables private or public operators to provide transportation services at the lowest possible cost, at a defined set level of quality.

1.5.4 M2M Applied to Cars

The automotive industry has been a laggard in terms of development of onboard IP technologies. In fact, many onboard network technologies are more than 20-years old. However, the industry has finally decided to work jointly on a common standard platform based on IP technologies.

The first cars using such an IP backbone are likely to hit the roads in 2015–2016. Furthermore, initiatives pushed by the US and EU Governments will force all car owners to equip their car with a safety related SIM card likely to be included in new cars, allowing automatic contact with emergency centers. By imposing SIM cards to all new cars, these Government initiatives will create the communication pipeline that will enable a new set of value added services such as predictive maintenance or automatic emergency calls as well as create some of the telecom infrastructure that will help push V2V and V2I telecommunication. By the same token, it will allow for the mobile advertising market to grow exponentially.

With the advent of SIM cards, an entire array of new services will become available to car owners. The same revolution that the railway industry is going through will hit the automotive industry. New added services, such as car failure anticipation, remote diagnostic will create thousands of new companies in a new market worth billions. It will in fact profoundly disrupt the automotive market as we know it. Other market trends and business models are already affecting the public and private transport markets. How quickly and how deeply they will affect transportation depends also on how governments can find financing and funding solutions to build new infrastructure.

1.6 New Business Models

Around one-third of all worldwide investments needed each year in infrastructure projects cannot be financed by private corporations or funded by governments. Moreover, with the growth of population and its urbanization, this gap is increasing. After the sub-prime crisis of the end of the past decade, many rich countries find it increasingly difficult to allocated funds for such big projects.

The reality is that most politicians find it difficult to justify big spending while at the same time imposing austerity measures on their citizens. This is unfortunate, as big infrastructure projects have the ability to create quickly thousands of jobs and ease the economy out of recession. This Keynesian approach is especially productive in the current situation of extremely low interest rates, resulting in cheap loans. The reality is that most countries are almost bankrupt and cannot easily increase their debt level. Furthermore, some countries like the USA prefer, for ideological reasons, to see their bridges collapse rather than refurbished them through tax hike.

In Chap. 6, we will propose solutions to bridge the $1 trillion gap, while reducing the environmental footprint.

Such solutions include:

- Creating favorable laws promoting the private sector's involvement in big transport projects;
- Asking the World Bank and global institutions to formally recognize that public transport is a green solution and whenever the carbon market recuperates, enable realistic numbers to be applied to metro and railway lines;
- Incentivize through subsidies or financial market institutions the transfer from conventional combustion cars to electric cars;
- Incentivize new business models that reduce car ownership, such as car pooling or car sharing, and thus decrease the pressure on new infrastructure;
- Allow rich people to cross-subsidize public transport by for instance, allocating urban fast lanes along metro lines, even though this could be perceived as socially unjust;
- Create financing instruments that capture for the project the anticipated real-estate speculation in a certain city area (i.e.: around metro stations); and
- Give corporations incentives to invest in big green infrastructure projects in parts of the word where no private companies would consider investing, through tax exoneration scheme.

1.6.1 Privatization

Most governments will continue not being able to bridge their financing gap for the foreseeable future. The private sector would have the resources to bridge that gap but has been incapable or not motivated enough to do it. In fact, many long-term investors find it hard to find attractive long-term projects and would be happy to become potential investors in infrastructure projects, if provided with the right opportunities. Thus, the easy way to solve the gap would be for governments to accept that the private sector needs a decent return on investment or reduce the level of risks transferred to the concessionaire.

The issue for many Governments is to assess what the acceptable profitability versus risk level should be when considering such risky business environments as a 30-year road or metro concession. Furthermore, even if this acceptable level can be found, it is often impossible for Governments to accept it for social and political reasons. Unfortunately for election purposes, Governments will still prefer to push through projects that cannot find serious investors, rather than adapting the profitability and risk levels. As a consequence and in the best case scenario, the project will be killed. In the worst scenario, the construction will start with adventurous companies and then become paralyzed for lack of liquidities, leaving a no-man's land in the middle of the city.

The reality is that Public Transport needs subsidies and privatization of such service requires a sound legal framework. It is a utopia to expect financial institutions or big corporations to accept 30-year concession risks and to finance the project from their own assets and the concession's operational profits, without any form of government financial transfer.

One of the first key factors to successful transport privatization is to accept that countries' legal foundation must allow for "Project Finance" rather than only "Corporate Finance". Because of the amounts and risks involved, companies need indeed to limit their potential losses to the equity strictly required by banks to get the necessary loans. The aftermath of the sub-prime crisis and the "Basel 3" banking rules consequence have already made banks reluctant to finance by debt a big share of such projects. Asking corporations to increase their equity commitment, while at the same time imposing unlimited liability exposure for 30 years, is definitely a no go for many institutions.

A second key success factor for privatization has been not to oversell the privatization benefits. Higher efficiencies that can be brought by the private sector, resulting from lower operating and maintenance costs, are usually an important reason invoked to justify privatization. In many countries, privatization of transportation infrastructure or services has been perceived favorably, especially when the level of financial commitment or risks transferred to the private sector was high. However, in some countries where only the operations were tendered to the Public Sector, resentment within population has made difficult the privatization of something perceived as of the Public domain. The route cause of unsuccessful privatization has often been linked to the creation of false expectations. Privatization cannot be the panacea and solve all problems. First of all and as mentioned, it is illusory to expect public transport concession to break-even in mass transit operations (the exception is when the operator also owns land, which it can resell such as in Hong Kong). When heavy investments are required along the operations (i.e.: doubling of autoroute lane or maintaining rail infrastructure), there is also a need for funding from Public Authorities. This Government financial help can come under different forms: cheap financing, direct subsidies, minimum revenue guarantee, etc. As Public entities need to pass along funding to the private sector, a clear legal framework is required.

The third success factor is thus a clear legal framework to enable the contracting of consortiums to take over the concession responsibilities. These laws, which come under the name of Public Private Partnership (PPP), have been pushed through legislature in many countries. Though there can be as many types of PPP as types of concessions, all PPPs have in common that they define the risks that will be shared by the different Parties.

1.6.2 Financing Transportation Projects

There is an appetite for financing big infrastructure projects, especially for markets where profitability can be achieved without Government payment transfer. Unfortunately in most transportation systems, governments need to fund partially the project. This makes transportation project more risky and more difficult to bank, especially after the sub-prime crisis. Indeed, the main banks that used to do project finance before the crisis have reduced the debt to asset ratio of the Special Purpose Enterprises (temporary entities created with the sole mission of managing the project). Other actors must put more equity, pushing more risk toward investors.

Short term investors, such as private equity funds, have stayed shy from public transportation markets, though some recently have entered PPPs in the energy sector. If the profitability could be increased, the more aggressive ones could probably be convinced of entering such opportunities. Long term investors such as insurers and pension funds have in the past entered as equity partners in big infrastructure projects. However they usually hate risks associated with the construction period and prefer to transfer this risk altogether to railway manufacturers or civil construction companies. Very often they are eager to buy back the shares of these companies after a few years of proven profitable operations. Sovereign wealth funds, another type of long term investors, have not been really active in the financing of big infrastructure market. With the change of its by-law to allow investment in energy projects, Norway's Sovereign Wealth Fund, could maybe set an example and enable transportation financing by these funds.

Private companies are being asked to invest more of their money to foot in the bill. There are few real concessionaires, which main business purpose is to take concession risk over a thirty years period. This is why Governments expect companies that have an interest in delivering the civil construction works or provide the system to participate with equity in the projects. The problem for these companies is that the financial markets don't assess favorably such investment, preferring financial ratios that show that money is better used in other areas.

As few players are really interested in such long-term risky investments, new financial incentives must be found to bridge the gap.

1.6.3 Financial Instruments and Incentives

We will present in Chap. 6 a few of the financial instruments that have been used or could be adapted for financing transport projects.

Most have in common that they are geared towards reducing the impact of human activities on the environment. Though they could in theory apply to other areas, their methodology is biased towards forestation or renewable energy projects. Green bonds, Carbon Credit and Voluntary Carbon markets have added useful financing, allowing several small infrastructure projects in poor countries to be implemented.

The big institutional players such as the World Bank or important private donors need to recognize that taking out thousands of polluting cars of the inner cities is as much a priority of the Greens as saving the green forest, or as important as fighting diseases in Africa.

Governments from the Rich World should also recognize the benefits of creating a BRT or a commuter line in their Development Aid Policy.

We suggest that investing in Public Transport projects in poor countries, should give right to corporate or personal tax exemptions. Providing transport to millions of poor citizens from the city outskirt will bring job opportunities, reduce journey time, accidents and health problems of the poorest of the poorest. This is in our view in line with the definition of eligible donation matters.

1.6.4 Capturing Future Wealth Increase

One of the most interesting financial instruments that has been created to finance infrastructure project comes from Brazil. The certificate of potential increase in construction (CEPAC in Portuguese) works on the principle that real estate speculation that would inherently result from the increased attractiveness of a region due to work improvements can be anticipated and promoted to finance public work.

CEPAC is a financial instrument emitted by a municipality (i.e.: Rio, São Paulo), used as a counterpart for the right to increase the construction level in a dedicated area of the city and is equivalent to a value of m^2 for construction area. This financial instrument has been used successfully to rebuild entire degraded Brazilian neighborhoods, allowing for the financing of private and public transport, such as tramway lines. However, to be successfully replicated in other countries, CEPAC would need a similar legal framework, sound financial controlling authorities and growing real estate market like the one of Brazil's megacities. This would obviously be complicated to implement in poor countries.

We believe out-of-the-box ideas based on e-mobility solutions could partially substitute such financial instrument, based also on improvement to the quality of infrastructure.

For instance, some public transport such as BRT or tramway lines underestimate their CAPEX, by not taking into consideration the land occupation cost. Building a monorail or a metro actually allows for road re-appropriation. By bundling with a private concession the future benefits of road appropriation, circulation improvement to a large avenue resulting from these new mass transit system could be used to cross subsidize the infrastructure works and mass transit operations. Rich people or even taxis would be willing to pay twice the price of a bus ticket to avoid spending hours in traffic jams. The same e-mobility technologies that exist on highways to pay tolls without stopping, would allow people to get in and out quickly from these exclusive lanes. For those who wouldn't want to create cross subsidies by mainly improving the life of the Rich, there would be the possibility of using congestion charges within the new line's neighborhood to subsidize the mass transit operation. The same charges used to discourage residents from driving within certain area could be applied. In fact, there is no technological limitation to installing the same video technology that reads your driver's plate and checks if you have paid electronically your charges in smaller area alongside the network's line.

The extra revenues resulting from such ideas could help reduce public transport reliance on subsidies. Another new source of revenues could be coming from online research, while passengers are commuting.

1.6.5 Mobile Advertising

Citizens of the rich or developing world spend hours in traffic jams or stuck in buses and metros. In fact, Americans will spend around 5 years of their life in their car or roughly 5–7 % of their daily time. If one would only consider available free time and if we were able to turn the lost time spent in transport into productive time, this could generate around 20 % more opportunity for people to buy or do research online.

Half the consumers based in Rich Countries, spends 75 % or more of their total shopping time conducting online research. This situation explains greatly the growth of companies such as Google and Amazon. With the advent of smart phones, product searches are now done increasingly on the go. For instance, European mobile users are spending more than one fifth of their time doing online research on the go. This trend is likely to continue and even accelerate with 4G telecom services.

When we add the craving for new mobile online searches and the available extra time provided by transport, we can all understand easily why the IT companies of this world are trying to get their technology into cars or in public transport.

1.6.6 Geo-localization Advertising

Location-based information or advertising uses GPS capabilities to know the position of a mobile user and deliver an ad for a product or service that can be found nearby. These ads or messages are triggered whenever a mobile user crosses a predefined area, which can be extremely narrow or cover an area of a few kilometers. Obviously areas such as train station or bus hubs are extremely pertinent for such specialized advertisement, as millions of people cross them every day. Public transport with their systematic financial losses could benefit tremendously from such extra income. Based on a pay-per-click of a little less than 1$, a great chunk of the deficits could be eliminated, if only passengers were to do a few daily online searches and the money be transferred directly to the Public operator.

The reality is ownership of geo-localized advertising isn't clear from a legal stand-point. Two fundamental legal principles are playing against each other: intellectual property and land owner rights. The ideal situation would be that passengers use public transport portals but this is unlikely to happen.

1.6.7 *Portal of Choice*

In the past few years, many transport applications have been developed.

"Waze" one of the most well-known Apps has been bought by Google. Other applications such as Moovit are also likely to be eventually snapped up by IT companies. By providing real time passenger information without having to go through the Public transport operator data base and intellectual property, these applications make the million passengers aware of any network disturbances and the best alternative routes. It also creates a strong motivation for thousands of passengers to go through the portal of these applications.

Having gone through these Apps' portals, the most likely next step for passengers is to do online research through related applications or to be exposed through geo-localized advertising. Big incumbent operators must fear to see their brand being commoditized in the future.

1.7 Changing the Face of Transportation

E-mobility technologies can bring fundamental changes to transport as we know it. In order to unleash the power of these technologies, Governments must make the right decisions and commit resources to enable their quick implementation. The first and easy decision to make is to promote home employment and flexible hours. Government should incentivize companies to organize shifts that would limit peak hour commuting. Out-of-the box ideas and easy measures can also be found.

For instance, incentivizing companies to ask their employees to start work twice a week outside of peak hours, would have in our view a more positive influence on traffic jams than reducing car owner's right to drive certain days, based on their driving plates. A simple accompanying measure such as allowing companies to account for carbon reduction resulting from their employees' avoided commuting trips could go a long way in reducing their carbon footprint.

1.7.1 Electrifying Transport

If governments are really serious about fighting global warming, they need to be coherent with their declared policies. One of the best ways of meeting global warming commitments is to electrify transport.

The UK Governments has made that choice and will be investing in nuclear power plants to support that shift in the midterm, as well as in new mass transit systems and high speed trains. In the long term, renewable energy should pick up the extra demand but in the mean time, transport will significantly reduce CO_2 emission.

Though the initial work and investments might be high, we will prove that the benefits of such policy on the economy will be huge, shrinking the commercial balance deficit of most Rich Countries.

1.7.2 Encouraging New Business Models

Public authorities should promote the new business models that e-mobility technologies are enabling. These business models revolve around the fact that in megacities and big cities of this world, owning and using a car is expensive and complicated.

Indeed, finding any free parking is increasingly difficult, as many city planners are intentionally reducing areas devoted to parking spaces. Buying or renting a parking place is becoming out of reach for most families. Furthermore, driving in inner cities is becoming increasingly nerve racking as city planners are favoring pedestrian zones, bicycle lanes and public transport over private transport. Driving is also much more expensive as congestion taxes or toll charges are applied in specific urban areas to dissuade suburbans to come enjoy the inner city's pleasures by car. In the past, citizens would have grumbled and just lived on with such annoyance but a megatrend is also favoring these new business models.

The younger generation doesn't crave to own a car, as the older generations did. For the baby boomers' grandchildren, holding a tablet or a phone is more valuable than driving a Porsche at 20 km/h. It makes increasingly more economical sense in inner cities to rent rather than buy cars, especially when we consider that cars stay idle on average more than 90 % of the time and lose 30 % of their value the day they are driven out of the car dealership.

For transport authorities wishing to find a quick win against traffic jams, the best solution is to promote usage of different car pooling services managed by several new Apps. Because any extra car sharing passenger has no impact on road capacity and very little on carbon emission, transport authorities should treat car pooling as a priority all over the world. We highly recommend investing in road infrastructure adaptation to facilitate boarding in safe areas near main arteries running into the main working areas.

The following business models have also the potentiality to reduce automotive footprint in inner cities:

- Peer-to-peer car sharing, which basically consists of renting private cars for a short period of time through Apps or car rental by the hours. City authorities should provide ample parking access in densely populated areas, to encourage people to abandon their second car for these more practical solutions.
- Peer-to-peer taxi or hailing services is democratizing taxi usage by reducing significantly the fee per kilometer charged. Moreover, these new service providers are giving credentials to an industry that lacks such credibility by increasing timely offer and providing traceability on the provided service.

Not surprisingly, these service providers, of which Uber is the most well known, are turning the industry upside down and meeting fierce resistance from taxi drivers. By offering better services for less, they are challenging the highly regulated and protected taxi market, which so often over charges passengers. Governments might decide to continue protecting such an industry for fear of massive retaliation through strikes and road blocking, but we believe it would be better off buying back licenses whenever applicable and liberalizing such an industry.

The fact is, the taxi business, as we know it, will disappear first through these hailing services and in the future through the use of unmanned cars. In most cities, unmanned taxi service will cost 70–80 % less and there is no way taxi drivers can resist such a price difference. The taxi drivers' fate will be the same as the elevator attendants; they just don't know it yet!

1.7.3 *Creating the Legal Framework for Unmanned Vehicles*

Government shouldn't resist but promote the advent of the unmanned vehicle era. The benefits for society in terms of accident reduction alone are such that transport authorities should embrace the enabling technologies. Job losses like the ones in the taxi industry will happen but thousands of new jobs and new companies will be created. Benefits to society in terms of health cost reduction, better quality of living for the elderly and the handicapped, reduced pollution are all additional benefits that should motivate governments to push through legislation enabling self-driving cars.

As we will see in Chap. 2, there will be a progression in the level of car autonomy. At first, passengers will stay in command and at the last stage humans won't be involved at all in driving. At each and every step, governments will need to support the initiative by creating the necessary laws but also by investing in the supporting infrastructure, till the private sector takes over.

Driverless cars will happen. The question is how soon? The answer to this question will depend on how fast barriers to adoption can be successfully broken down.

1.7.4 *Barriers to Adoption*

One of the main barriers to adoption is cost.

> The 2014 price of a self-driving system from Google that reached around $150,000 will need to be lowered to a range between $10,000 and $20,000 to be installed in the high end automotive market. To be installed in mid size cars, the extra cost will need to be lower than $5000. The big car manufacturers are working together with their OEMs to find ways to lower cost by miniaturization and mass producing equipment that is still custom-made, such as Lidar or lasers. According to IHS Automotive forecasts[4], the $3000 threshold should be achieved by 2030. At that price, self-driving technologies will permeate the entire automotive supply chain allowing for its adoption by the global mass market.

Another barrier to adoption is legal.

> We believe that driverless vehicles are much safer than driven vehicles. The proof can be found in the railway industry that has been able to reduce

---

[4]IHS Automotive forecasts, January 5, 2015 IHS is a registered trademark of IHS Inc.

significantly the number of accidents by getting rid of the drivers or attendants in the control centre. After the elevator and more recently the train environment, the automotive industry will go driver free. This will create fundamental changes in the way the courts judge accidents. Today the likely culprit is one of the drivers involved in the crash. When there will be no driver and since judges cannot charge a robot-car, the logical culpable will most likely be thousands of kilometers from the accident.

This book will give a view on the likely penal responsibilities based on a system approach. It is indeed our view that the likely technological and business model that will prevail for driverless cars will follow the same paths as the train approach, with a separation between signaling, vehicle manufacturing, infrastructure provider, car owner, and operator.

The last barrier for adoption will be the attitude that people have toward robot-cars. Though unmanned trains could already be defined as robots, people don't have such a close relationship as they have with their cars. Thus, we can say that for the first time, humanity will be confronted with machines able to circulate in the open, by their own means.

Many science fiction novels have depicted the resentment of human beings for robot. Luckily for the industry, cars don't look like human beings. We believe that for the older generation, it will come as a shock to hop on a car without any steering wheel or brake pedal. As for unmanned elevators or trains, people will at first feel strange but will quickly get use to their robot-car.

Acknowledgments and disclaimer The use of pictures, or references made to studies or companies and their brand does not in any way suggest that the authors of such studies or the mentioned companies endorse in any way this book or its content. The author endeavors in respecting the copyright subsisting any of the graphics, and texts that he uses, to use graphics and texts he has himself created or to use graphics and texts not covered by copyright. All trademarks and brand names quoted in the book including those protected by copyright of third parties are subject unreservedly to the provisions of current copyright law and the rights of ownership of the registered copyright holders.

Companies and Brands Stated in the Chapter

- Twitter Inc.
- Facebook Inc.
- Ferrari S.p.A.

- BMW AG (Bayerische Motoren Werke AG)
- Nissan Motor Company Ltd
- Renault S.A.
- Toyota Motor Corporation
- Tesla Motors Inc.
- Otis elevators company
- Google Inc.
- Amazon.com Inc.
- Yahoo Inc
- Transmilenio SA
- Waze™ of Google
- Moovit
- Porsche AG
- Uber Inc.

Chapter 2
Risk Adverse Society

Acronyms

AC and DC	Alternating current and direct current
APM	Automated people mover
ATC	Automatic train control
ATO	Automatic train operation
ATP	Automatic train protection
ATS	Automatic train supervision
CAPEX and OPEX	Capital expenses and operational expenses
CCTV	Closed-circuit television
CBTC	Communication-based train control
CALM	Continuous air-interface, long and medium range
DSRC	Dedicated short-range communications
DVR	Digital video recorder
DTO	Driverless train operation
EMU	Electrical multiple unit
ETMS	Electronic train management system
ETD	End of train device
EGNOS	European Geostationary Navigation Overlay Service
ERTMS	European Rail Traffic Management System
ETCS	European Train Control System
EDR	Event data recorder
GPS	Global positioning system
GSM	Global system for mobile communication
ITCS	Incremental train control system
IP	Internet protocol
IMS	IP multimedia subsystem
JPEG	Joint photographic experts group
LDW	Lane departure warning
LMA	Limit of movement authority
LTE/4G	Long-term evolution/fourth generation
MADD	Mother Against Drunk Driving
MPEG	Moving Picture Experts Group

S. Van Themsche, *The Advent of Unmanned Electric Vehicles*,
DOI 10.1007/978-3-319-20666-0_2

MSAS	Multi-functional Satellite Augmentation System
NDGPS	National Differential GPS
NHTSA	National Highway Traffic Safety Administration
NTSC/PAL	National Television System Committee/Phase Alternating Line
OCC	Operational Control Centers
OCR	Optical Character Recognition
PRT	Personal rapid transit
PTC	Positive train control
PFD	Probability of failure on demand
RRF	Risk reduction factor
SIL	Safety integrity level
STO	Semi-automatic train operation
SDP	Service delivery platform
SOA	Service oriented architecture
SIP	Session Initiation Protocol
UTO	Unattended train operation
UMT	Universal Mobile Telecommunication
UPnP	Universal Plug and Play
VCC	Vehicle control center
VCU	Vehicle control unit
VII	Vehicle infrastructure integration
VOBC	Vehicle on-board controller
V2C	Vehicle-to-cloud
V2I	Vehicle-to-infrastructure
V2V	Vehicle-to-vehicle
VANET	Vehicular ad hoc network
VoIP	Voice over IP
WAAS	Wide Area Augmentation System
WAVE	Wireless Access in Vehicular Environments
WLAN	Wireless Land Area Network
WHO	World Health Organization
Wi-MAX	Worldwide Interoperability for Microwave Access

2.1 Introduction

We've used the expression "risk adverse society" to describe several elements and factors of modern society that are growing in importance and creating a durable impact on the way people see things and how they help shape laws. Although these megatrends are seen by many as favorable per se, they have consequences that can create issues for society and will have a strong impact on transportation. Two mega trends play a key role in this new behavior: the graying of society and a society prone to litigation.

On the other hand, an important mega trend which is the increasing role of women in modern society cannot be associated as a key factor for a more risk adverse society. Indeed, two studies have tended to show that women don't seem more risk adverse than men.

A 2008 study[1] on the Kasai tribe of India showed that it is untrue that the average female avoids risky behavior more than the average male. Additionally a 2009 study[2] "Gender differences in risk behavior: Does nurture matter" shows that an "*average girl from single-sex schools are found in their experiment to be as likely as boys to choose the risky behavior. This suggests that observed gender differences in behavior under uncertainty found in previous studies might reflect social learning rather than inherent gender traits*".

Maybe can we only say that women bring more caring to society? Definitely the creation of associations such as Mother Against Drunk Driving (MADD) is a good example of such attitude. Such a program, summed up hereafter has a great impact on transportation policies in the USA:

- Education about the dangers of drunk driving, advocacy and victim assistance;
- Strict policy on illegal blood alcohol content (0.8 % or lower) and using stronger sanctions for offenders;
- Helping victims of drunk driving;
- Maintaining the minimum legal drinking age at 21 years; and
- Mandating alcohol breath-testing ignition interlock devices for everyone convicted of driving while legally impaired.

2.1.1 *Graying of Society*

Steady advances in medical technology, increase in wealth, better diets as well as a wide range of other factors have caused populations around the world to reach old age in growing numbers. Life expectancy of a woman in Japan is now of more than 82 years, while the average world age expectancy is about 67 years according to the

[1]Kasai tribe of India; Author studied by Gneezy, Leonard and List (2008).

[2]Gender differences in risk behavior: Does nurture matter. Author Alison L. Booth and Patrick j. Nolen (University of Essex/Australia) 2009 study.

World Factbook of the CIA[3]. Global population aged 65 and over is set to double by 2050.

In wealthy nations, older people now constitute 15 % of the population but will account for 26 % by 2050. Poor nations historically have had very low percentages of older people, but this percentage is expected to increase from about 6 % now to 15 % in 2050.

At the same time, citizens of all continents, except Africa, are deciding to have fewer children. Several countries are now experiencing a fertility index of less than 2.1, level at which deaths are replaced by births. This growth in old age population cannot compensate for the decline in growth in many rich countries, and thus population in countries such as Japan and Germany is shrinking.

While we all know senior citizens that can be more risk prone than many teens, the reality is that they are on average more cautious, conservative and risk adverse. A study published in 2013[4], by the Yale School of Medicine showed that with increased age comes decreased risk-taking in decision-making.

To add to this scientific analysis, we will use the thoughts of Isaac Asimov, probably the most renowned science fiction novelist. In his first library success in the 1950s, "The Caves of Steel"[5] an earth police officer Elijah Baley, was sent to resolve a murder mystery on Aurora, a planet colonized by explorers whose life expectancy was between 300 and 350 years. Hans Fastolfe, an Auroran politician who was a colleague of the murdered victim expressed his wary of the drawbacks of such a long life to Baley:

"*If you were to die now, you would lose perhaps forty years of your life, probably less. If I were to die, I would lose a hundred fifty years, probably more*".

Horrified by all of this, Baley thought Aurorans were unable to collaborate with one another and too risk-averse, because of their longevity.

We believe that the graying of society will give a formidable push on e-mobility technologies and especially self-driving cars. We know that older generations still have a love affair with their cars, especially men. For those who live in the downtown area, grocery shops, hair dressers, banks, doctors, and all the other services that we take for granted when we are young can still be easily available for

---

[3]World Factbook of the CIA (Public domain). https://www.cia.gov/library/publications/the-world-factbook/index.html.

[4]Study published in 2013 by the Yale School of Medicine (If at Levy, assistant professor in comparative medicine and neurobiology at Yale, and colleagues).

[5]The novel "The Caves of Steel"; Author Isaac Asimov, (1954).

people with eyesight or reflex problems. However, for senior citizens living in suburbs or on the countryside, not being able to drive is a synonym of having to rely on family or friends to get by.

Transport is freedom and not being able to commute means very often having to move into old age residences or at least settling into a more urbanized environment. Having driverless cars would allow senior citizens to delay or even avoid the day when they need to go through such a traumatizing experience. Furthermore, and as we will see, unmanned technology is much safer than driven technology. The baby boomers will plebiscite safer and more reliable transportation means.

2.1.2 Society of Litigation

Already in 1981, A.E. Dick Howard an American specialist in constitutional law suggested in the Wilson quarterly report[6]:

> *we may be well on our way to becoming a "litigation society". The courts have often served as a useful "safety valve" – they led the way in ending de jure racial segregation. But of late, they have tried to resolve an increasing number of social questions that are less susceptible to judicial remedy. The real difficulty, Howard says, may be the breakdown of old sense of community and compromise that led Americans to settle political disputes out of court – in legislatures and party conventions.*

2.1.3 Impact of These Trends on Transportation

The graying of society and the increasing importance of litigation throughout the world have three immediate impacts on transportation:

Safety In Public transport, this has been an important investment driver, creating new technologies, which have reduced drastically the number of fatalities and injuries. Society doesn't accept fate in this regulated environment. Although accident statistics have improved throughout the years in private transport, there are still too many killed and injured people. With the emergence of e-mobility technologies, road fatality numbers will need to drop or there will be massive class actions against car manufacturers or road infrastructure operators.

Security Senior citizens who are very often the most vulnerable people in public transport need to feel comfortable and secured while traveling. Under terrorism threats and other security issues, more and more specific security technologies are

[6]A.E. Dick Howard; Wilson quarterly report (1981).

being developed for public transport, especially CCTV. Here, also society doesn't accept inaction, and public transport authorities will therefore cover all stations and onboard vehicles with IP cameras, which is able through intelligent software to detect abnormal behavior or situations.

System homologation A litigation society always needs someone to blame. Homologating bodies' role is perceived as being the system's ultimate guarantor, creating extra hurdles on system approvals, with huge impact on delivery schedule and cost of all new systems.

We will present in the next sections, the worldwide issues that transportation is facing in regards to safety, homologation and security, and how they are impacted by the e-mobility revolution.

2.1.4 Safety Facts and Figures

Safety is the condition of being protected against physical consequences as a result of failure, damage, error, accidents, harm, or any other non-desirable event. It integrates the control of recognized hazardous situations to achieve an acceptable risk level. It also takes into account protection measures against exposure to an event that could cause health or economical losses. It considers protection of people and assets.

Worldwide accident statistics According to the World Health Organization (WHO)[7], each year nearly 1.3 million people die as a result of a road traffic collision. Worst, more than half of these fatalities aren't even on board the car. The vast majority of these victims die in poor or middle-income countries (less than 8 % die in rich ones). Twenty to fifty million more people sustain non-fatal injuries from a collision, and these injuries are an important cause of disability worldwide. To bring that to a country level, 32,885 Americans died in 2010 of car accidents, while over 2 million people were injured!

Due to a quick increase in car use in developing countries, the WHO is estimating that the death number will skyrocket, more than doubling by 2030. On top of the terrible consequences for the families and the injured themselves, economic losses due to car crashes are catastrophic. Based on an estimated negative impact on the economy between 1 and 3 % of the respective GNP of the world countries, these crashes can generate a total yearly cost for society of over $500 billion (Tables 2.1 and 2.2).

[7]Global Plan for the Decade of Action for Road Safety 2011–2020, produced by the World Health Organization (WHO).

Table 2.1 Description of fatalities according to Private and Commercial Transportation type

Transport mode	Private transportation		
	Crashes solely involving Private users	Crashes with commercial highway carriers	Crashes with trains and rail transit
Cars and light trucks	26.678	3.766	245
Pedestrian and bicycles	4.930	545	592
Motorcycles	3.989	156	2
Other	1.252	–	−2
Total	**36.849**	**4.467**	**837**
Large trucks	Commercial transportation		
	Passengers	Employees	Bystanders
Buses	30	9	
Rail Road	7	27	4
Rail Transit	22	3	–
Other			15
Total	**182**	**883**	**19**

Source Comparing the Fatality Risks in United States Transportation Across Modes and Over Time; Author: Ian Savage; White paper published in Research in Transportation Economics: The Economics of Transportation Safety, volume 43(1), 2013

Table 2.2 US fatalities per billion passenger miles and km (2000–2009)

Passenger fatalities per	Billion passenger	
	Miles	km
Riding a motorcycle	212.57	342.03
Driver or passenger in a car or light truck	7.28	11.71
Passenger on commuter rail and Amtrak	0.43	0.69
Passenger on urban mass transit rail	0.24	0.39
Passenger on a bus	0.11	0.18

Source Comparing the Fatality Risks in United States Transportation Across Modes and Over Time; Author: Ian Savage; White paper published in Research in Transportation Economics: The Economics of Transportation Safety, volume 43 (1), 2013

The white paper[8] from Mr. Ian Savage describes the fatalities from the various transportation modes during the period of 2000–2009. The most important finding is that 94 % of the US total deaths happened on the

[8]Comparing the Fatality Risks in United States Transportation Across Modes and Over Time; Author: Ian Savage; White paper published in Research in Transportation Economics: The Economics of Transportation Safety, volume 43(1), 2013.

Table 2.3 Fatality comparison between the USA and the UK, IRTAD 2012[a]

Measure (2009)	Country	Highways	Rural roads	Urban street	Total
Fatalities	USA	4.122	17.264	12.497	**33.883**
	UK	132	1.423	782	**2.337**
Distance driven (billion km)	USA	1.154	1.191	2.414	**4.759**
	UK	101	224	191	**516**
Distance driven (km) per thousand licensed drivers	USA	5.5	5.7	11.5	**22.7**
	UK	2.8	6.3	5.4	**14.4**
Distance driven (km) per thousand licensed drivers	USA	3.8	3.9	7.9	**15.5**
	UK	1.6	3.6	3.1	**8.3**

[a]IRTAD: Database including accident and traffic data and other safety indicators for 29 countries (2012); http://internationaltransportforum.org/irtadpublic/about.html

American road network. Of these road fatalities, 74 % were car and light truck passenger related. 55 % of the fatalities were involved in crashes with no other vehicle, but occurred when a vehicle rolled-over without a prior collision, stroked a fixed object at the side of the road, an animal or debris in the roadway, or caught fire. Almost 10 % of all fatalities were motobikers, an unreasonably high number considering the low biker percentage. About 15 % of total road fatalities were not occupants of motorized vehicles, but mainly pedestrians.

Car accidents Even though there are several campaigns against drunk driving and for buckling up, about a third of highway fatalities involved in the US at least one car or motorcycle driver impaired by alcohol and almost half the fatalities concerned occupants not wearing a seat belt or using a child safety seat at the time of death. The type of road also has a significant effect on the fatality risk. The following chart, extracted from the IRTAD 2012 study[9], shows fatality values for the USA and the UK (Table 2.3).

Based on this data, roads in rural areas have a fatality risk that is 2.7 times greater than that in urban areas. In general, the lower average speeds, greater provision of lighting, greater deployment of traffic control devices and fewer curves in urban areas more than compensate for factors such as the greater number of intersections and the presence of pedestrians. The safest functional class of roads is the Interstate Highway System. This type of highway has a fatality rate per vehicle km that is about half the US average for all roads.

[9]IRTAD: Database including accident and traffic data and other safety indicators for 29 countries (2012). http://internationaltransportforum.org/irtadpublic/about.html.

Table 2.4 Fatality comparison between the US and a few European countries; IRTAD 2012[a]

Measure	Year	US	Sweden	UK	Netherlands	S.UK.N Total
Fatalities	2006	42,708	445	3298	730	4473
	2007	41,259	471	3058	709	4239
	2008	37,423	397	2645	677	3719
	2009	33,883	358	2337	644	3339
	2010	32,885	266	1905	537	2708
	Average	37,632	387	2649	659	3696
Population (in thousands)	2006–10	303,965	9.188	61,408	16,436	87,032
Vehicles (in thousands)	2006–10	256,944	5.351	34,850	9045	49,246
Vehicle distance driven (million km)	2009	4,758,450	81,444	516,007	126,966	724,417
Vehicle distance driven/person (million km)	2009	15,511	8799	8345	7701	8272
Fatality rate per million people	2009	123.8	42.2	43.1	40.1	42.5
Fatality rate per million motorized vehicles	2009	146.6	72.4	76	72.9	75
Fatality rate per billion km driven	2009	7.1	44.4	4.5	5.1	4.6

[a]IRTAD: Database including accident and traffic data and other safety indicators for 29 countries (2012); http://internationaltransportforum.org/irtadpublic/about.html

> Safety improvement to infrastructure such as elimination of intersections and integration of space or concrete blocks between opposite lanes reduce the frequency of crashes, albeit crashes when they do occur tend to be more severe due to higher speeds. In general, the riskiest types of roads are those in rural areas that do not have a middle division between oncoming traffic.

Table 2.4, extracted from the same study, gives the number of fatalities, population, motor vehicles (excluding mopeds), and vehicle distance driven by country. It shows that the fatality rate is substantially lower for each studied European countries than for the USA.

Compared with road transport, other modes have considerably lower annual fatality counts, even though the totals are still substantial.

Railway and mass transit accidents Public transport figures are much better though not perfect. Fatal train collisions and derailments command most media attention because they are usually spectacular, even though they are infrequent and account for only a small minority of railway fatalities.

Table 2.5 Main cause of accidents on Indian Railways, per type of users; Indian Railway yearbook[a]

Year	Collisions	Derailments	Level crossing accidents	Fire	Other	Total
2001	20	344	83	15	2	**464**
2002	30	279	88	9	8	**414**
2003	16	216	96	14	7	**349**
2004	9	197	95	14	5	**320**
2005	13	136	70	10	3	**232**
2006	9	130	75	15	4	**233**
2007	8	96	79	4	8	**195**
2008	8	100	77	5	4	**194**
2009	13	85	69	3	7	**177**
2010	9	80	70	2	4	**165**

[a]Indian Railway Yearbooks; (2012 study)

In the USA, railroads (which includes in that country a very large network of freight lines) and mass transit claimed an average of 63 lives a year to which we would need to add 65 fatalities involving pedestrian trespassers (2013 Study with data considering more than 10 years of statistics) (see Footnote 8).

The yearly average in this study shows that there were on the American rail transit 22 passenger deaths and 3 employees' death.

To give the reader a rough order of magnitude of casualties in developing countries, we decided to select India. The following charts show the improving evolution of accidents in this country with a huge train network but still with deficient safety systems. Tables 2.5 and 2.6 were extracted from the Indian Railway Yearbooks[10]. It shows that the main cause of accidents is derailments followed by level crossing accidents and collisions.

The table 2.6 from the same yearbooks gives an idea of what are the main causes of the accidents on the Indian network. As expected it clearly indicates that at the origin of an accident, 75 % of the time, a human being was involved, be it is a staff member, a passenger or another person, such as a trespasser or a car or lorry driver.

Tables 2.7 and 2.8 from the same yearbooks show the favorable trend in the number of accidents and the number of casualties and injuries per type of people. If we look at the 2010 figures, we can see that the reduction achieved was significant with only 0.17 accidents per million train × km.

[10]Indian Railway Yearbooks; (2012 study).

Table 2.6 Main causes of accidents on Indian Railways; Indian Railway Yearbook[a]

Cause	2001	2002	2003	2004	2005	2006	2007	2008	2009	2010
Failure of railway staff	284	248	184	161	119	120	85	86	76	63
Failure of other person	109	103	118	107	78	86	84	81	75	75
Rolling stock problem	16	11	6	6	5	1	4	4		3
Track problem	17	13	11	7	7	6	6	3		3
Electrical problem			1	1	2					
Signaling				2		1		1		14
Sabotage	19	14	10	18	4	5	8	7	13	1
Combination of factors	4		2		1		1		4	4
Incidental	11	20	15	15	14	11	6	8	5	2
Other	4	5	2	3	2	3	1	4	4	
Total	**464**	**414**	**349**	**320**	**232**	**233**	**195**	**194**	**177**	**165**

[a]Indian Railway Yearbooks; (2012 study)

Table 2.7 Train accident evolution on Indian Railway (2001–2010): Indian Railway Yearbook[a]

Year	Accidents per million train km s in %
2001	0.65
2002	0.55
2003	0.44
2004	0.41
2005	0.29
2006	0.28
2007	0.22
2008	0.21
2009	0.19
2010	0.17

[a]Indian Railway Yearbooks; (2012 study)

Bus accidents The following table 2.9 from the US department of transport shows that buses have a great safety track record. It also shows that the trend in fatalities is favorable but might have hit a threshold at about 275 fatalities per year.

Public versus Private transport track record The clear conclusion from all these numbers is that public transport is much safer than private transport. In fact, any American has a 15 times higher probability of dying in his car than on board a train or a bus. The odds of dying in a motorcycle accident are just staggering and it is surprising that in a country prone to litigation, no class action has yet been done against motorcycle manufacturers.

Table 2.8 Number of accidents on Indian Railways, per type of users; Indian Railway Yearbook[a]

Year	Passengers		Railway employees		Others	
	Killed	Injured	Killed	Injured	Killed	Injured
2001	55	281	8	27	153	175
2002	114	595	14	38	168	175
2003	157	658	29	45	232	279
2004	84	279	3	28	155	159
2005	35	86	5	8	181	209
2006	168	483	9	31	138	113
2007	38	227	6	24	164	151
2008	9	245	10	26	172	135
2009	52	257	12	22	145	165
2010	67	253	4	9	167	135

[a]Indian Railway Yearbooks; (2012 study)

With such a poor track record, what could the automotive industry learn from the railway or bus industries? Well, as we will see they can go driverless. By doing this, the number of accidents and fatalities will drop drastically, probably to a level comparable to the other transportation means.

2.1.5 Security

Security definition It is the condition resulting from the protection against deliberate action of another human being to harm someone else or damage intentionally property. It integrates the control of hazardous situations involving malevolent acts to achieve an acceptable risk level. This takes into account the protection measures against exposure to someone who causes health or economical losses, including protection of people and assets.

Nature of criminality The nature of crimes to be committed within public transport is quite diverse. It includes various threat categories to which several scenarios can be associated:

- **Crime**: pick-pocket, violence, prostitution, drug trafficking;
- **Terrorism**: conventional bombs, dirty bombs (biological, chemical, nuclear), stabbing, poisoning;
- **Fire**: Arson, person on fire;
- **Sabotage**: infrastructure, switching equipment, cable theft;
- **Vandalism**: graffiti, destruction; and
- **Disorder**: Hooligans, aggressive behavior, racial harassment.

Fear of crime Fear of crime is an important issue for the rail and bus industry. Passenger growth, and the general health of the industry, could be undermined if

Table 2.9 Analysis of bus crashes; US Department of Transport 2010[a]

Year	Number of buses registered	Fatal crashes involving buses	Buses involved in fatal crashes	Occupant fatalities	Total fatalities in bus crashes	Million vehicle miles traveled	Rates per 100 million vehicle miles traveled by buses		
							Number of buses registered	Fatal crashes involving buses	Fatalities in bus crashes
2000	746,125	323	325	22	357.00	7590	4.26	4.28	4.70
2001	749,548	289	292	34	331.00	7070	4.09	4.13	4.84
2002	760,717	274	274	45	331.00	6845	4.00	4.00	4.84
2003	776,550	288	291	41	337.00	6782	4.25	4.29	4.97
2004	795,274	276	279	42	315.00	6801	4.06	4.10	4.63
2005	807,053	278	280	58	340.00	6980	3.98	4.01	4.87
2006	821,959	303	305	27	337.00	6783	4.47	4.50	4.97
2007	834,436	280	281	36	325.00	14,516	1.93	1.94	2.24
2008	843,308	251	251	67	311.00	14,823	1.69	1.69	2.10
2009	841,993	221	221	26	254.00	14,387	1.54	1.54	1.77
2010	846,051	245	249	44	276.00	13,789	1.78	1.81	2.00

[a]US Department of Transport's 2010 analysis of bus crashes; FMCSA Analysis Division/Large Truck and Bus Crash Facts 2010

stations were to become places which people would rather avoid. Security measures such as implementing CCTV systems clearly reduce malevolent acts.
On the contrary, in private transport, the type of crime is rather limited to carjacking and auto theft. With self-driving technology, we will see that cyber terrorism will become possible in the future.

2.1.6 Homologation

Homologation definition It is the granting by an authorized organization to operate or sell a given product or system, according to safety and other technical requirements.

Homologation bodies play mainly two important roles:

- Making sure that a product or system meets the specific technical requirements or the general technical requirements, as indicated in a standard; and
- Giving the authorization to operate under safety conditions.

The first role can be considered a traditional role of homologating bodies, whereas the second is quite new and becoming increasingly important.

Railway standards The railway market is highly regulated. The vast majority of products and equipment applied in such environments must meet many stringent standards. To give the reader an idea of these norms, we've indicated some of the most important railway standards:

- EN 15 227: Requirements of crash safety;
- EN 15 085: Welding of railway vehicles;
- EN 45 545/DIN 5510 EBA: Guideline for fire protection requirements;
- EN 50 155: Electrical equipment, requirements on hardware;
- EN 61 508: Functional safety, electrical systems;
- EN 50 121: Electromagnetic compatibility (EMC);
- EN 50 126: RAMS LCC Management; and
- EN 50 128/EN 50 129: Software/safety proof.

Authorization to operate In a risk adverse society, governments and administrative bodies want to be sure that a metro or train will operate safely under any condition. More importantly, they want to be sure that if any accidents happen, there will be someone to blame. Till a few years ago, these homologating bodies were there mainly to review the information supplied by the engineering or manufacturing companies. However, they are now being seen as the guarantor of the system safety. Not only have these companies been prosecuted in case of accidents, but also their employees have been personally liable to criminal inquiries.

This simple change of status has created huge bottlenecks in the approval process, as their employees now request much more details before approving personally safety related solutions. This has also had a drastic impact in the volume of documents that must be supplied before approval. As a consequence, delivery schedule of European train suppliers haven't been met in the last years. In fact, this movement towards more stringent homologation has been estimated by some specialists as adding as much as one to two years to deliveries. A consequential damage of this change in status and role is that important extra costs have been added to manufacturers or system integrators in an industry known for very low margins. Unless the industry is able to change its processes to adapt to these new demands, the extra costs will be passed to society through higher taxes or more expensive fares.

Safety improvements do not come cheap but whenever new accidents happen, there will always be public outcry for more safety measures. As one of our nine principles is that all transportation means compete against each other, all these measures are pricing out trains in regards to cars or buses. Luckily two important elements should help the railway industry. First and as we will see in the next section, the railway industry has shown resilience to new safety regulations throughout its history and has a strong embedded safety culture. Second and probably more importantly, in a litigation society, no industry can go on forever allowing millions to be killed or injured. The day all vehicles are unmanned, the same safety concepts, and stringent regulation will be applied to the automotive and bus industries, leveling the playing field.

2.2 Safety Concepts

Safety of passengers has always been a key concern of public transport authorities. No politician or operator's President in his or her right mind can show complacency when the life of passengers is involved. On the other hand, car and motorcycle manufacturers have mostly been able to escape costly legal battles, except when class actions were able to prove that the deaths or accidents were due to equipment malfunction. In cases where accidents were due to poor road state or bad signaling, infrastructure owners have been able to escape such litigation, because of the difficulty of proving that such accidents were caused by malfunctioning infrastructure. How can someone pushing his or her way into a train through partially closed doors successfully sue the metro operator for any injuries suffered by his or her act, when someone hitting an icy patch on the road can't?

The reality is that there is a double standard measure. Private transport industry can invoke the act of God principle or the "he took his chance" syndrome. Public transport and especially the train industry must on the contrary plan for all types of conditions, even if this means increasing tremendously the cost of trains or its railway infrastructure, even though the probability of such event occurring is negligible. As we've just seen, the safety homologation process is becoming increasingly complex, lengthy, and thus expensive, reducing the competitiveness of the railway transport means in regards to private transport. Imagine if car manufacturers had to build in fail-safe systems to ensure braking under any conditions, or equipment redundancy to ensure that any equipment malfunction would automatically be taken over by substituting equipment.

Everyone within the private transport chain, be it manufacturers, infrastructure operators as well as private car owners can still invoke the "fate" excuse to avoid being prosecuted. On the other hand, under the megatrend of a risk adverse society and under the influence of a litigation society, laws have been getting tougher for railway or bus employees responsible for gross negligence involved in accidents. Recently, railway employees responsible for accidents have been trialed and arrested if found guilty. It has not always been that way. At the beginning of the era of train transportation, risks were part of train travel. Today even a small derailment without any onboard passenger is likely to make the newspaper headlines. Is the private transportation going that way?

The first signs of going in that direction are there. Under the assaults of the litigation society, drunk or over-speeding drivers must now respond rightfully so for their acts criminally. Things are changing and we believe that with the new e-mobility technologies, the day isn't far where any car accident involving casualties will be investigated and the responsible party be it manufacturer, infrastructure, driver or car owner, prosecuted. With the emergence of new technologies allowing for driverless cars, this trend will accelerate. In order to understand how e-mobility is likely to influence safety in the automotive industry, let's take a look at the evolution of safety in the railway industry. We could ideally learn how railway safety could help shape a safer automotive environment or at least see how the automotive industry is likely to evolve from a safety perspective in the coming years.

2.2.1 Railway Safety Concepts

Trains being guided by fixed rails are uniquely susceptible to collision, unlike cars or buses. Furthermore, trains cannot stop as quickly as automobiles, and frequently operate at speeds that do not enable them to stop within the driver's sighting distance. Thus safety was from the start one of the top priorities of any new train operation, though fate was at the start an inherent part of the journey.

In the early days of the railway adventure, horse mounted flagmen preceded trains. Hand and arm signals were then used to direct train drivers. To compensate for poor visibility conditions (i.e.: fog, rain, snow), wayside attendants introduced in the 1830s elevated flags and lanterns, which could be seen from far away. However, many accidents still occurred. So much so, that in 1889, England introduced the first safety regulation (Regulation of Railways Act), which forced the application of interlocked block signaling. Almost simultaneously, the USA introduced similar regulation.

Regulation evolved and new technologies were developed, but the three main safety principles that were established at the end of that century are still applied today:

- Locking a section of the railway;
- Making sure that no trains occupy such section; and
- Insuring that the system's integrity is maintained throughout operations.

Block interlocking It is a very simple concept: under normal circumstances, railway sections—known as blocks—can allow only one train at a time. Thus, trains cannot collide with each other if they are not permitted to occupy simultaneously the same section of tracks. However, this is only a certainty if the trains are spaced far enough apart to ensure that they cannot collide, taking into consideration factors such as speed, braking time, etc.

In the early railway days, signalmen were responsible for ensuring that any switch was set correctly before allowing a train to proceed. Early interlocking systems used mechanical devices both to operate the signaling appliances and to ensure their safe operation. Beginning around the 1930s, electrical relay interlocking was used and since the late 1980s, new interlocking systems have tended to be electronic based.

Block signaling This is the second safety principle: signaling to the next train driver (or an unmanned train today) that no train is in that section.

At the beginning of the railway adventure, employees were standing at intervals along the line with a chronometer and used hand signals to inform train drivers that a previous train had passed, as well as when it did so. However, the signalmen couldn't know whether the train had cleared the line ahead, so if this preceding train had stopped for whatever reason, the following train driver would have no way of knowing it before it was too late. Even though drivers were expected to slow down when a train had passed very recently, accidents were frequent then. With the use of telegraph, it suddenly became possible for signalmen in a station or at an interlocking tower to send a message to confirm that a train had passed and that a specific block was cleared. This safety enhancement became known as the "absolute block system". Fixed mechanical signals began replacing the hand signals from the 1830s. When a train passed into a block section, signalmen would protect that block by setting mechanically the signal to "occupied". When an unoccupied message was received, signalmen would move the signal into the "clear" position.

With the 1989 legislation, Great Britain made mandatory the use of block signaling together with interlocking for all passenger railways. Based on these two principles, railway engineers came up afterward with new developments to increase safety on the wayside, on board the train and at the controllers' level in the station, and more recently within the control centers.

Integrity The next safety issue railway engineers tried to solve was that even though a train might have left a block section, there was always the possibility that a few wagons had separated from the front locomotive. This safety problem which can be described as the "carriage integrity" principle was especially critical for long freight convoy. The other integrity issue that needed to be solved was the integrity of the railway tracks. In long distance freight operations like the ones in North America, freight trains could travel through forest for thousands of miles without meeting anyone. Being able to identify that the tracks weren't broken was a key issue.

2.2.2 Safety Procedures

Things don't always go as planned. To address issues coming from abnormal or emergency situations, the railway engineers came up with safety operational mode, also known as degraded modes.

Degraded mode safety principles: Under normal weather conditions (not under misty or rainy conditions) trains were allowed at the low speed of 30 km/h, a velocity judged sufficiently low to enable safe braking during sight driving, to overpass signals indicating the line ahead was occupied. This operating mode was called "permissive block". In order to split or join trains together or even rescue failed trains, the operator had to allow multiple trains to enter in an absolute block. In giving authorization, signalmen needed to ensure that drivers knew precisely what to expect ahead, and act in a safe manner. In such situation, the signal would remain at "danger" and the driver could pass that signal at low speed.

2.2.3 Interoperability

Interoperability is a newer concept not related to safety but more to economics or to a political vision. Convergence of rail networks is promoted by governments. Countries seek to build interoperable networks, and international organizations seek to build macro-regional and continental networks. For instance, the EU has set out to develop interoperable freight and passenger rail networks across its entire area, and is seeking to standardize gauge, signaling, and electrical power systems.

Although interoperability decision isn't made for safety reasons, it has implication for safety. Once the decision is made to integrate various elements of a system, these elements need to work together in a safe way. Physical but also conceptual interoperability needs to be sought to integrate various rail networks.

The most obvious railway interoperability parameter is the gauge similarity. Rolling stock on the network must have wheel sets that are compatible with the gauge, and therefore the gauge is a key parameter in determining interoperability between two lines. There are many other parameters, such as electro-magnetic compatibility, voltage, compliance with control system parameters such as the signaling system, axle load, loading envelope, etc.

2.2.4 WaySide Safety Technologies

Railway engineers came up with several technologies to solve the issues of signaling, interlocking, and integrity. On most railways, physical signals are erected at the line side to indicate the drivers whether the line ahead is occupied and to ensure that sufficient space exists between trains to allow them to stop.

Mechanical signal Older forms of signal displayed their different aspects by their physical position.

The earliest types comprised a board that was either turned face-on and fully visible to the driver, or rotated so as to be practically invisible. While this type of signal is still in use in some countries (e.g. France and Germany), by far the most common form of mechanical signal worldwide is the semaphore signal (introduced as way back as 1842). This comprises a pivoted arm that can be inclined at different angles. Each angle would define a level of risk.

Color light signal On most modern railways, color lights have largely replaced mechanical signals as they can display the same aspects at nights and days and require less maintenance.

Although signals vary widely between countries, and even between railways within a given country, a typical aspect would be (as on roads):

- Green: run to the next line speed;
- Yellow: Prepare to find next signal displaying red; and
- Red: Stop.

The following two technologies were used mainly to help identified cleared blocks:

Axle counter Using axle counting is a simple method of determining the occupied status of a block. These devices located at the block's beginning and end count the number of axles entering and leaving. If the numbers accounting for axles leaving and entering the block are the same, then the block is assumed to be clear.

Track circuit Its principle is a relatively simple alternative: the rails at either end of each section are electrically isolated from the next section and an electrical current runs through both running rails at one end. A fail-safe relay at the other end is connected to both rails. When the section is unoccupied, the relay coil closes the electrical circuit, and is thus energized. However, when a train enters the section, it short circuits the current in the rails, and de-energizes the relay, indicating the same token that the block is occupied. This method does not explicitly need to check that the entire train has left the section. If part of the train is left in the section, this part will continue to be detected by the track circuit.

This type of circuit is also used to detect the absence of trains, both for the purpose of setting the signal indication and for providing various interlocking functions—for example, not permitting switching points to be moved when a train is standing over them. Electrical circuits are also used to prove that points are in the appropriate position before a signal supervising them may be cleared.

Track integrity check Rail breaks are a serious threat to operation and to passengers. Another interesting feature of track circuit is that it can automatically detect some types of track defect such as a clear broken rail, though cracked rails cannot be identified, as current can still pass through. Many companies are working on alternative technologies such as ultrasonic or electromagnetic broken rail detector or similar other technologies. All these technologies have in common that they send a flow in the rail which measures any variation in regards to a normal response time or pattern. These systems are installed either on the trains or on maintenance machines.

2.2.5 Fixed, Semi-Fixed, and Moving Block Principles

As we've seen, railway engineers came up with the concept of block which forbids the entry of a new train before the previous train exits such block. Without changing this safety principle, the engineers worked throughout the years to shorten the spacing between the blocks. Until recently, this spacing was fixed along the tracks and depending on the operational mode, its length was more or less important. With the need to transport thousands of commuters during peak hours, spacing became critical because reducing it meant increasing significantly the operational capacity.

Headway Space between two vehicles, running on the rails, influences directly the system's capacity. It is a direct consequence of the minimum stopping distance required to bring a train to a complete stop. The stopping distance is the sum of the distances a train will run due to the driver's perception and reaction time, and the distance a vehicle will travel from the point when its brakes to when it comes to a complete stop. It is primarily affected by the vehicle's speed and the coefficient of friction between wheel and rail surface.

The railway industry came up with the concept of headway, which translates distance into time. The minimum safe headway measured tip-to-tail (from the front to the end of a train) defined by the braking performance is on a flat section (on a slope, we would need to take into consideration the gravity impact):

$$T_{\text{min}} = L/V + t_{\text{r}} + P_{\text{r}} + kV/2\left(1/a_{\text{f}} - 1/a_{1}\right);$$

where:

T_{min} minimum safe headway time in minute;
L length of the vehicle;
V speed of the vehicle;
t_{r} reaction time;
P_{r} perception time;
a_{f} maximum braking deceleration of the following train in m/s^2;

a_l maximum braking deceleration of the leading train in m/s^2; and
k arbitrary safety factor superior or equal to 1.

Let's calculate the minimum theoretical headway, using realistic values for a light metro operation, with a driver. We can select a 100 m long EMU composed of 4 cars train-set. A speed of 80 km/h (22 m/s) is typical of metro operation's maximum speed, running between two stations. 1.5 and 0.5 s for reaction time and perception respectively can be considered accurate values.

For a metro car of this length, the average braking deceleration would be around 1.2 m/s^2, but we could use an emergency braking value of around about 1.38 m/s^2. A higher value of 1.5 m/s^2 is possible but could cause injuries to passengers who would fall. As railway safety rules require that the minimum distance considers absolute distances (also called "brick wall" stopping), this means that the speed of the leading train shouldn't be considered. With a safety factor of 1.5, the minimum headway time would be

$$T_{\text{tot}} = 100/22 + 2 + 1.5 * 22/2 * (1/1.38) = 18.5\,\text{s}.$$

With an unmanned system, the 2 s due to perception and reaction time would be eliminated.

Based on this 18.5 s, the braking distance for such a train configuration would be around 230–250 m.

Calculating operational headway is much more complex than what we've just done, as it requires taking into consideration other safety factors such as overlapping distance, as well as integrating an intermediary signaling for safety reasons and the dwell time at stations. To complicate matters even more, some metro operations can consider up to three-signal spacing (the following graph (Fig. 2.1) shows two-signal spacing).

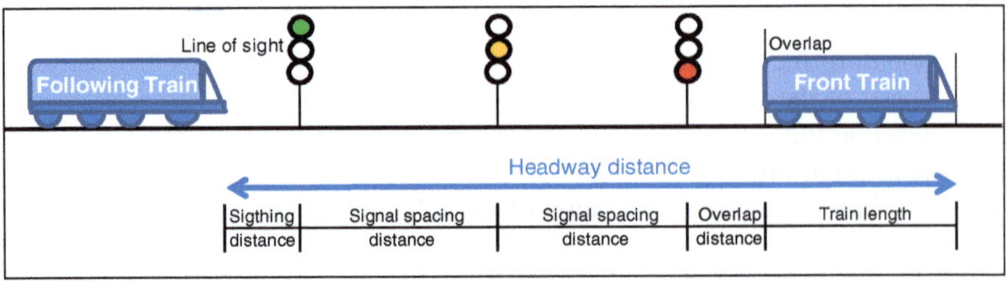

Fig. 2.1 Headway distance considering two-signal spacing. *Source* Author

Fixed block As seen, a fixed block signaling solution allows a single train to occupy a specific block of the railway line. As a train or metro moves along the track, it will occupy blocks which prevent another train from entering that area; mechanical or color light signals provide information to the driver on available blocks and routes. Railway engineers calculate the blocks' length using speed, grades, stopping areas, operational properties of the train, and location of physical elements such as switches and stations. Once the blocks' length has been calculated, signals are physically placed on the line to instruct drivers on scheduling and routing information.

For many lightly operated lines, fixed block length isn't an issue. For these non mass transit lines, railway designers usually favor positioning the blocks at the start and end of the stations. The following operational modes still use mainly such positioning:

- **Timetable**: trains following a schedule; and
- **Train order**: messages sent from a central dispatcher to line side station operator and train crews (regarded as immutable command orders), which control and protect train movements.

However, when system capacity starts to be an issue, as in mass transit operations, designers usually put the blocks with the start and end of the signal levels. The lengths of blocks are then designed to allow trains to operate as frequently and quickly as safely possible. A lightly used line might have blocks many kilometers long, but a busy commuter line might have blocks only a few hundred meters long.

As indicated, the previous headway formula would need to integrate the overlap, additional spacing distances, and the dwell time at stations (necessary time for passengers to hop on and off the metro). When all these distances and additional safety time are factored in, the minimum headway time, in which metros can operate safely with fixed block technology, is around 90 s. As we will see for many mass transit operators, this isn't good enough and this is why technology providers have been trying to introduce a newer technology.

Semi-moving block technology In the early twenty-first century, companies tried to adapt fixed block technology to enable more and more trains to operate on the heavy commuter and metro lines, on shorter spacing distance. With blocking spacing calculated to their bare minimum, operating systems could almost simulate the performance of moving block technologies. However, technology providers are now all proposing moving block technology, which changes the concept of spacing.

Moving block technology One disadvantage of having fixed blocks is that the faster trains are allowed to run, the longer the stopping distance, and therefore the longer the blocks need to be. This obviously decreases the line's capacity (Fig. 2.2).

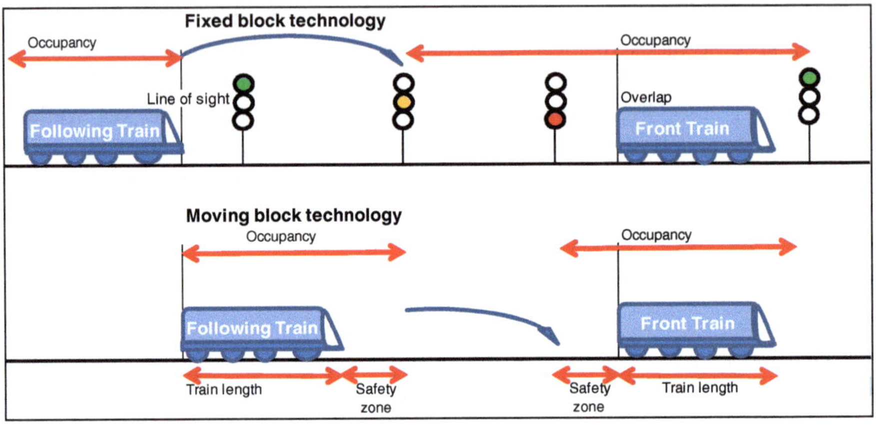

Fig. 2.2 Distance comparison between fixed and moving block technologies. *Source* Author

The spacing concept of moving block system doesn't take into consideration two fixed points on the line but rather a safety zone around each moving train that no other train is allowed to enter. Similar to the calculation of fixed block length, this safety zone is calculated using speed, grades, and the operational properties of the train. An additional buffer zone is provisioned around the train, which can vary depending on the location of the vehicle and its surrounding elements.

Normal operating and worst-case scenarios are simultaneously considered when calculating this safety zone. Normal braking rates are used for calculating safety stopping distances. Other factors integrated in the calculation are positional uncertainty, runaway propulsion, and delays in brake application. A guaranteed emergency braking rate is also applied.

This safety zone, which is calculated in real time by a vehicle on-board controller (VOBC) and centralized computers, depends on knowledge of the precise location, speed, and direction of each train, which is determined by a combination of several sensors:

- **Passive markers along the track**: Track-based transponders are activated by a low-frequency signal, which receives its energy from a passing train and transmits data to the train. Typical data transmitted by fixed balise typically includes the location of the balise, the geometry of the line, such as curves and gradients, as well as any speed restrictions;
- **Active markers along the track**: Track-based transponders are powered from the signaling supply that continuously sends packets of information to passing trains;
- **Train-borne tachometers**: This technology is based upon the principle of detecting rotational wheel speed by means of a pulse produced by a transducer and calculating any deviation of distance between two markers;

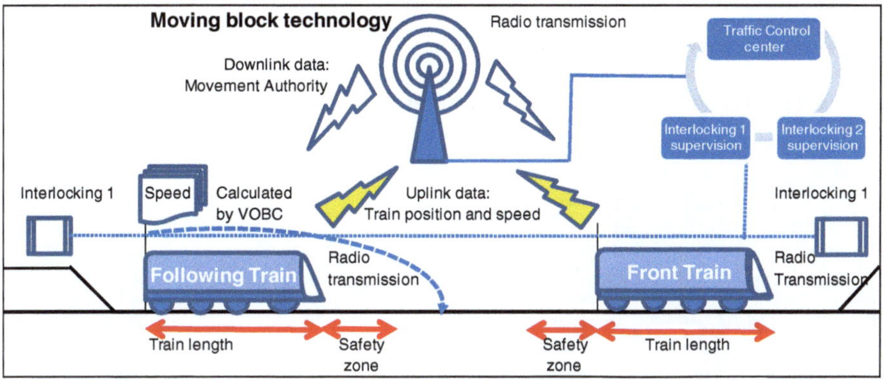

Fig. 2.3 Illustration of moving block transmission technology. *Source* Author

- **Speedometers such as GPS systems:** They cannot be used alone, because they don't work in tunnels (see section on GPS for more technical information); and
- **Wayside signals:** With moving block technology, they are unnecessary.

It is important to mention that the occupancy calculated in these systems must include a safety margin for location uncertainty added to the length of the train. This safety margin depends on the accuracy of the train's odometers and sensing devices.

On a moving block system, the line is usually divided into sections, managed on the wayside by one interlocking equipment. Each interlocking equipment is controlled in real time by a computer at the control center level. As shown in the following illustration (Fig. 2.3), this control center receives and downloads data to all fleet trains. In such a system, the train position and its braking curve are continuously calculated by the train's VOBC, and then communicated via radio to the wayside equipment.

Based on the speed and train position data received from the trains, the wayside equipment is able to establish protected areas, called Limit of Movement Authority (LMA), up to the nearest obstacle (i.e., the front train, or whatever obstacles, such as an end of line) (Fig. 2.4).

2.2.6 *WaySide Interoperability Technologies*

Things that aren't interoperable per nature cannot work together.

Track gauge This was one of the main reasons why track gauge was originally so different from one country to the other, as many countries built different systems in order to avoid invasion, limit freight competition from neighboring countries, or simply adopted the gauge system from their colonial power.

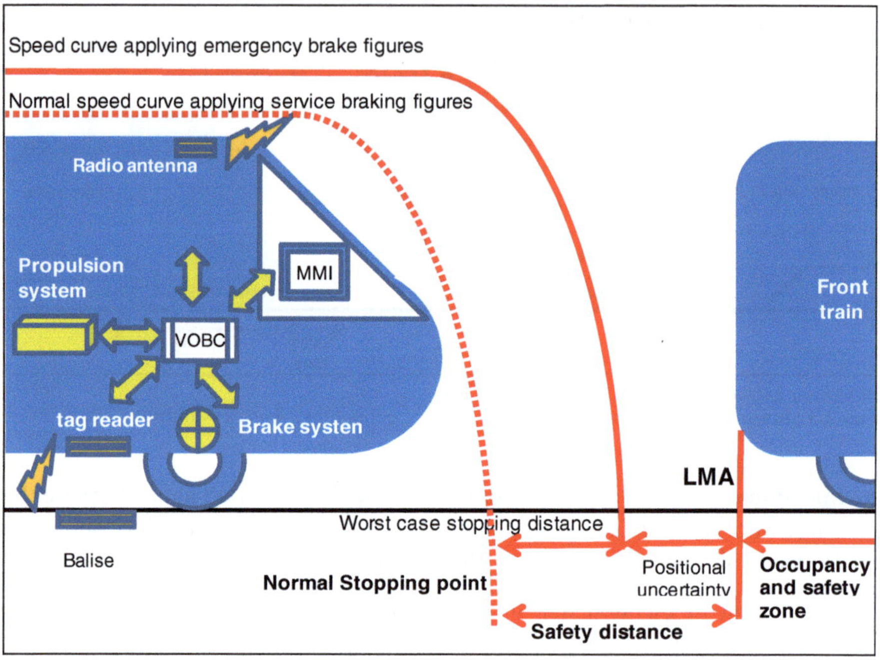

Fig. 2.4 The safety distance concepts for moving blocks. *Source* Author

As incredible as it seems, throughout history many different track gauges could be found, for instance 0.6, 1 m (small gauge), 1.6 m (large gauge). A 3 m gauge track was even contemplated (the Breitspurbahn) by Adolph Hitler's Nazi regime. Today all new built lines tend to adopt the standard gauge, which is 1.435 m system. Interoperability issues can be huge in some areas such as Spain, as one can find small, large and standard gauges alongside. No wonder that for interoperability purposes, Spanish manufacturers introduced in 1968 a bogie with variable gauge wheelsets (variable axle). Such variable gauge technology allows trains to travel across a break of gauge and as the train passes through the gauge changer, wheels are automatically unlocked, moved closer together, or further apart, and are then re-locked.

Overhead electric system Electric trains collect their current from an overhead line system by pressing a device called pantograph against the lowest wire of an overhead line system (contact wire). Metro lines use mainly a brush that is in contact with a connector called third rail. The current collectors are electrically

Fig. 2.5 Picture of a tramway's pantograph in contact with a catenary line. *Source* Author

conductive and allow current to flow through to the train or tramway and back to the feeder station through the steel wheels on one or both running rails (Fig. 2.5).

Alternative electrical power transmission schemes for trains include ground-level power supply, batteries, and electromagnetic induction.

There are two main electrical power transmission technologies: AC and DC. Each one may present different voltages and in the case of AC, different frequencies as well.

Example of AC and DC technologies

- 600 V DC;
- 750 V DC;
- 1500 V DC;
- 3 kV DC;
- 15 kV AC, $16^2/_3$ Hz (16.7 kz);
- 25 kV AC, 50 or 60 Hz.

Although there are always some exceptions, AC lines are mainly used for main line trains and DC for commuter and metro operations.

Signaling system interoperability Signaling interoperability is mainly an issue on main railway lines (but also sometimes on commuter lines of huge city commuter network). Indeed on metro networks, lines operate independently from one other.

For main lines, the EU specified the use of a standardized signaling system called European Train Control System (ETCS). It is a signaling, control and train protection system designed to replace the many incompatible safety systems currently used by European railways, especially on high-speed lines. ETCS requires standard trackside equipment and a standard controller within the train cab. In its final form, all wayside information is passed to the driver electronically, removing the need for signals which, at high speed, could be almost impossible to see. It is now a requirement that all high-speed trains within the EU adopt ETCS and many high-speed networks outside the EU have also adopted it.

ETCS is specified at four different levels:

Level 0: ETCS-compliant locomotives or rolling stock interact with wayside equipment that is non-ETCS compliant;
Level 1: ETCS and non-ETCS equipment co-exist on the same line;
Level 2: a dedicated system with wayside train integrity supervision; and
Level 3: works like Level 2, but train integrity is controlled by onboard equipment monitoring distances between trains constantly and fluidly.

Other wayside interoperability issues Most railway lines have different clearance requirements, as well as axle load issues (maximum weight per axle that the line can support). This complicates the life of railway manufacturers who need most of the time to build tailor made equipment for each railway and metro operations.

2.2.7 Train Integrity Technologies

Carriage integrity: In the early railway days, when a train left a block, the driver needed to inform the signalman controlling the block entry, but had no way of knowing if the last convoy's wagon was still attached. This is why railway engineers came up with a procedure: even if the signalman received the message that the previous train had left a block, the signalman of the next block had to make a visual contact to check if he could see the 'end-of-train marker' on the last vehicle, confirming the train's integrity and allowing the next train to enter the block. The end of train marker was usually a red colored disc by day and a red electric lamp by night. In case he couldn't see it, the signalman would ask the next signal box to stop the train and investigate.

End of Train markers Today, the "end-of-train marker" used in freight convoys, usually called end of train device (ETD), is a self-energized electronic device. Introduced in the 1980s under the drive to lower operational cost, it can be divided into two categories: flashing red light and radio-based ETDs, sending back a message that the end wagon occupies the same position in regards to the head locomotive. Any significant change in distance will automatically give a message to the crew that there is a problem.

For passenger trains, the carriage integrity is today performed by logical sequencing done by the computer-based vehicle system called vehicle control unit (VCU). These VCUs, before entering in operation, will check the vehicles integrity and even establish the train's configuration.

2.2.8 Train Protection Technologies

The consequence of a train driver failing to respond to a signal's indication can be disastrous. As a result, various auxiliary safety systems have been developed. Any such system will necessitate the installation of onboard equipment to some degree. Some of these systems only intervene in the event of a signal being passed. Others include audible and/or visual indications inside the driver's cab to supplement the wayside signals. Some systems act intermittently (at each signal), but the most sophisticated systems provide continuous supervision.

Cab signaling This system communicates track status and driving position data to drivers present in the train cab. The simplest systems display trackside signal aspect, while more sophisticated systems also display allowable speed and dynamic information about the track ahead.

> Cab signaling systems range from simple coded track circuits, to transponders communicating with the cab, and communication-based train control (CBTC). In modern systems, a train protection system is usually overlaid on top of the cab signaling system to warn the driver of dangerous conditions, and to automatically apply brakes and bring the train to a stop if the driver ignores the dangerous condition.

Train protection systems All types of train protection systems are based on the desire to eliminate possible driver errors resulting from failing to obey to a visual displayed line side or to an in-cab signal instruction. The development of such technology on main line railways began with the introduction of warning systems and subsequently progressed to instruction enforcement issued by these systems. There are three main train protection systems that will work closely together to maintain a train within a defined tolerance of its timetable: ATO, ATC, and ATP. The combined systems will marginally adjust operating parameters such as the ratio of power to coast when moving and station dwell time, in order to bring a train back to the timetable slot defined for it.

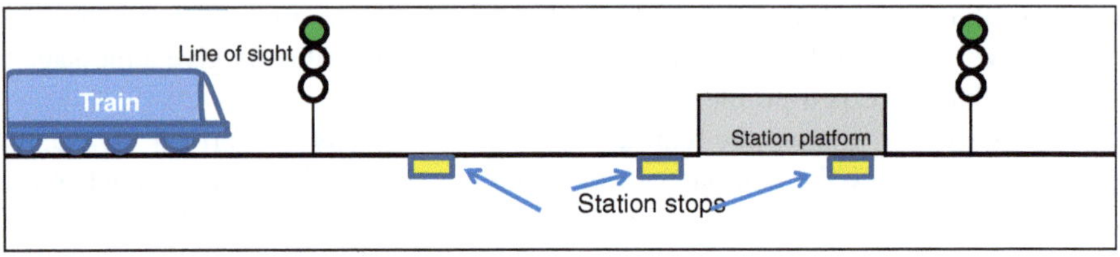

Fig. 2.6 ATO principles. *Source* Author

Automatic Train Operation (ATO) The earliest ATO system on a full metro line was introduced in 1961 on the Barcelona Metro line 2. ATO is the non-safety part of train operation related to station stops and starts. The basic requirement of ATO is to tell the train approaching a station where to stop, so that the complete train is in the platform, after receiving confirmation from the ATP that the line is clear. The sequence operates as shown in Fig. 2.6.

The train approaches the station under clear signals so it can do a normal run-in. When it reaches the first balise (originally a looped cable now mostly replaced by fixed transponders), a station brake command is received by the train. The VOBC calculates constantly and updates the braking curve to enable it to stop at the correct point.

Many operators using ATO systems decide to maintain drivers to mitigate risks associated with failures or emergencies.

Automatic Train Control (ATC) Many modern systems integrate ATC systems that carry out normal signaling operations such as route setting and train regulation. The ATC hardware is divided into two parts:

- **Vehicle on-board control** (VOBC) located on the vehicle and composed of a dual-processor computer that continually monitors the position, speed, and general status of the train; and
- **Vehicle control center** (VCC) located in the operations and maintenance center that directs the train movement, via the VOBCs of a portion of the fleet. The VCC is indeed limited to the control of about 125 trains and normally communicates with trains at least once every second. An important safety feature is that if the communications between a VOBC and the VCC were to be lost for more than 3 s, the VOBC fail-safe mechanism would immediately halt the train by applying emergency brakes.

Automatic Train Supervision (ATS) This system is usually integrated within the ATC system. Its main function is to monitor the system status and provide appropriate controls to direct train operation with the objective of maintaining the intended traffic patterns and minimizing the effect of train delays on the operating schedule.

Automatic Train Protection (ATP) ATP is the safety system which ensures that trains remain a safe distance apart and have sufficient warning to allow them to stop without colliding with each other. ATP systems were first introduced on metros in the late 1960s. Most metro applications use ATP in conjunction with ATO. ATP was also introduced in various forms on a number of main line railways, often in conjunction with high-speed train operation. There are mainly two types of ATP systems:

- **Intermittent systems**: They use electronic beacons (inductive or radio frequency) or short electrical loops positioned within a meter; and
- **Continuous systems**: They use permanently active data transmission and monitoring systems, either through electrical inductive coupling by means of track loops or coded track circuits or by means of radio transmission.

ATP is a predictive enforcement system, which continuously monitors the speed of a train in relation to either a target speed or distance. It intervenes when a train is prevented from passing a Limit of Authority or exceeds a speed limit. This speed limitation will be defined by the line's profile or signal indication. If the allowable speed is exceeded, braking will be applied until the speed is brought within the required limit or the train stopped.

As an ATP system recognizes track condition and driving information of a corresponding train, it permits shorter headway and increases track capacity, by ensuring a minimum braking distance.

Braking system The interface between braking and ATP systems has been designed to automatically authorize trains to brake if they consider that the driver does exceed predefined speed or if the train isn't in a safe mode.

2.2.9 Onboard Operational and Safety Procedures

Operating rules, policies, and procedures are used by railroads to enhance safety. Specific operating rules may differ from country to country and even from railroad to railroad within the same country. Most railway systems around the world use what is known as "speed signaling." Under this speed signaling principle, the signaling elements inform the driver of the speed at which he or she may proceed, but not necessarily the route the train must take. Speed signaling requires a far greater range of signal aspects than "route signaling," but places less dependence on the drivers' route knowledge. This usually takes out the human error factor, which is far safer.

A notable exception is Great Britain, which generally conforms to the "route signaling" principle. Under such procedure, the driver is informed about which route the train must take beyond each signal. This is achieved through route indicator attached to the signal. The driver uses his route knowledge, reinforced by wayside speed restriction signals to drive the train at the correct speed throughout the portion of the route. This principle requires that drivers be familiar with the route, and in emergency situations drivers might have problem when diverted to other routes. As several accidents have been caused by such situation, UK drivers are only allowed to drive on routes for which they have been trained on and must regularly drive on these alternative routes to keep their knowledge up to date.

2.2.10 *Positive Train Control (PTC)*

In the USA, PTC is a system of functional requirements for monitoring and controlling train movements to provide increased safety. The American Railway Engineering and Maintenance-of-Way Association (AREMA) describes positive train control (PTC) as having these following primary characteristics:

- Train separation or collision avoidance;
- Line speed enforcement;
- Temporary speed restrictions; and
- Rail worker wayside safety.

The main concept in PTC (as defined for North American Class I freight railroads) is that the train receives information about its location and where it is allowed to safely travel (LMA). Equipment on board the train then enforces this, preventing unsafe movements. PTC systems may work in either dark territory or signaled territory, and may use GPS navigation to track train movements.

2.2.11 *System Interoperability Procedures*

One of the fascinating differences between the North American and European railway systems is in the priorities given to freights and passenger cars. Actually, if we want to generalize, one could say that priorities are inverted across the two sides of the Atlantic. In Europe, freight cars must wait on the sideline for the passage of the passenger cars, whereas in North America, passengers must wait several minutes for the long wagon convoy to pass through.

This significant difference (together with the difference in distances, the electrification of the lines, and the type of transported goods) has had consequences on the design of the locomotives which favored more powerful locomotives in the USA running at lower speed, whereas European locomotives can typically go at speeds of 160 km/h but are usually limited in power output.

2.2.12 Grade Crossing

A railway crossing is an intersection between a railway line and a road or path, at the same level. Many accidents occur at these level crossings with cars or even pedestrians. Grade crossing safety systems use either passive solutions, such as signs, or use safer active warnings, such as lights and warning tone. Radar sensor systems or empty space detection by video analytics are some of the new e-mobility technologies implemented to improve safety of level crossings.

2.2.13 Safety Integrity Level (SIL)

SIL is defined as a relative level of risk reduction provided by a safety function. It is defined by a number of quantitative and qualitative factors (i.e., development process and safety life-cycle management). In the European Functional Safety standards, based on the IEC 61,508 standard, four different SIL levels are defined. This standard defines two broad categories: hardware safety integrity and systematic safety integrity. A device or system must meet the requirements for both categories to achieve a given SIL level.

The SIL requirements for hardware safety integrity are based on a probabilistic analysis of the device. In order to achieve a given SIL, the device must meet targets for the maximum probability of dangerous failure and a minimum Safe Failure Fraction. The concept of "dangerous failure" must be rigorously defined for the system in question, normally in the form of requirement constraints whose integrity is verified throughout system

Table 2.10 SIL probabilistic requirements

SIL level	PFD	RRF	PFH	RRF
1	0.1–0.01	10–100	0.00001–0.000001	100,000–1,000,000
2	0.01–0.001	100–1000	0.000001–0.0000001	1,000,000–10,000,000
3	0.001–0.0001	1000–10,000	0.0000001–0.00000001	10,000,000–100,000,000
4	0.0001–0.00001	10,000–100,000	0.00000001–0.000000001	100,000,000–1,000,000,000

Source Standard IEC 61508

development. The actual targets required vary depending on the likelihood of demand, the complexity of the device(s), and types of redundancy used.

The probability of failure on demand (PFD) and the risk reduction factor (RRF) of operational demand for different SIL levels are as in Table 2.10.

For continuous operation, it is necessary to use the Probability of Failure per Hour, which we've indicated in Table 2.10.

Hazards of a control system must be identified then analyzed through risk assessment. Mitigation of these risks continues until their overall contribution to the hazard is considered acceptable. Certification schemes are used to establish whether a device meets a particular SIL level. Requirements of these schemes can be met either by defining a rigorous development process or by establishing that the device has sufficient operating history to argue that it has been proven in use.

2.3 Communication-Based Train Control (CBTC)

CBTC technology is behind the concept of various technologies, such as

- PTC,
- European Train Control (ETCS),
- electronic train management system (ETMS), and
- incremental train control system (ITCS).

The IEEE defines CBTC as follows:

- Continuous, ATC system utilizing high-resolution train location system, independent of track circuits;

- Continuous, high-capacity, bidirectional train-to-wayside data communications; and
- Train-borne and wayside processors capable of implementing ATP, ATO, and ATS functions.

The advent of digital radio communication technology during the early 90s encouraged the signaling industry to replace transmission loop by radio telecom-based systems for wayside to train communication. CBTC makes use of telecommunications between train and track equipment for traffic management and infrastructure control.

CBTC technology gives a more accurate train position than traditional signaling systems. This results in a more efficient and safe way to manage railway traffic. Metros and other railway systems are able to improve headways while improving safety. Improving headway as we have already seen, allows for capacity increase. Furthermore, CBTC allows for reduced signaling system CAPEX and OPEX by eliminating all mechanical and electromechanical signals along the line.

CBTC systems are designed to be vital standalone systems, which may also have an overlay capability on fixed blocks for migration purposes. A CBTC system must be implemented in a fully fail-safe manner in order to be able to provide vital moving block functionality (Fig. 2.7).

Fig. 2.7 View from a monorail's front car, showing heavy concrete switch beams managed automatically by a CBTC system. *Source* Author

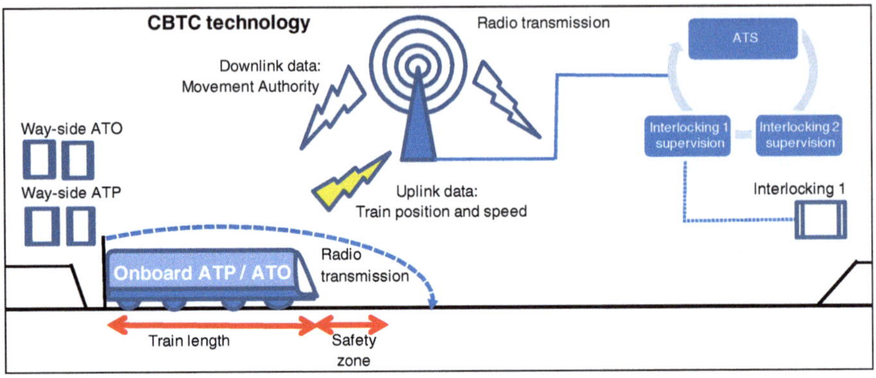

Fig. 2.8 A typical radio-based CBTC architecture. Technical solution may differ from one supplier to another. *Source* Author

2.3.1 CBTC and Moving Block

The CBTC technology was first introduced in 2003, at SFO AirTrain, in San Francisco Airport, and on Singapore's North East Line. In the modern CBTC systems, trains continuously calculate and communicate their status via radio to the wayside equipment distributed along the line. As seen, this status includes, among others parameters, the exact position, speed, travel direction, and braking distance. This information allows calculation of the area potentially occupied by the train. It also enables wayside equipment to define the points on the line that must never be passed by other trains on the same track. These points are communicated to make the trains automatically and continuously adjust their speed while maintaining safety and comfort requirements. So trains continuously receive information regarding the distance to the preceding train and are then able to adjust their safety distance accordingly (Fig. 2.8).

Typical architecture of a modern CBTC system comprises the following main subsystems:
Wayside equipment, which includes interlocking and the subsystems controlling every zone in the line or network (typically containing wayside ATP and ATO functionalities).
CBTC onboard equipment.
Train to wayside communication sub-system.
Onboard ATP system: Subsystem in charge of continuous train speed control and communication with the wayside ATP subsystem;

Onboard ATO system: Responsible for automatic control of traction and braking effort;
Wayside ATP system: Manages all communications with the trains in its area and calculates the limits of movement authority;
Wayside ATO system: Controls the destination and regulation targets of every train.
Communication system: Digital networked radio system using antennas or leaky feeder cable (especially used in tunnels) for bidirectional communication between the track equipment and the trains. The 2.4 GHz band is most commonly used in these systems (same as WiFi);
ATS system: Interface between the operator and the system, managing the traffic schedule according to the specific regulation criteria.

2.3.2 Metro Evolution Toward Unmanned Railway Systems

The evolution of autonomy in the railway environment can be seen as a continuous movement toward more sophisticated systems. Based on the Brussels located International association of public transport UITP, there are five Grades of Automation for trains:

- **GoA 0** is on-sight train operation;
- **GoA 1** is manual train operation where the driver controls many operational functions;
- **GoA 2** is a semi-automatic train operation (STO) where starting and stopping are automated but the driver in the cab can intervene;
- **GoA 3** is a driverless train operation (DTO) where starting and stopping are automated but a train attendant operates the doors and drives the train in case of emergencies; and
- **GoA 4** is an unattended train operation (UTO) where all operations are fully automated without any on-train staff.

Completely man managed As expected, railway systems started as entirely man-aged by personnel. In the Grade of Operation 0 and 1 environments, all operational functions are performed by human being. On the wayside, tracks are switched mechanically by the train dispatchers and the trains are obviously driven by people. The safety functions were firstly ensured by mechanical devices and logic. With electromechanical devices, relay-based logics started to be implemented and safety functions on the wayside started to be controlled by the signaling control center. Similarly, onboard safety was improved with more intelligent electromechanical devices.

Semi-autonomous As more and more intelligence was built into the signaling systems and the train, the track switches started to be operated remotely from the control centers or even from the moving train. In systems, with Grade of Automation 2, drivers were and are still there mainly to make decisions in emergency situations. These systems are still the norm in most regional and intercity lines. Vast majority of metro lines are now going to driverless applications.

Driverless Driverless trains (Grade of Automation 3) were introduced in the 80s. It is probably of no surprise to anyone that the introduction of driverless systems initially happened in closed environments such as airport or small metro applications.

The paternity of such introduction is however still in doubt. The first light driverless metro was built in the North of France in the city of Lille, by the French company called Matra (later purchased by Siemens). The VAL (véhicule automatique léger), which is a light rubber tire metro, was originally built in 1983. However, the first automated people mover (APM) commercial operation opened for service in September of 1980 at Hartsfield-Jackson Atlanta International Airport with an unmanned system. The system was originally built by the American company Westinghouse. As a result of acquisitions and mergers with other companies, the system has been operated and maintained under several brands.

The next technologies that evolved toward autonomous driving were the Light metro, APM, and Monorails. Metros are also going unmanned in the big cities of this world.

Fig. 2.9 A view of a monorail GOA 4 operation front end, without any driver or attendant in the city of Sao Paulo in Brazil. *Source* Author

Today, we see new systems being implemented without the need for any onboard staff to attend the train (Grade of Automation 4). What people should realize is that for the first time in history, they are confronted to a completely unmanned operation, which could also be described as a robotized application (Fig. 2.9).

2.4 Applying Railway Safety Principle to Cars

Does it make sense to apply railway technologies or principles to future automobile design? To answer this question, let's look at Great Britain. There were 1754 road deaths in 2012, which means that around five people died that year on Britain's roads every day. As seen hereafter, the variety of route causes are such that merely maintaining the status quo isn't acceptable.

The following are the most common causes of road accidents due to human factor and must be considered jointly (there is often more than one cause of accidents):
Speeding: Around 400 people (22 %) a year are killed in crashes in which someone exceeds the speed limit or drives too fast for the conditions;
Drink driving: Around 280 people (16 %) die a year in crashes in which someone was over the legal drink driving limit;
Seat belt wearing: Around 300 lives each year (17 %) could be saved if passengers always wore their seat belt;
Careless driving: Around 300 deaths a year (17 %) involve someone being "careless, reckless or in a hurry";
Aggressive driving: a further 125 deaths were involved (7 %);
Fatigue: involves around 20 % of all car accidents;
At-work: Around one third of fatal and serious road crashes involve someone who was working (33 %);
Inexperience: More than 400 people are killed in crashes involving young car drivers aged 17–24 years, every year, including over 150 young drivers, 90 passengers and more than 170 other road users (22 %)
Failed to look properly: 40 % of road crashes involve someone who failed to look properly;
Loss of control: One third of fatal crashes involved loss of control of a vehicle (33 %); and
Misjudge other person or car Path/Speed: One in five crashes involve a road user failing to judge another person's path or speed (20 %).

Source of information[11]: web page of Royal Society for Prevention of Accidents

[11] Web page of Royal Society for Prevention of Accident (ROSPA) updated on 15/072013). http://www.rospa.com/faqs/detail.aspx?faq=298.

Obviously, car accident statistics are typical of any given country and might differ from one region to the other in relation to each government's attitude toward such cause of accidents, as well as in function of the media's crucial role in promoting safe driving behaviors. The point is that many of these fatality causes could have been eliminated with unmanned operations, especially had railway safety principles and technology been applied in the UK or anywhere else. The following paragraphs will give ideas about how to adapt railway principles to the car environment.

2.4.1 Automotive Block Interlocking Concept

Block interlocking concept, which forbids vehicles to occupy the same space, at the same moment, will need to be applied to unmanned cars. The issue to solve for automotive system engineers will be how to space cars far enough apart to ensure that they cannot collide.

Fixed block The automotive environment will not use the fixed block concept as road environments simply don't allow for fixed sections.

Moving block We strongly believe that the concepts behind the moving block technology could be applied to the unmanned car environments. Firstly, its spacing concept is well suited for cars as it doesn't take into consideration two fixed points but a safety zone around each moving vehicle in which no other vehicle would be allowed to enter. This safety zone would be calculated using speed, grades, and the car's operational properties. Secondly, an additional buffer zone would be provisioned around the vehicles, which would vary depending on location and surrounding elements. Thirdly, the car's normal braking rates would be used for calculating safety stopping distances, with probably a guaranteed emergency braking rate that could be used in critical situations. The big question will be whether the concept of brick wall stopping will be applied or not, and under which conditions.

This safety zone, which would be calculated by an onboard computer like the one used within the Google cars, would use the define location, speed, and direction of each car through an accurate GPS. In some critical areas, passive or active markers could be used to indicate dangerous curves or gradients, or even in tunnels to give more accurate positioning information (Fig. 2.10).

The block notion will need to evolve because of the fluidity of the road environment, from an exclusively horizontal dimensional concept toward a tri-dimensional space, which will take into consideration length, width and even height parameters. Because of the need to overpass cars or run on parallel lanes, virtual moving blocks will also be calculated to account for the safe zone needed to overtake front cars (Fig. 2.11).

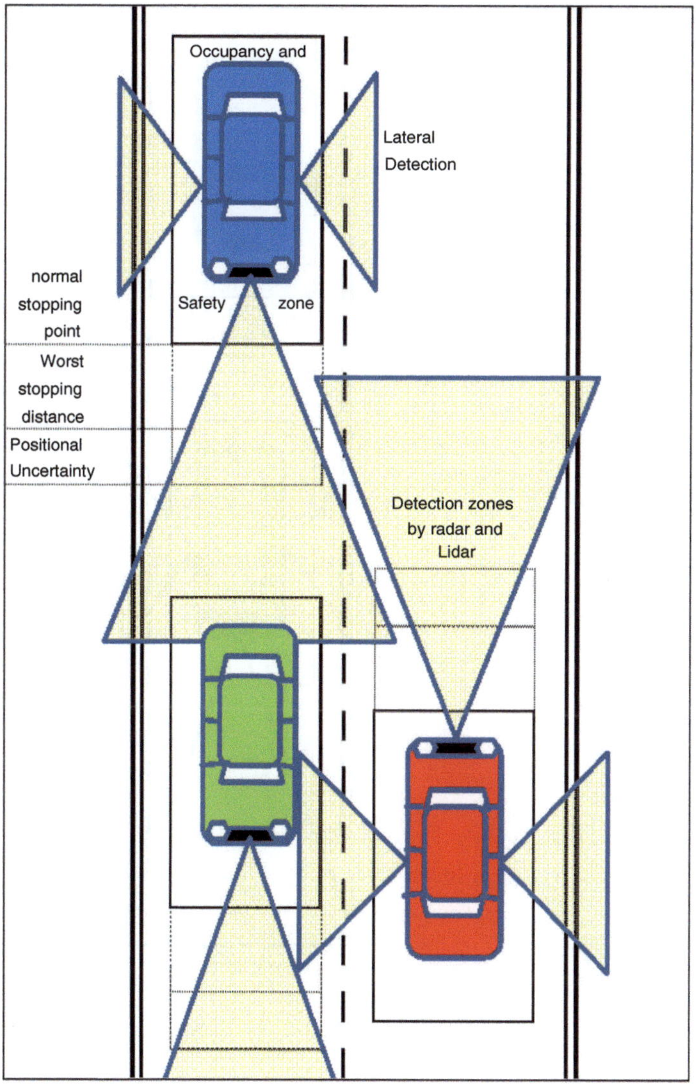

Fig. 2.10 Automotive moving block interlocking with safety zone calculation and detection zones. *Source* Author

2.4.2 Automotive Block Signaling Concept

The second railway safety principle of signaling to the next vehicle that no train is in that block will also need to be applied.

In a railway moving block application, the track line is usually divided into sections that are each controlled in real time by a computer at the control center

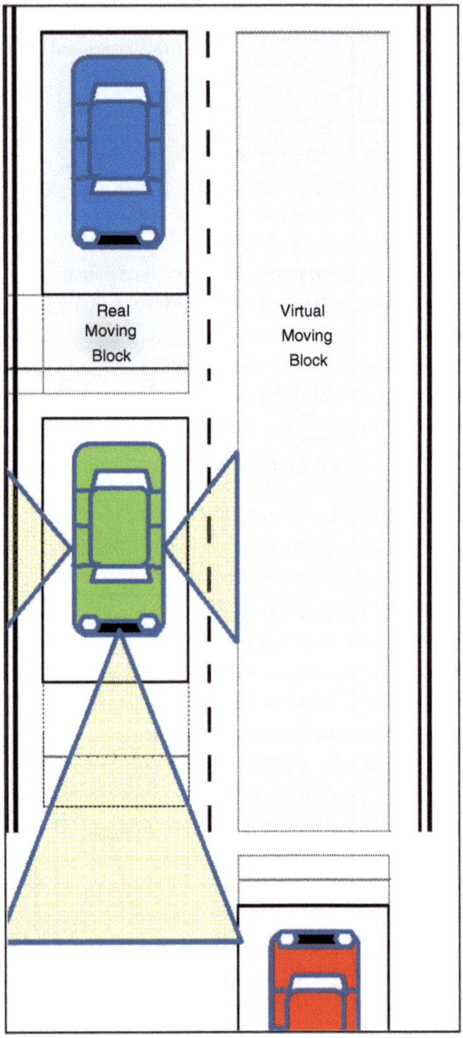

Fig. 2.11 Real and virtual moving blocks. *Source* Author

level, with its own telecom transmission system. Would that architecture be replicable in a road environment? This would be important because in a moving block environment, the continuously calculated car position and its braking curve must be communicated to the wayside equipment to establish protected areas (LMA). In a highway environment, we could imagine having such dedicated computers per lane sections. However, in an urban environment besides the complexity and cost of having computers throughout the city areas, the major issue to apply such a moving block concept would be to maintain constant connection of vital functions (the absence of communications between the VOBC and the VCC stops the trains).

Obviously and as we will see in the next sections, engineers are working on a blend of various communication technologies, as well as a mesh network architecture, which would provide fallback solutions in case of technical problems. With cloud computing, information would be centralized without having to dedicate fail-safe computers. However, major differences between railway and automotive operational environments will need to be taken into consideration.

- **Vehicle-to-vehicle (V2V) communication**: Because in a road environment cars can come from all directions, communications between all cars will be required within a calculated perimeter, which will be accounted for in real time according to the cars safety braking distance.
- **Obstacle detection**: Most metro networks are closed environments, reducing uncertainties to the bare minimum. On the contrary, cars (bus or even tramways) do face issues with obstacles such as pedestrian, animals, cars, or even falling objects such as rocks. It is true that long distance detection technology is being tested in the railway environment to prevent accidents but the difficulty is that trains at great speed need very often a kilometer to brake, eliminating the benefits of such technology. Unmanned cars, on the other hand, can brake within meters and as they will need mandatorily to avoid obstacles, they must use sensor technology to avoid crashing on pedestrians, animals, cars equipped with failing V2V equipment, or even working cruises. As we will see in the next sections, sensor technology such as lidar, radar, and cameras will be integrated within unmanned car environments.
- **No dispatching**: Unlike trains, car destination won't be centralized as it goes against the main transportation principle: freedom to go wherever we want.

With these major differences needing to be accounted for, it is unlikely in our view that the automotive concept of LMA will be centralized. Indeed the safety spacing calculation, which is given by a centralized computer in the railway environment, will be calculated by each car individually and shared directly between all cars within a certain perimeter. Additional safety parameters will need to be accounted for, such as direction, safe braking distances of each car, distance and time to reach another car's LMA (especially when overtaking), safety priorities, all non-car detected object, animal or human being, fixed or removable beacons or balises, etc.

Geo-localized messages (see Chap. 5 for more information) will be pushed to the road from the V2C to indicate the passing cars that an obstacle or a special situation, such as a working crew, can be found within vicinity of the location (Fig. 2.12).

2.4.3 Automotive Integrity Concept

Is the integrity concept relevant on roads? After all, cars unlike trains are made of a single body shelve and no wagon or section can depart from the motorized section. Roads also unlike tracks cannot brake suddenly and derail a car.

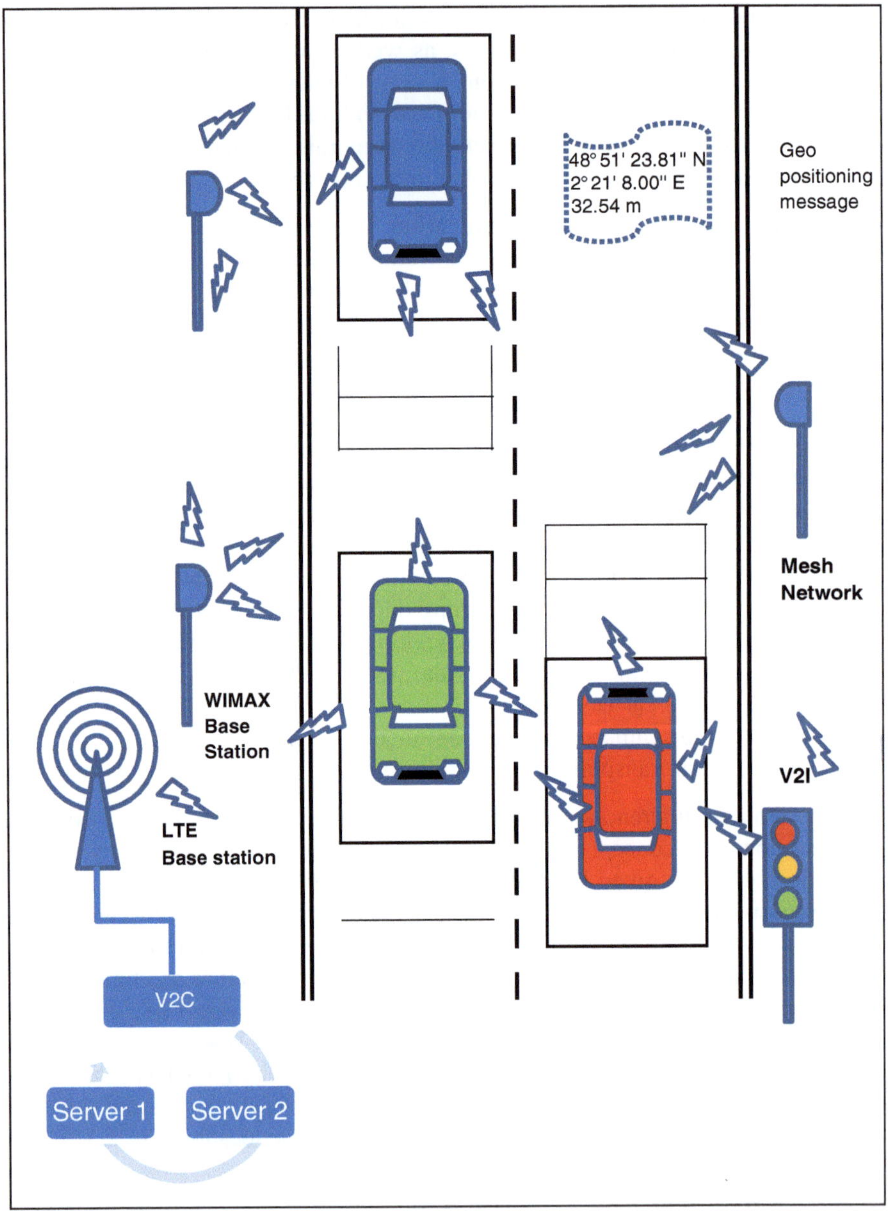

Fig. 2.12 Automotive block signaling with V2V, V2C, and V2I communication through LTE, Wi-MAX, and mesh networks. *Source* Author

Well in our view it is. First, cars don't run alone. Trucks can lose their charge. Cars also can tow a motor home or a trailer, which can separate from the cars. These would need to be accounted for. Specialists could always argue that any separated trailer would be treated as any other road obstacle and be detected by the onboard sensors. True, but the car's VOBC would need to calculate that such element could change lane, roll over, decelerate or accelerate depending on the slope, etc. Secondly and more importantly, in a platooning environment (a convoy of cars that follow each other a 5 m apart), no detection system would give the braking system enough time to avoid such separated trailer. In fact, the integrity concept is extremely relevant in such platooning operation, as the platoon itself could be considered a convoy of attached parts, which integrity would need to be checked regularly (Fig. 2.13).

2.4.4 Automotive Protection Technologies

Unmanned cars will need to integrate some of the protection technologies we've described. Obviously, the equipment and the software will be different, but the basic functionalities will be similar.

ATC The Vehicle On-Board Control (VOBC) continually monitoring the position, speed, and general status of the car will be mandatory. On the other hand, the VCC portion of the ATC will be superfluous as the safety braking distance will be calculated within the VOBC.

ATS It will be needed to monitor the system status and provide appropriate controls with infrastructure equipment, with the objective of maintaining the intended traffic patterns. In fact, the ATS functions will be integrated within the V2C (vehicle-to-cloud) environment.

ATO This non-safety part of train operation related to station stops and starts won't really be necessary in a road environment. Exact stopping distances will be calculated by the VOBC with the support of sensors.

ATP This predictive enforcement system, which continuously monitors the train's speed in relation to either a target speed or a distance, will be required. It will intervene whenever a car exceeds a speed limit (as speed restrainers already do) or when the car infringes the Limit of Authority, calculated in the VOBC, taking into account other cars moving block and safety braking distance.

Braking system If the allowable speed is exceeded, braking will be applied until the speed is brought within the required limit or the car stopped.

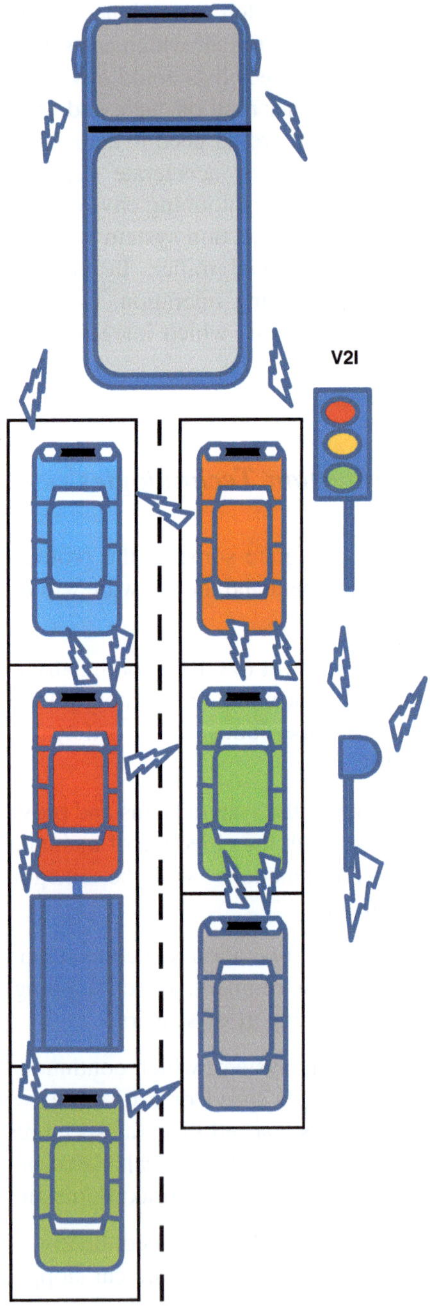

Fig. 2.13 Platooning with illustration of integrity issue. *Source* Author

2.4.5 *Automotive System Interoperability*

Interoperability for any system is always a key issue to address. Cars will need to communicate with each other, with the infrastructure and the cloud in ways that can be understood by each other, independently from the cars' brand. This means that at least continental standards will need to be applied for cars to be able to travel from one country to the other. As we will see, it is unlikely that a worldwide standard will be adopted, as for instance, we already know that the technology behind the V2V Japanese, American, and European system won't be interoperable.

2.4.6 *Other Relevant Automotive Safety Concept*

In the previous section, we've seen other concepts such as degraded mode procedures, safety integrity level (SIL). There is no doubt in our mind that these safety concepts and other technology from the railway environment will be applied to the automotive unmanned environment. In fact, in the next section, we will present the car automation level. As the reader will see, the similarities are just striking!

2.5 Automation Level in the Automotive Environment

For every 30 s, someone somewhere dies of a car crash, and ten more are seriously injured. This hideous number doesn't show the complete picture. In the rich world, the number of crashes per driver has been constantly dropping. Higher vehicle standards are a big reason for falling death rates in the rich world. Restraints on drivers and investment in safety programs, be it through better infrastructure, awareness campaigns on seatbelts or new technologies, have also helped slashed road deaths in rich countries. Newer technologies such as alcolocks, which prevent drunk-driving and self-driving cars, will make roads in the rich world even safer.

However, additional safety measures must be taken in developing and poor countries, if we want to avoid WHO's hecatomb prevision that expects deaths to more than triple in the very poor countries.

> Experience in the Rich World has shown that roads can be made safer cheaply and simply. Some of the recent safety measures such as dedicated cycle roads, count-down lights at crossings and strict vehicle standards are pricey, but if the road designers and investors would only earmarked a small fraction of the road cost for safety improvement, it would dramatically improve the fatality statistics. For instance, roads need pedestrian paths but 84 % of the worldwide roads have none. This measure is easy to address.

Roads with fast traffic need well designed junctions and crossings. For very poor countries that can't invest in central barriers of concrete or metal, allocating space between the opposite lanes can dramatically help stop head-on collision. Installing rumble strips on hard shoulders can also be a good way of alerting drivers that they are moving from their lanes. Ensuring that courts and police enforce laws against speeding and driving under alcoholic or drug substances as well as bringing ticket money also can reduce significantly all the road tragedies.

All these measures make a big difference and should be contemplated everywhere. However, we will maintain the focus of this section on the e-mobility technologies that can also reduce significantly the number of fatalities. We will try to present these technologies that can address some of the main causes of road accident as described in the previous section.

2.5.1 Reducing or Eliminating the Human Factor in Driving

One of the more interesting lessons we can learn from the evolution of safety in the railway industry is that the human factor is one of the main causes of accidents and eliminating it altogether will improve accidents statistics. As the car manufacturer Volvo puts it, most accidents are caused by the four D's: distraction, drowsiness, drunkenness, and driver's error. Another crucial lesson learned from the railway industry is that safety improvements can come from three different areas: system, vehicle, or infrastructure. In railway industry, three industries have originally pushed these three options: signaling companies, train manufacturers, and railway operators. Governments, through regulations and even more essentially through norms and standards, have also played a key role in improving safety. Let's understand what the concepts behind automated cars are.

2.5.2 Level of Car Automation

We've seen that there are different levels of automation within the metro or train for CBTC signaling technologies (as well as safety levels). As for the train industry, the automotive industry has come up with an almost similar classification. In the United States, the National Highway Traffic Safety Administration (NHTSA) has established an official classification system. However, there is still no worldwide standard and thus the following is an explanation based mainly on this classification.

Level (0) Non-Automated The driver completely controls the vehicle at all times. There is no automation at all. While most current vehicles can be included in this category, vehicles with warning systems that assist drivers also fall into this category. Vehicles equipped with these technologies will not assume control for any driving tasks, but will provide additional information to the driver and/or warn the driver of situations requiring immediate attention.

> Navigational global positioning system (GPS) is an example of a currently available e-mobility technology, which provides information useful to the overall task of driving. V2V communications and Lane Departure Warning (LDW) are also two examples of warning technology, alerting the driver when the vehicle begins to drift out of the lane of travel.

Level (1) Automation-Assisted Individual vehicle controls are automated. This category still leaves the driving authority with the driver. However, under limited normal driving or crash imminent circumstances, technologies in this category will take control away from the driver (Fig. 2.14).

Fig. 2.14 Picture of an active lane-keeping shown on the dash board. *Source* Author

An example of this type of technology is the Electronic Stability Control. This system uses automatic computer controlled braking of individual wheels to assist a driver in maintaining control in critical driving scenarios in which the vehicle is beginning to lose directional stability. Another advanced example of automation-assisted driving is a lane-keeping system that will actively steer a vehicle back toward the center of its lane when the system detects that the vehicle is drifting into an adjacent lane or is on a collision course with a vehicle in an adjacent lane.

Level (2) Monitored Automation This category is the first in which technology will share the driving responsibility with the driver. However, in this category, the human driver is expected to intervene at any moment (e.g., lane markings disappear and the vehicle can no longer position itself in the center of the lane). Thus, the autonomous technology is only able to assume the responsibility of driving when the conditions permit.

For example, some vehicles on the market today are available with automatic parallel parking systems. This type of technology differs from automation-assisted driving technologies because the driver gives a general command to the vehicle (e.g., "park in this space") and the vehicle performs that command by assuming entirely the steering control and making the necessary steering calculations. Another potential example is the combination of adaptive cruise control with lane-keeping. The combination of these two technologies would potentially enable vehicles to proceed on the highway with little or no input from drivers.

Level (3) Conditional Automation In this category, the driver can fully cede control of all safety-critical functions in certain conditions to the onboard computer. The driver doesn't need to keep constantly an eye on the road. The car senses when conditions require the driver to retake control and provides a "sufficiently comfortable transition time" for the driver to do so. Level 3 is full automation in certain situations (most likely for highway driving).

In 2014, several automakers, like Audi, BMW, Nissan, and Mercedes targeted self-driving capabilities by roughly 2020. Others are more bullish. Tesla and Ford for example, targets respectively 2016 and 2017 to market such cars. For speed lower than 60 km/h (38 mph), Mercedes Benz's "intelligent drive" kit can already allow drivers to relax and enjoy, while their car accelerate, steer and brake autonomously.

Table 2.11 Comparison between car and rail automation level according to UITP and NHTSA classification

Level of train automation (UITP)		Level of car automation (NHTSA)	
Level	Description	Level	Description
GoA 0	On-sight train operation	Level 0	Non-automated
GoA 1	Manual train operation	Level 1	Automation-assisted
GoA 2	Semi-automatic train operation	Level 2	Monitored automation
		Level 3	Conditional automation
GoA 3	Driverless train operation	Level 4	Full automation
GoA 4	Unattended train operation	Level 5	Unmanned

Source Author

Level (4) Full Automation The car performs all safety-critical functions for the entire journey, with the driver not expected to control the vehicle at any time. However, level 4 will maintain the possibility to shift control to a human driver. This vehicle would integrate various technologies from the previous three categories to perform all driving tasks.

Level (5) Unmanned This is an unmanned system where human control won't be allowed at all. The only driver input would be the destination. Level 5 driving encompasses all of the systems necessary for the vehicle to perform automatically and independently all driving tasks in all driving scenarios.

2.5.3 *Similarities Between Level of Car and Train Automation*

Table 2.11 shows how similar the principle of automation level for cars and trains are.

2.6 Personal Rapid Transit (PRT)

Before being accused by Nay-Sayers of being over-optimistic about the future of unmanned vehicles, let's present a car technology which is already completely driverless: the Personal Rapid Transit system.

PRT is a public transport mode based on small automated vehicles operating on a network of specially built and segregated guideway, arranged around a network of small stations. The principle of a PRT is very similar to the concept of a network of taxis running on dedicated and exclusive lanes. Like taxis, they offer relative privacy, though one could expect to share the PRT car with a few other passengers. They also allow for relative fast access to the specific final destination station.

2.6.1 PRT References

PRT technology has been around for a few years but isn't well known from the general public. The main reason is that the industry has a record of failure due to various project problems, such as financing issues, cost overrun, regulatory conflicts, flawed designs, etc (Fig. 2.15).

As of 2014, two PRT systems were already operational.

- Since 2010, 13 pod cars continue to shuttle students along an 800 m stretch between a station and the university in Masdar, UAE; the air-conditioned vehicles have a maximum speed of 40 km/h. The entire system was originally designed to run up to 5000 trips per day, with each of the 810 vehicles.
- Since May 2011, a 21-vehicle Ultra PRT system at London Heathrow Airport links the business car park, by a 3.8 km route to the terminal. The driverless cars can reach speed of 40 km/h and carry up to four passengers and their luggage. Since its opening, the PRT system has carried in around 2 years more than 700,000 passengers from the Terminal 5 Business Car Park across to Terminal 5 itself.

Unfortunately, it is true that there is still today no city-wide deployment, but this is due more to the fact that no Public authorities (except for Masdar city) have already committed to building PRT, because of the risks associated with being the guinea pig. But these two applications have shown that PRT can work. With the right political support for financing the project and provided a large technological railway or car manufacturer would embrace such market, things could be very different. In the mean time, the small PRT developers are facing a bleak future as they don't have the financial strength required to sustain the R&D and marketing costs to build such a market.

However, if this market were to integrate technologies being developed for driverless cars, this market could pick up and grow tremendously. The fundamentals are there. To use an image, PRT is the convergence of smart infrastructure with smart cars (which could become much smarter with the integration of driverless car technology), as well as the convergence of automotive and railway technologies and concepts.

Fig. 2.15 Heathrow Airport car park operational PRT system. *Source* Picture of ULTra PRT POD at Heathrow Airport car park During a trip to the airport 2012-02-16; Author: Moshrunners; file is licensed under the Creative Commons Attribution-Share Alike 3.0 Unported license

2.6.2 Smart Infrastructure

All PRT suppliers have been highly influenced by rail transit technologies and concepts. For instance, technologies like fare systems, guideway, track and station design, as well as concepts of system safety, reliability, operations, and maintenance are originated mainly from the railway environment.

PRT station PRT stations are similar in many aspects to subway stations. The fundamental difference is scale. Indeed, PRT stations must have the same functionality than mass transit but must be designed for only a small numbers of passengers. In order to make the PRT network attractive, the system usually positioned stations close together. However, and unlike metro stations, the station's vehicle access isn't positioned directly on the main lane. The station uses one- or two-way section connected to the main lane by the same junction (vehicles come back and forth through the same junction) or by two junctions positioned in parallel to the main lane. Each station might have multiple berths, for several passengers drop-off or hop on, or for vehicles storage. When user demand is low, surplus vehicles can be configured to stop at empty stations at strategically placed points around the network. This enables an empty vehicle to quickly be dispatched to wherever it is required, with minimal waiting time for passengers.

Switching As mentioned, many concepts from the railway industry have been adopted by the PRT system designers. Switching, which is one of the main railway features, has been used to ensure guideway clearance for most PRT systems. However, and as we will see on the section dedicated to self-driving cars, conventional steering technology with mounted sensors could become an interesting alternative to switching. There are two PRT switching concepts:

- Track switching like in railway system; and
- Vehicle mounted switching.

Each approach has its advantage and inconvenient.

Track switching increases the vehicle reliability by reducing the number of small moving parts in each car and weight. It also increases the overall system safety using successful tested rules from the metro industry. On the other hand, vehicle switching allows faster switching, reducing the headway time and distance between vehicles, as well as obviously simplifying the guideway infrastructure. Because of differences in switch setting between two cars, it increases the minimum distances required between consecutive junctions.

Dedicated guideway As for trains, the notion of dedicated lanes is extremely important in PRTs. It allows control of the environment like in the railway industry, eliminating potential hazardous situation, resulting from mixed traffic.

Furthermore, all PRT systems use segregated lanes with no possibility for pedestrian or other cars to cross the guideway. Many use over-pass to avoid this potential conflict with pedestrians, animals or car crossing. Masdar's system ambition has been limited because it attempted to dedicate ground-level to its PRT system, which led to expensive buildings and roads. Many guideways have a loop shape for easier reinjection of cars into the system.

Power and telecom equipment Most designs use the guideway to distribute power and data communications, including the vehicles.

Signaling system Unlike railway signaling system, which uses fixed traveling concept, PRTs use a unique flexible signaling concept. In most metro operations, the vehicle rides from the first station on one side of the city to the last station, stopping at all operating stations on the way. Even when there are exceptions, such as when metro cars make shorter shuttle trips between two intermediate stations of the more busy section of the metro, all the cars defined as such will have the same fixed sequence of stopping.

On the following (Fig. 2.16) railway design, a metro vehicle would usually go from station 1–7 and back from station from 7 to 1. In case of shorter carousel (planned trip), the metro could in principle use the switch situated at the end of station 3 and 5 to short circuit stations at the extremities of the line. We say in principle, because this configuration would create operational problems and not be favored by operators and passengers.

PRT systems apply an operational principle based on point-to-multipoint travel. For instance, passengers might decide to go directly from the first to the last station, by passing all intermediary stations (Fig. 2.17).

On the design here above we show two configurations where the vehicles can use 3 park lanes either with no parallel line to the main guideway or with a parallel line to inject more easily the parked cars (at station 3 and 5).

Thus in theory, if there was a network with sufficient guideway length and closely spaced stations, passengers could easily board in a station and take relatively direct routes to their destination without stops. Once again, this is only

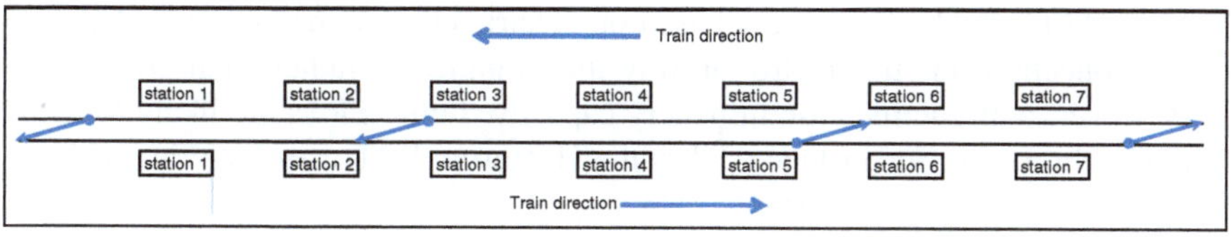

Fig. 2.16 Train carousel with intermediary switching possibility. *Source* Author

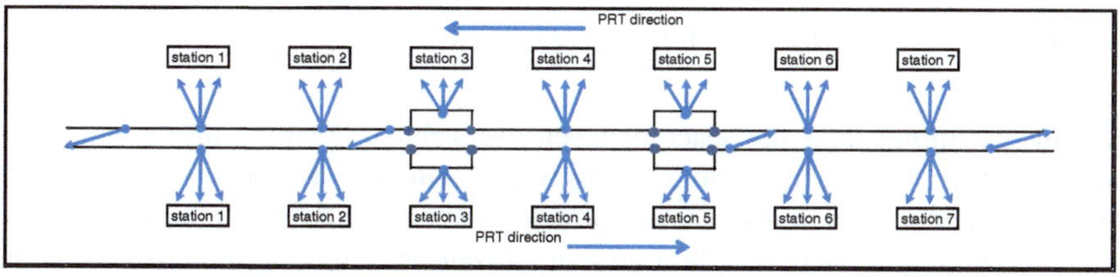

Fig. 2.17 PRT carousel with intermediary and station switching possibility. *Source* Author

possible because the stations aren't directly on the main lane, thus vehicles are never blocked by other vehicles stopped on the main guideway at stations.

Running gear Most designs enclose the running gear in the guideway to prevent derailments.

Fare system As for trains, PRT would be working with reliable ticketing system.

Supervision and monitoring system Onboard computers communicating with the control center main computers eliminate the need for human control. Errors from human drivers are eliminated increasing the public's safety. Other public transit safety engineering approaches, such as redundancy and self-diagnosis of critical systems, are also included in PRT designs. The monitoring PRT system places the cars in moving Slots (block) that go around the guideway loops. Real vehicles are allocated a slot by the trackside controllers. Traffic jams are prevented by placing north/south cars in even slots and east/west vehicles in odd slots. At intersections, traffic in the systems can interpenetrate without slowing.

2.6.3 (Reasonably) Smart Cars

PRTs have adapted automobile technology to carry comfortably from 2 to 4 passengers. Using many standard components of automobiles such as electric motors, batteries, tires, axles, wheels, and body components, the automotive sector possesses all of the needed competencies to efficiently design and manufacture these types of vehicles.

> Most PRT vehicle designs are based on electric cars. Some systems adapted the concept from the metro of wayside conductors rather than batteries. However, with the increase in battery capacity, future PRTs are likely to use that technology. For instance, Ultra the Heathrow PRT uses batteries which are recharged when at stops.

2.6.4 PRT Operational Characteristics

Headway distance It is the space between two vehicles running on the guideway and influences directly the system's capacity. Existing rail regulations usually apply to PRT systems. Thus, railway safety principles are typically applied and headways are calculated in terms of absolute stopping distances. This stopping distance is the

Table 2.12 Braking distance in meters of PRT cars (with 0.5 s perception and reaction time), according to speed

Speed in km/h		10	20	30	40	50	60
Braking distance	With 0.5 s perception + Reaction	2.0	5.0	9.2	14.5	21.0	28.6

Source Author

sum of the distance a car will run due to the driver's perception and reaction time and the distance necessary for stopping the car (Table 2.12).

The braking distance is one of two principal components of the total stopping distance and is defined as the distance a vehicle will travel from the point when its brakes to when it comes to a complete stop. As we've seen in an earlier section, it is primarily affected by the vehicle's speed, the coefficient of friction between tires and road surface and negligibly by the tires' rolling resistance and vehicle's air drag.

The second element is the perception-reaction time required to hit the brake. In cars involving a driver, 1.5 s is considered for legal purposes. In PRT system 0.5 s can be used, as this process is linked to communication between control centers and cars and don't involve human beings. A coefficient of kinetic friction of 0.7 can be used to calculate such distance, but this is under normal condition (under rain it would need to be lowered). The following table gives the braking distance in meters required for different speeds under normal road conditions.

The minimum safe headway measured tip-to-tail defined by the braking performance is as follows:

$$T_{\min} = L/V + t_r + kV/2(1/a_f - 1/a_l)$$

where

- $T_{\min}$ minimum safe headway time in minute;
- L length of the vehicle;
- V speed of the vehicle;
- t_r reaction time in minutes;
- a_f maximum braking deceleration of the follower;
- a_l maximum braking deceleration of the leader; and
- k arbitrary safety factor superior or equal to 1.

However, using an absolute stopping distance (also called brick wall stopping) for safety reasons and thus not taking into consideration the speed of the leading

Table 2.13 Safety time PRT cars (with 0.5 s perception and reaction time and safety factor of 1.5), according to speed

Speed in km/h		10	20	30	40	50	60
Braking distance	With 0.5 s perception + Reaction	2.0	5.0	9.2	14.5	21.0	28.6
Safety time in second	With safety factor of 1	2.8	3.0	3.6	4.3	5.1	5.9

vehicle, as well as considering a safety factor of 1 for a 4 m car, we would have the following minimum safe headway in seconds (Table 2.13).

PRT system engineers have calculated that two-second headway is sufficient. It is true that computerized control theoretically allows for simultaneously multiple vehicles braking. However, such short headway is controversial, because it would consider the following:

- A system reaction time of around only 0.01 s without any perception time;
- A brake failure of around 1 m/s of the lead vehicle; and
- An average speed of 30 km/h.

The problem from a safety perspective of this approach is that usually, the authorities making the safety case use the most pessimistic scenario. This pessimistic scenario would need to consider rainy conditions, a leading car or object at complete stop at full speed of around 60 km/h (and not the average speed of 30 km/h). Indeed most PRTs' speeds are in the range of 40–70 km/h though a few PRT designs have operating speeds of 100 km/h. Having said that, the UK Railway regulating body has evaluated the Heathrow design and accepted in principle one-second headways, pending successful completion of initial operational tests at more than 2 s. The jury is out there to conclude on this required safety time.

PRT system Capacity This headway calculation is fundamental because system capacity is inversely proportional to headway. System capacity in terms of vehicle and in terms of passenger is, respectively,

$$N_{\mathrm{veh}} = 3600/T_{\min} \quad \text{and} \quad N_{\mathrm{pas}} = P \times 3600/T_{\min}$$

where

N_{veh} the number of vehicles per hour;
P the maximum passenger capacity per vehicle; and
$T_{\min}$ minimum headway

Using a four-passenger per car vehicle capacity and 2 s of headway, the theoretical PRT capacity with one lane would be 1800 vehicles and 7200 passenger/car. However, most estimates assume that vehicles will not generally be filled to capacity, due to the point-to-multipoint nature of PRT. At more typical average car occupancy of 1.5 persons per vehicle, the maximum capacity is 2700 passengers per hour (see Chap. 4 on megacities for more information on capacity calculation).

PRT promoters argue that the theoretical minimum PRT headways should be based on the mechanical time to engage brakes, and these are much less than half a second. Moving from two to one-second headways or half a second would respectively double and quadruple PRT capacity, making it more attractive and financially interesting. However, one shouldn't forget that in simulations of rush hour or high-traffic events, about one-third of vehicles on the guideway need to travel empty to resupply stations with vehicles in order to minimize response time. This would obviously reduce capacity proportionally.

Travel speed PRTs have an advantage in terms of journey time in regards to normal buses or metros. Given their point-to-multipoint travel nature, nonstop journeys are about three times as fast as those with intermediate stops. This is not just due to time savings for accelerating and stopping. Scheduled vehicles are also slowed by boarding and exiting.

2.6.5 Cost Characteristics

Most of the PRT initial investment is in the guideway, with most estimates falling in the \$10–\$15 m range per kilometer line, excluding rights of way or system infrastructure, such as storage and maintenance yards and control centers.

A design with many modular components, mass production, driverless operation and redundant systems should in theory result in low operating costs and high reliability. However, this isn't the case currently and will only be known once a full scale PRT is in operation. According to Ultra, the Heathrow pod integrator, it took six years to develop this PRT and costed £30 m (50 M US\$), which at a little over 13 M US\$ per kilometer corroborate this costs per kilometer.

2.6.6 PRT Versus Unmanned Cab

In chap. 6, we will see the new business model of hailing companies such as Uber against which most European taxi drivers were protesting in June 2014. Furthermore, we describe a future where there will be no taxi drivers left but where PRTs will become predominant. Will it be through unmanned taxis with a multipoint-to-multipoint model (current taxi model) or point-to-multipoint model (current PRT model) is still a question mark. What is also sure is that because of the investments made by the big IT players and car manufacturers, unmanned cabs are likely to have more intelligence on board the car than on the wayside. The main reason for this is that unmanned cars need to ride in non-segregated lanes and PRTs don't have this capacity today, as they rely on switching technology just like the train industry. We believe that the automotive industry cannot reduce this driving flexibility if unmanned cars are to prevail in the future.

2.7 E-Mobility Technologies Reducing Fatalities

The human factor is the main cause of accidents; drunk driving, seat belt wearing, speeding, fatigue, in-car distraction (eating, not fixing the road, phone, and e-mail), and wrong perception or judgment are a few of these causes. The human factor isn't, however, the only cause. Bad weather conditions (i.e., rain, fog, and ice) and poor road states (pot holes, dirt) also kill several people and injure many more. We will analyze the e-mobility technologies that are being developed to try to reduce such problems. Let's start by presenting a technology that actually allows recording of all bad driving behaviors or equipment malfunctions—the black box—that has been already installed in our cars for a few years.

2.7.1 Black Box

People are mostly familiar with black boxes in airplane. In the last decade, event data recorders (EDRs), as they are called in the railway industry, have been introduced and are now recording safety-related information such as speed, acceleration, braking, door closing, etc. Black boxes are now gaining momentum in the automotive market as well. Despite obvious privacy issues, black boxes have been used by car accident investigators and insurance companies to reconstruct the events before accidents occurred. Countries such as Canada have used, as early as October 2003, black box information to charge drivers responsible of dangerous driving, involved in-car crashes with fatalities. The reality is, and although most people aren't aware of their onboard presence, event recorders are already installed on around 80 % of all new cars. Originally, car black boxes were installed to determine the cause of airbags activation, but now their use has changed to enable recording of metadata such as speed, acceleration, braking events, and other safety-related measures. Fortunately, they can also identify if a malfunction happened and caused an accident.

Lately, new types of black boxes coupled with video cameras installed on the windshield have made their market introduction. These digital video recorders (DVRs) can associate image, with positioning through GPS, as well as performance of the car through the event recorder. Thus on top of the safety-related measure of the event recorder, these DVRs also record the car's time, location, and direction, as well as the driver's view which makes it helpful for a number of hard-to-prove situation involving accidents (i.e., someone crossing a lane on a red light).

How does DVRs work? It basically records constantly images on which metadata (I.e.: speed, positioning, braking data, etc.) is inserted for a specific period of time. There is yet no rule for how much time they can record but for instance in the Public Transport a rule of 20–32 days is usually applied to maintain some kind of privacy protection. Video footing isn't destroyed but overwritten by newer video and sound streams and metadata on a First In First Out basis. If a car hit is felt by the system, the last 30 s are tagged and secured in a safe area where it cannot be erased.

If cars have already been equipped for so long with black box under the form of EDR with limited information recording, why do people suddenly want to add more information? There are several reasons to this.

The obvious reason is that DVR technology has evolved tremendously and prices have gone down allowing its diffusion on the mass market. Digital technology has allowed camera miniaturization as well as a significant reduction in mechanical parts always prone to problems linked to vibration. Being cheap, easy to install and use is a great argument for potential consumers.

Secondly, GPS have now become generalized in many new cars. Thus adding such information is simple.

The third reason is legal. In litigation society, having information that the other party doesn't have is clearly an advantage and a powerful motive to install such DVR. In the USA especially this is a strong selling argument. In Russia, where law enforcement is more fragile, DVRs have also been extremely popular as they enable fighting false accident claims. Even countries like France or Belgium, where black box video information can't be used as element of proof in courts are changing their views and starting to follow countries like America or Russia.

The fourth reason is financial. Many insurance companies are pushing for car black box installation. In many countries insurers are still imposing their own black box system, as there isn't yet a global standard.

Though DVR technology might be different, all motivations are identical: reward the good driving behaviors by reducing the insurance costs. These new insurance policies are still at an early stage with big insurance companies like AXA still testing their own black box and deciding on its commercial policy.

The last reason, which also explains greatly why insurance companies are reducing their insurance premium, is that it works. In the United States, studies on the use of black box have shown a reduction of more than 50 % in the accident rates of cars equipped with such systems.

For all these reasons and considering that car fatalities are under scrutiny of an increasingly risk adverse and pro-litigation society, the question isn't if these DVRs will be installed on each car but when. Civil right privacy concerns will be again a big issue.

Privacy advocates need to be wary, though they won't be able to impede, in our view, such inevitable trend. By installing black boxes, insurance companies will be entitled to have access to many details of the car owner's private life and will be able to use this information against their clients. But things will get worst for privacy advocates, the day DVRs become mandatory. Indeed today this information is used as forensic evidence. The justice uses it in case of accidents involving fatalities. People use it in court when it suits them. However with the event of more powerful telecommunication network such as 4G, any car will become a supplier of real time information about the driver's behavior. Ford's CEO Alan Mulally said in 2013:

"*We have GPS in your car, so we know what you're doing*".

Navigational systems can now transmit real time information about a car's location, as well as details about the vehicle's performance, such as speed, acceleration, etc. The US Government Accountability Office reported in 2011 that all carmakers collected location data and shared it with third parties that provide services to the car owner. Many automobilists are happy to provide such information to service tracking providers such as General Motors' OnStar™ system.

But the question is how long it will be before the companies that own such system decide granting access to advertising companies selling geo-positioned promotions? As we will see in the connected city chapter, companies such as Google aren't investigate in driverless technologies for the sake of it because they see a huge market opportunity for ads. Furthermore, how long will it be before vehicles' speed is fed to local law enforcement?

In our view, in a couple of years, all car speed and safety-related driving patterns will be fed back into central systems. This is the only way that driverless cars can dominate the road, and dominate they will.

2.7.2 *Drink Driving: Alcohol Ignition Interlock*

More than 10,000 people die in the US each year in alcohol-impaired driving crashes, accounting for about a third of all deaths.

An alcohol ignition interlock is an equipment, which measures the level of alcohol in the body. It requires a driver to blow into the device and register a blood alcohol reading that is below a predetermined level. If the driver exceeds the level, the vehicle will not start. In some countries such as the US, such equipment must be installed on frequent drunk driver's car. In case of offense, the data recorded by the device can be uploaded to the driver's Department of Motor Vehicles or even to a court system, depending on the State's law. Some groups, such as (MADD) support ignition locks for first-time offenders.

2.7.3 *Seat Belt Wearing*

In 95 % of accidents, wearing your seat belt will save your life. All newer cars are equipped with alarms that indicate when front drivers don't buckle up.

A Seat Belt reminder system, which is based on weight sensed by the front seats, usually sends a noise signal but doesn't stop the car. This system is usually not connected to back seat belts and rear passengers can still decide to buckle or not, even if legally they are obliged to do so. In the UK, one third or rear passengers don't wear their seat belt. The problem with backseat passengers is that backseats are also used to carry things. Thus alarms for unbuckled passengers are more complicated to manage.

Would it be reasonable to envision as for drunk driving, locking the ignition if not wearing your seat belt? From a technical perspective, this would be extremely simple. The same alarm could be used to stop the ignition, though the issue for back seat passengers would still remain. As we've seen, air bags are connected to the EDR, which records all safety-related measures. Thus using this information to stop ignition would be simple. An alternative choice would be to send the information through the EDR to the police force to inform them that car occupants haven't buckled up.

However, sending this information would be in most cases useless. Indeed, not wearing seat belts is a choice. People can always buckle up their seat belt without actually being secured by it. No technical solution could solve this problem easily.

2.7.4 Real-Time Limitation on Over-Speeding

Newer GPS give more than geo-positioning and time information. They can calculate speed and overlay this speed on a map where legal upper limits can be dispelled. Integrating a physical speed restrainer in the car, which would automatically be triggered after an acceptable over passing tolerance of let's say 20 %, would be extremely easy to do.

2.7.5 Real-Time Information on Over-Speeding

As we've just seen, it would be extremely easy for car manufacturers to supply through GPS all speeding information with an acceptable level of accuracy to the nearby police radio receiver. Integrating the acceptable tolerance of 20 % for over passing, any driver could be fined at the control center directly without even having to be arrested.

2.7.6 Automatic Car Parking

Some cars already have systems that assist with parking, but these are not completely autonomous. These systems can identify empty parking space and steer into it, while the driver uses the brakes. The Volvo system, however, lets the driver get out and uses a smart phone application to instruct the vehicle to park. The car then drives off, maneuvers into a parking place, and sends an information message to the driver on where it is. The driver can collect the car in person or use his phone to call it back to where he dropped it off.

In the past, designs for automatic car parking relied on car parks being fitted with buried guide wires that vehicle needed to follow. That, though, creates a chicken-and-egg problem: car park operators will not invest in such infrastructure until there is a sufficient number of suitably equipped cars and conversely drivers will not want to buy self-parking cars if there is nowhere to use them. This means that for autonomous parking to work most of the technology will have to be in the car itself.

2.7.7 *Fatigue*

Some of the luxury brands newest safety systems try to address each of the four D problems that cause most accidents, especially trying to keep the driver alert.

Volvo uses cameras, lasers, and radar to monitor the car's progress. If the car crosses a lane line without a signal from the blinker, an alarm sounds. If a pattern of tiredness emerges, an icon on the dashboard flashes and the words "Time for a break" is showed. To instill better habits, the car rates the driver's attentiveness as it goes, with bars like those on a cell phone. Mercedes goes a step further: its advanced cruise control won't work unless at least one of the driver's hands is on the wheel.

2.7.8 *Wrong Perception or Judgment*

With all of the distractions inside of a car today, rear-end collisions have become far too common. Nearly, a third of all car accidents where someone was killed or injured in the United States involved a rear-end collision. Most often, the driver in the rear is at fault for tailgating, following too closely, or not paying enough attention to see that the driver in the front car has stopped or is slowing. Most of these accidents could be avoided if drivers would systematically respect the minimum inter car distance. This distance which we transform into time is around 2 s under normal weather conditions. Unfortunately, many drivers don't use such precautionary measures nor do they add the recommended additional 20 % reduction under rainy or misty conditions. Without such measures, drivers are often caught hitting the brakes at the last moment, creating a possible chain of accidents with nearby cars. Companies are working on technology which would alert other drivers and prevent most of these accidents.

For instance, Ford in 2013 tested equipment that could provide early warning to other motorists when brakes were applied hard. The broadcast signal illuminated a warning light in the dashboard of following vehicles, even if they were out of sight or not directly behind the braking vehicle.

Obviously, receiving in advance an alarm informing us of a nearby potentially dangerous situation such as hitting the brakes or abruptly changing lanes would significantly reduce this cause of accidents.

2.8 The Advent of Vehicle-to-Vehicle Communication Technology

In a move that signals a great technological leap forward for the automotive industry, the US Government announced in February 2014 that it will take steps to require all new cars to communicate with each other. New regulation is expected before President Obama leaves office in 2017. This is the first step on a series of automotive evolution that will ultimately lead to unmanned vehicles, resulting from internal electronics integration with external infrastructure, communications, and database systems.

Similar to trains with radio technology, cars that would communicate with each other need some basic wireless technology. The communication blends multiple networking technologies including dedicated short-range communication (DSRC), 4G/LTE cellular wireless broadband, mesh networking, and accurate geo-positioning. All these technologies that we will describe in more detail later on form a system that goes under the name of VANET (for vehicular ad hoc network).

2.8.1 VANET

As the word VANET describes it, this system is composed of elements (nodes) that spontaneously join and self-organize to form a network in an Ad Hoc mode. These elements can be vehicles, communication base stations, traffic lights, security panels, etc. This system is characterized by rapid but predictable network topology changes, one-time interactions between these elements, and partially available connectivity. Intermittent communication can be a real concern, especially in rural areas, where economical reality won't allow for the systematic implementation of telecom access points. Because of this connectivity issue, several telecom technologies are likely to be involved and integrated within the cars: 4G, GSM, UMT, WI-MAX, and a specific technology to VANET called WAVE (Wireless Access in Vehicular Environments) also known under its standard name: IEEE 802.11p.

We will only describe in more details this WAVE technology, as the other ones are already well known and not specific to transportation (check study[12] for more information on WAVE technology). We will also introduce the draft standards being introduced to ensure that all these heterogeneous technologies work together.

[12]Smart Vehicles, Technologies, and Main Applications in Vehicular Ad hoc Networks; Author Anna Maria Vegni, Mauro Biagi and Roberto Cusani.

2.8.2 *Wave Technology*

For those familiar with wireless technology, the WAVE technology is derived from the IEEE standard 802.11 (i.e., similar to the wireless technology people use in their home). The IEEE 802.11p standard uses channels of 10 MHz bandwidth in the 5.9 Gigahertz (GHz) band in the USA. In Europe, the EU has allocated the 30 MHz of spectrum in also the 5.9 GHz band to road safety applications. Unfortunately, these European and American (but also Japanese) dedicated short-range radio (DSRC) networks aren't compatible.

IEEE 802.11p requirements:
Communication distance: These radio emitters need to be able to communicate over a distance allowing for sufficient safety braking distance, preventing dangerous events. The IEEE 802.11p range can reach up to 300 m, which gives around 10 s of go-ahead at highway speed (sufficient as we've seen in the section on PRT headway).
Communication connections: between cars themselves or with the roadside infrastructure, communication must be extremely short. Thus, the technology must work without lengthy association and authentication procedures. IEEE 802.11p amendment defines a way to exchange data through that link without the need to establish a basic service set, such as an access point and associated stations. For that purpose, IEEE 802.11p enabled telecom stations may start sending and receiving data frames as soon as they arrive on the communication channel.

As several technologies will need to work together seamlessly, an international body working group (ISO TC204/WG16) produced a series of draft standards, known as CALM (continuous air-interface, long and medium range).

2.8.3 *CALM Technology*

The objective of the CALM standards is to develop a network terminal ensuring seamless connectivity between cars and roadside systems for several of these technologies we've already mentioned. This is an EU initiative. The USA has its own initiative known as vehicle infrastructure integration (VII).

2.8.4 *LTE Technology in VANET*

LTE or 4G is a standard for wireless communication of high-speed data for mobile phones and data terminals. LTE's would be well suited for periodical communication with cloud computing servers to store information from cars and their environment.

Indeed, 4G provides downlink peak rates of 300 Mbit/s, uplink peak rates of 75 Mbit/s and Quality Of Service provisions permitting a transfer latency of less than 5 ms in the radio access network. Furthermore, LTE has the ability to manage fast-moving mobiles and supports multi-cast and broadcast streams as well as scalable carrier bandwidth from 1.4 to 20 MHz.

Many experts believe that when driverless cars start hitting the road, 5G will also be implemented (around 2020). With 5G communication will come 100 times faster than 4G with much lower latency and ability to connect 10 bn devices around the world.

2.8.5 *Mesh Network Infrastructure*

Network infrastructure will need to sustain various mobile communications between information sources and transaction stations on the roadside and mobile radio units, as well as between portable and mobile units. The envisioned network is based on mesh technology (for more information see Chap. 5). Wireless mesh architecture infrastructure is, in effect, a router network minus the cabling between nodes. It's built of peer radio devices that don't have to be cabled to a wired port, like traditional WLAN access points do. Mesh architecture sustains signal strength by breaking long distances into a series of shorter hops. Every mesh node (small radio transmitters that function in the same way as a wireless router) can send, capture, and retransmit signals to intermediate nodes (other cars, smart traffic signal, etc.) that not only boost the signal, but cooperatively make forwarding decisions based on their knowledge of the network, thus performing routing functionalities (Figs. 2.18and 2.19).

Such architecture may provide high bandwidth, spectral efficiency, and economic advantage over the coverage area. Mesh network is important because, most access points would be in locations with no easy cable/fiber access. Moreover, fast deployment in rapidly changing urban landscape is desirable and mesh structure can easily be fixed on traffic lights, light poles, highway bridge structures, without the needed physical interconnections.

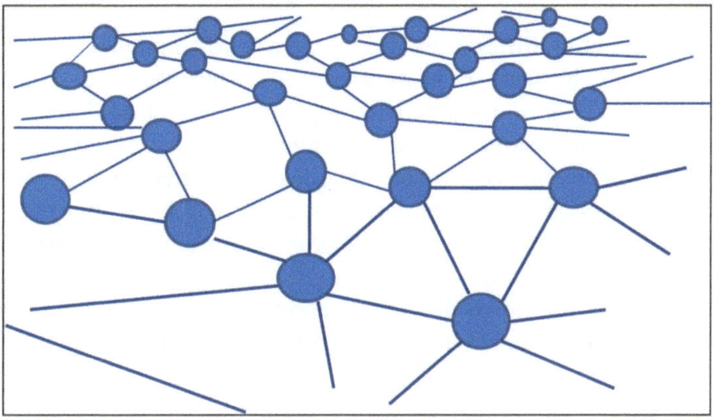

Fig. 2.18 A mesh network routing architecture. *Source* Author

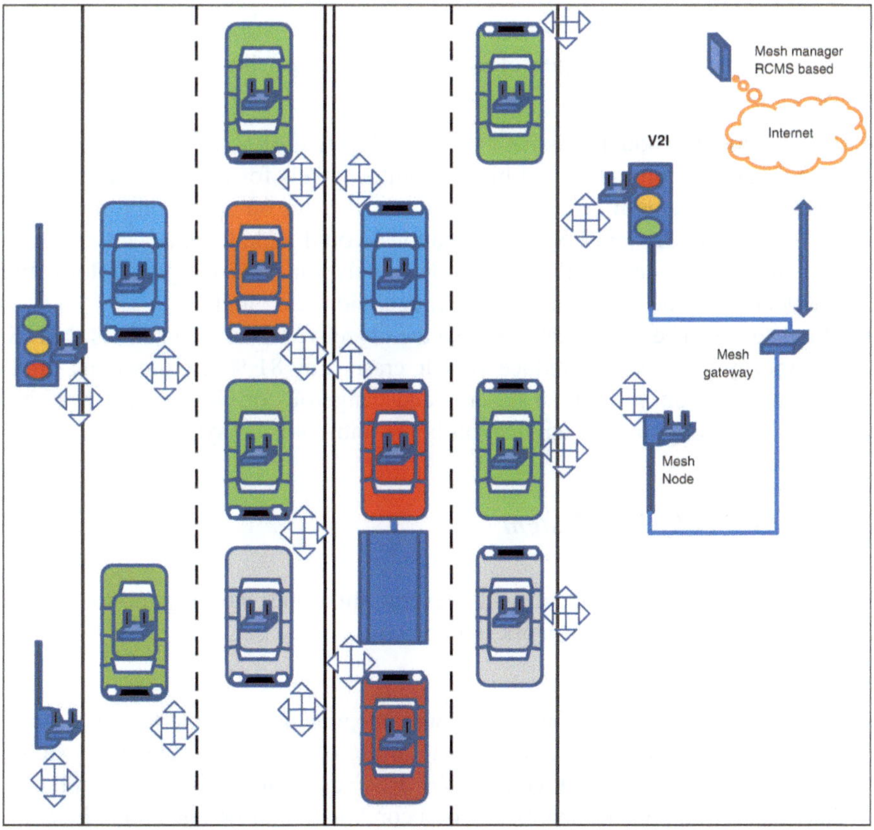

Fig. 2.19 Vehicle Ad Hoc Network composed of mesh nodes communicating through the 802.11p standard and interfacing through a mesh gateway with managing system in the cloud. *Source* Author

- **By-pass functionalities**: Cars need to be able to communicate even when intermittently blocked by passing vehicles. Mesh networks can use the hundreds of other nodes around to adjust to find a clear signal.
- **Fallback options and robust message networks**: If one network is down, alternatives need to be identified and strengthened to reliably propagate messages between networks. For example, if an accident were to cause V2C communications to be broken, a car may still have access to a V2V communication network. An emergency signal message could potentially be sent through V2V to a vehicle nearby, and then between cars and infrastructures until reaching its destination.

2.8.6 *Vehicular Application*

There are typically three types of application that can be regrouped within the VANET systems:

- **Infotainment**: Though this isn't the real objective of VANET, it should bring commercial opportunities and higher comfort levels for passengers;
- **Traffic management**: These applications will focus on optimizing the car flow by reducing traffic jams. Such applications would include enhanced navigation and mapping system, traffic light flow regulation, lane merging information, accident real-time information, speed regulation, etc.; and
- **Road safety**: These applications have the objective to reduce fatalities (NHTSA reports say that it would reduce vehicle crashes by 81 %) by providing drivers with information about road hazards and dangerous driving behavior or situations. The most obvious application is the anti-collision system.

2.8.7 *Anti-collision System*

Since V2V is still a concept with several thousand working prototypes, all description of how V2V will work is uncertain.

Most likely the anti-collision systems will integrate a visual interface included within the front board, forewarning the driver of a potential dangerous situation. V2V warnings might come to the driver as an alert, perhaps a red light that flashes in the instrument panel, or a combination of orange and red alerts for escalating problems? The direction of the risk might also be identified on the front board. In a second generation, for which prototypes already exist, cars are likely to brake or even steer around hazards. All the basic information required for providing this anti-collision system with the required input already exists in most cars.

As seen for trains, cars will need to provide other cars with basic safety information. The EDR, though much simpler than a Train Control Unit, can already provide most of the mandatory information, such as speed, position, direction of travel, car steering angle, car weight, braking conditions, and even loss of stability. All this information will be provided by the onboard equipment except one: the position of the car.

2.8.8 Accurate Geo-Positioning

Everyone has already been aboard a car with GPS. The GPS is a satellite navigation system that gives geo-localization and indicates time. In order to work properly, it needs a direct access to at least four satellites. Three satellites can work but data isn't accurate enough for precise 3-D location (latitude and longitude are still available).

How do GPS work?
Using information from satellites, it basically triangulates its position in regards to the satellites. The satellites transmit a signal of their own location in orbit, with a time signature. The receiver compares the time a signal was transmitted by the different satellites with the time it was received, which gives its location. The receiver uses the signals overlap to do this triangulation. However, as the size of the waves transmitting the signal can be large, their overlap can be several meters, thus limiting geo-positioning accuracy. Furthermore and as most drivers have experienced, GPS can stop emitting geo-positioning information. In fact, GPS satellites transmit a low-powered radio signal that travels by line-of-sight. This means it will pass through clouds, glass or plastic but will not go through most solid objects such as buildings or trees. Furthermore, it can be impacted by atmospheric turbulences or electrical interferences. This is especially problematic during storms and when passing through tunnels.

So there are two fundamental issues to solve with GPS for V2V communication to work properly: geo-localization precision and satellite losses.

Geo-localization precision GPS accuracy was on average around 15 m. While this could be sufficient to indicate where you are in relation to a road junction, it isn't accurate enough to tell you to stop especially if you are in a traffic jam, 3 m apart from the neighboring car's bumper. Furthermore, certain atmospheric factors and other sources of error can also affect the GPS receiver's accuracy. This is why GPS haven't really been used for signaling by railway industry so far. Less than 5 m can separate two tracks and thus, with this level of accuracy, it was impossible to detect if the train was on the same track heading toward a collision or if the trains were just passing by normally.

GPS receivers with WAAS Newer generation of GPS and new safety concept can solve most of these problems. Newer GPS receivers with Wide Area Augmentation System (WAAS) capability can improve accuracy to less than 3 m on average, though the nominal value is 7.6 m. In North America, measurements of the system at specific locations have shown that GPS can provide accuracy level of 1.0 m laterally and 1.5 m vertically. Similar correcting systems exist in Europe (EGNOS) and Japan (MSAS).

PTC initiatives: In order to see how this problem could be solved for cars, let's look at the GPS based railway initiatives. The American signaling program called (PTC) manages to overcome these limitations by using the Nationwide Differential GPS (NDGPS), which comprises of a network of ground-based reference stations, in order to serve as a closer point of reference. The concept behind PTC works by the train receiving information about its location and along which lines it can safely travel. Onboard computers receive the GPS data and can control the movement of the train automatically, preventing any movement that is unsafe according to up-to-date travel information. Using the NDGPS also prevents areas of poor signal, such as tunnels, from significantly impacting the system.

Galileo GPS Europe's answer to the American GPS is called Galileo. When completed by 2019, 30 satellites will be able to give an accuracy level of around 4 m for horizontal and 8 m for vertical positioning. Furthermore, GPS and Galileo will be interoperable.

DGPS Users can also get better accuracy with Differential GPS (DGPS), which corrects GPS signals within an average of 3 to 5 m.

When looking at initiatives in the railway industry, we can conclude that the GPS level of accuracy will enable in the future (around 2017 -2018) a sufficient level of precision for V2V communication to work.

Satellite loss As we've all experienced, GPS can lose satellite signal. In a V2V system, this can be complicated as the exact car's position in regards to the other cars is geo-located. However, in such case, an alarm is likely to indicate the driver that her car's position isn't accurate enough and should be more careful. This loss will, however, be more complicated in an unmanned vehicle environment.

Dead reckoning It is a GPS feature available on many high-end GPS models that help cars keep track of their position after losing their satellite signal. It works by still being able to monitor a vehicle's speed and course of direction, though less accurately than through true satellite guidance. Some systems measure the average speed for a stretch of road based on recent car data and calculate the position based on projections. Other systems may use a digital compass and a connection to the car's sensors to help define the car's speed and direction.

Active markers We've seen that these tools are often used in railways to give basic information about geo-positioning. Most likely in critical area such as tunnels, they will be used to compensate for the absence of satellite access.

2.8.9 V2V Operational Mode

V2V operational mode relies on trust. For experienced drivers, putting their life into the hands of their car's onboard computer won't be easy. They should. After all, millions already do with unmanned trains and as we've seen reducing the human error factor in transportation can only increase safety. However, it is true that passengers will need to trust their car but also the nearby vehicles in a few different ways:

- Firstly, the driver must trust that the received signal comes from the car directly in front or on its side;
- Secondly, he is trusting that this car is accurately reporting its state, and not sending wrong messages; and
- Thirdly, she is trusting that the received signal isn't trying to hack into the car's system or being sent by malicious people.

Operationally, the following six components that would be deployed in vehicles equipped with V2V would need to be working together:

1. An appropriate blend of telecom technologies, such as DSRC that receives and transmits data through antenna, other telecommunication technology, and a mesh network;
2. GPS receivers providing vehicle position and time to DSRC radio and supplying timekeeping signal for applications, as well as accurate and updated maps;
3. An onboard communication network that incorporates the existing network that interconnects components in the vehicle;
4. An electronic control unit that runs safety applications;
5. A driver–vehicle interface that generates warning issued to driver; and
6. A memory that stores security certificates, application data, and other information.

A ruggedized communication security system, which will check the proper functioning of these six components, will need to provide and verify V2V security certificates to ensure this trust between vehicles.

2.9 Intelligent Wayside Technologies

Infrastructure may sound boring in regards to smart cars, but these new e-mobility technologies will demand that Public Authorities decide how to promote such technologies and define the role they should play in their implementation. Unfortunately, and as we will see in Chap. 6, there is a huge deficit in public spending for infrastructure. In many countries, Governments, States, and municipalities are broke and can barely afford to maintain the transportation infrastructure they already have. No public authorities will look favorably at having to invest in new infrastructure and pay the maintenance of such new equipment. However, without clever infrastructure, the concept of V2V and unmanned cars can run into trouble. In fact, we could probably say that smart vehicles are only as intelligent as the infrastructure that surrounds them is.

In this section, we will look at which key infrastructure should be made more intelligent and who could pay for it. However, in the last instance, if no one can pay for it, there is always the possibility to ensure that the required functionality be performed directly by the car and paid by the owners themselves.

In order to make complex concepts easy to remember, marketers have come up with additional acronyms. V2I and V2C are used to indicate Vehicle-to-Infrastructure (V2I) and V2C communication, respectively. Under V2C, we include all the telecommunication backbone and network. Under V2I, we include all the road equipments that will provide real-time information.

2.9.1 Vehicle-to-Cloud (V2C)

We've just seen that continuous wireless component is a key success factor of V2V. V2C, which incorporates this wireless system, includes two elements:

- **Cloud computing**: This portion can be hosted by servers from big IT providers. As we all know, this hosting capacity is growing exponentially and this is now a service which can be sold easily; and
- **Wireless communication link**: We've seen that V2C will be based on a mesh network with base stations or LTE tower positioned in key areas of cities or roads (i.e., junctions).

The interesting aspect of mesh networks that we've seen is their ability to reassemble themselves to fit changing environments. Moreover, the more connection points, the better. Indeed, one of the driving forces behind mesh technology is Metcalfe's Law, (quote from Bob Metcalfe one of the Ethernet inventor) that says that the value of a network grows as its number of connection points increases. Imagine wireless traffic relaying information from car to car until reaching its destination. More cars would just mean more capacity to transmit data. It should also be noted that some network companies such as Cisco are working on newer network technology, which would allow for decentralized data processing, directly within the networks, increasing reaction responses.

Some infrastructure providers would also need to support this V2V initiative. One area which will be critical is parking lots.

2.9.2 Intelligent Parking

Some companies have already developed small wireless sensors that can indicate if a parking space is occupied or not. Like balises for the railway industry, these magnetometers that are glued to the road or buried a few millimeters under the pavement can detect when a car arrives or leaves.

In 2014, San Francisco will implement the largest mesh network for monitoring 6,000 parking spaces. The city hopes that displaying information from the sensors on Web maps, smart phones, and signs on the street will reduce the traffic and pollution caused by circling cars.

2.9.3 Intelligent Traffic Systems

Intelligent lighting systems based on sensors or cameras as well as centralized computer making adjustment to help car flow can be found throughout the world. V2I, however, will enhance this capacity by giving additional information that ordinary sensors cannot.

By leveraging and installing wireless transponders and smart embedded sensors, cars will feed safety information into highways and roads. Such information would include static road hazards like curvy roads, flooding areas, low bridges, changing road conditions such as construction, minimum and maximum speed, as well as traffic input such as density, flow, volume, and speed. Furthermore, information at cross sections will also be available under V2I, such as average car speed and time wasted at such junction. However, the main advantage of such technology is that the information flowing from the lighting system to the car could synchronize automatically the vehicle's speed to reduce the braking needs.

One could also imagine that in areas with low car density, information flow from the car could reduce the lighting's stopping period, enabling the vehicle to avoid stopping altogether. It could even eliminate altogether the need for lighting systems, in areas with no pedestrian. As a matter of fact since 2012, computer scientists from the University of Texas have been developing smart intersections designed for self-driving cars. If successfully applied, their technology will insure that onboard and intersection computers communicate directly and synchronize safely the required car speed at intersections without the use of any traffic lights or stop signs.

2.9.4 Distributed Intelligence

How many people have been stuck for hours in a traffic jam, wishing someone or something would have given it an advanced warning of an accident or works lying ahead? On many toll highways of this world, dedicated radio services now give such information. However, not everyone wants to listen to elevator music or to what happens 300 km from their current car location. Furthermore, getting alternative route information is often difficult. With new GPS technology, things have improved as they often provide real-time road status and alternative roads. However, such services are often limited and alternative routes often end up taking more time than just staying on the road.

V2V intelligent traffic management should help. Traffic information already exists on many highways and main city arteries, centralized in control centers. Often traffic reports are relayed to radios or to GPS real-time service provider, and even sometimes, displayed simple electric panels on the road. However, V2C technology will extend these services to all types of road. Cars will be able to send messages to cloud servers informing everyone about the road condition and potential hazardous situations they've encountered, through the means of geo-localized messages (giving a latitude, longitude, and vertical position). Any cars passing by will be able to get such information in quasi real-time and intelligent mapping will then be able to select and display the best alternative route.

2.10 Driverless Cars

Marketers have come up with several names to characterize cars that can run without the need of a driver: self-driving, driverless, unmanned, autonomous, robot car, etc. All these names describe the level 4 or 5 of automated cars. Independently from how we name them, these cars rely on four basic principles: Data acquisition through sensors, data treatment through onboard computers and software programs, two-way communication, and interface with the car's drive train equipment.

2.10.1 Data Acquisition

Unmanned cars need to sense their environment in all potential conditions:

- Exterior and interior environments, such as tunnel;
- In all spectrum of light, from daylight to night; and
- In all weather conditions, be it fog, rain, snow, or even sand in the desert.

In order to sense all these conditions, a mix of several technologies will be required. These are a few of the technologies currently used by Google and that could equip these cars, if their unit cost can be reduced significantly.

Lidar It is a laser radar system, which measures the distance by illuminating elements positioned in front of the laser beams. Lidar uses a spectrum of ultraviolet, visible, or near infrared lights, to picture objects in all conditions. It can target a wide range of materials, including metallic and non-metallic objects, rocks, rain, human beings, etc. In fact everything that can be found on a road can be picked up by a lidar.

> Google driverless car used till 2014 for testing a 64-beam laser mounted on the top of a car, which unit price was $70,000. The laser provided three-dimensional depth. Each beam flashed around ten times per second, scanning more than 1 million points in concentric waves. This laser could detect a 30 cm long object crossing a road 50 m ahead of the car.

Radar Google's car integrates also radar technology, which is an object-detection system that uses radio waves to determine the range, altitude, direction, or speed of objects. The radar has twice the range of the lidar but doesn't have the same precision.

Cameras Finally, it also integrates vision through cameras. Cameras are used for identifying road signs, signals, colors, and lights.

Data fusion All views from these three technologies are then integrated, combined, and overlaid by the digital maps and Street Views™ that Google collects throughout the world. The result is a clear 3 D model of the road.

Once the system gets a clear view of its environment, it still needs to understand what it is seeing. The difficulty is that roads, unlike metro lines, aren't controlled environments. Kids, animals, and other cars can rush from the side or run in front of the car. Luckily, driving is mostly simple: follow the front car within the two delimitating lines, with a defined interspacing. This is especially true on highways. The complication is managing the remaining stuff, especially in urban environments.

2.10.2 Data Treatment

How can computer treat such complicated stuff? The answer is by writing appropriate algorithms. The difficulty is what is an appropriate algorithm and how do you make sure you aren't missing any critical set of rules or interpreting it erroneously? To illustrate such risk, let's take an example from the science fiction writer Isaac Asimov, who set the three laws of robotic[13], a very appropriate example for a robot car:

- *A robot may not injure a human being or, through inaction, allow a human being to come to harm*;
- *A robot must obey any orders given to it by human beings, except where such orders would conflict with the First Law*; and
- *A robot must protect its own existence as long as such protection does not conflict with the First or Second Law.*

Mr. Asimov described in his book Robots and Empire, a 0 law created by a robot which modifies the three first laws:

- *A robot may not harm humanity, or by inaction, allow humanity to come to harm.*

Obviously, this is pure science fiction but it shows that if software engineers rely too much on self-learning programs, it can create rules that are difficult to validate through scenarios or testing.

Let's take a more practical example to illustrate how testing and self-learning engines can be useful. Anyone who has accelerated with a rear wheel drive car, when hitting an icy patch may have experienced its car rear end trying to

[13]The novel Robots and Empire from Isaac Asimov; published by Doubleday Books in 1985.

come around. In order to redirect the car in the road axle, one must turn the wheels in the direction of the spin. This move is counter-intuitive and anybody who hasn't gone through this experience will turn the wheel in the opposite direction of the spin, ending crashing in snow banks, hopefully without injury. A self learning system would do the error once but would either define its action plan or identify the issue for the software engineer to solve.

If we take the example of the spinning car rule, we could create an algorithm replicating the following scenario: "if"

- the car is rear wheel drive (x1);
- Sensors have identified a quick spin (x2);
- ice has been detected by vision (x3);
- temperature is lower than 0°C (x4); etc.,

"than do":

- turn the driver's wheel in the same direction as the spin(y1) if there is any wheel;
- use engine braking to decelerate (y2);
- if using ABS brakes, maintain a steady pressure on the brakes (y3);
- if not using ABS brakes, do not maintain steady pressure, apply light stabs, hold, release and press again (y4);
- Brake at speed under 10 km (y3);
- etc.

After setting such rules and the necessary interface with sensors and drivetrain, one needs to drive the car and test by trial and error that the rules don't make the car crash. This is a slow and demanding process. Luckily, software engineers can take advantage of self-learning software programs. Other logic models can be used, such as fuzzy logic to integrate and understand uncertainty. In other words, software engineers can combine learning and rule setting to come up with the best algorithms.

Many set of rules can be tested according to several scenarios. For instance, simulations based on accidents documented by the US National Highway Traffic Safety Administration can be applied. Many accidents in rural areas are due to collision with wild life. If a dear crosses in front of the car what rules should be applied? Should the car try to avoid it or hit it? Is the previous rule true in all conditions or it needs to integrate speed? How much advance warning does it need?

As we see, driving hours and experience with self-teaching systems will optimize these algorithms. Having worked on these rules and having been able to fine-tune them will give companies with an early start, a clear advantage. For instance, we all heard that the IT company Google's unmanned cars have been accident-free for a few years now. This proves that their algorithms are working. It also proves that they are safer than normal vehicles, as statistically after more than 830,000 km, and human drivers would have already had an average of two crashes.

2.10.3 Financial Barrier to Adoption

One of the barriers to adoption is cost. In 2010, the cost of Google's self-driving technology was $150,000, of which as we've seen $70,000 was just for the lidar. Obviously, such price is dissuasive but some suppliers believe that mass production could reduce quickly the lidar to less than 1000$. Computational processing, which is still another large component of the overall price, will experience the usual exponential cost reduction.

Many specialists in the automobile industry believe that Google's idealistic approach to the driverless cars won't bring the price of all these technologies down far enough to make it attractive to the mass market. So many car manufacturers together with their OEMs are looking at ways to consolidate and simplify the hardware. Miniaturization, sensor fusion, and integration of controllers are a few approaches considered. Better optical systems that could replace Lidar and radar technologies are also envisioned.

Whatever the solutions that will prevail, one thing is certain, cost will go down. How quickly is still a question mark, but according to a recent study[14], price for the self-driving technology will add between $7000 and $10,000 to a car's price in 2025, a figure that will drop to around $5000 in 2030 and about $3000 in 2035, the year when the report says most self-driving vehicles will be operated completely independent from a human occupant's control. Some other reports suggest that by 2030, the price of the additional technologies will fall under $1000, at which point the autonomous option will cost probably less than the annual savings in insurance.

If costs come down, the next most important barrier is legal. In our opinion, and though these legal issues are real, they will be tackled in time to allow the advent of unmanned cars. The fact is that some US States such as Nevada, Florida, California and Michigan have already legalized self-driving cars. The UK has also legalized their use on public roads in 2013. As economical interest and proof of safety will prevail, more and more countries and States

[14]Emerging Technologies: Autonomous Cars—Not If, But When; Source IHS Automotive forecasts; Date January 2, 2014. http://press.ihs.com/press-release/automotive/self-driving-cars-moving-industrys-drivers-seat.

will legalize self-driving cars. Though legal concerns are legitimate, we see them more as a resistance to change than a road blocker. After all, the railway industry as gone unmanned and this is a market with stringent standards and heavily regulated. In our view, legal issues won't stay in the way of all the forces advocating for autonomous vehicles, such as:

- Fewer accidents, reducing overall risk and liability, which will cause insurance companies to favor self-driving cars;
- Significant reduction in the number of deaths and injuries, bringing undeniable social benefits;
- Greater convenience and possibility for elderly citizens to continue using their cars, leading to strong consumer demand;
- Market differentiator, giving a huge incentive to manufacturers to sell profitable new options; and
- Drunk driving reduction and increase in alcohol consumption, making alcohol companies, bars and restaurants strong supporters, especially in country with 0 alcohol tolerance for drivers.

Simply put, the money is with the forces for autonomous vehicles. Insurance companies, liquor companies, vehicle manufacturers, customers, and governments will all want the benefits of self-driving cars.

2.10.4 Legal Barrier to Adoption

As mentioned, a few states and countries have accepted the principle of driverless cars. Unfortunately, this is more a permission to test vehicles rather than the real legal foundation for a society composed of a fleet of driverless vehicles. In fact, the efforts by the Nevada State to pass the first law authorizing the use of autonomous cars in June 2011 were reported to be highly supported by Google, in order to legally conduct further testing of its driverless technology. However, this law still limits the use of unmanned systems. Although the law acknowledges that the vehicle "operator" (this will replace world driver, at least till there is no operator at all) doesn't need to pay attention during operation, he/she must not have disturbing activities such as sending text messages and requires a person behind the wheel at all time. In California, the law demands for a car operator behind the steering wheel to take over in case of sudden car failure.

2.10.5 Legal Responsibilities

With all these legal limitations, what is required for the advent of unmanned system? First, let's look at what are the implications of a car crash for ordinary car drivers.

Current liabilities There are three potential causes to an accident: issues linked to cars, problems with the infrastructure, or an act of God. Let's focus on the two first causes, as the latest is actually invoked in cases where there can't be any proven liability. In case of car accidents, there are five potential liable Parties:

- One or both drivers;
- Car owner of one of the cars who didn't maintain the car's physical integrity;
- A mechanic who had recently serviced one of the vehicles badly;
- One of the car dealers; or
- One of the car manufacturers.

According to David Chung, a partner of an American law firm (April 2013[15]).

> *There are numerous factors which would be taken into account in order for the Court to determine who is liable. Depending on circumstances, multiple parties may be liable where more than one party is found to have materially contributed to the incident resulting in the injury or damage. Where multiple parties are culpable, they would share the liability in the proportions determined by the Court... Who is liable is a question of fact.*

People or entities can also be liable for infrastructure issues. Bad road conditions due to failure to maintain decent driving conditions, or unexpected events causing fatalities, such as construction work not being properly announced and trucks or people being where they shouldn't be, are also potential liable parties. Obviously, in an era without drivers, one of the most frequent liable parties—the driver—will disappear, but new actors will be introduced.

Liability for Driverless Cars According to Mr. Chung's view on principle of law: "*relationships of trust give rise to duties of care.*"

In unmanned trains, passengers trust that the system will perform correctly. Similarly, passengers using self-driving cars will trust that all elements will operate flawlessly to avoid any accidents. Thus, in this e-mobility era, passengers will be off the hook for any liability, as they would trust that someone or something else is in control.

> Once again comparison with trains will be useful to understand litigation changes brought by e-mobility technologies. The segmentation, we've used between, cars, infrastructure and even act of God isn't appropriate anymore.

---

[15]David Chung, Driverless cars, whose liability; March 4, 2013.http://www.bennettphilp.com.au/content/driverless-cars-whose-liability.

> We need to think in terms of system integration and the different elements that are likely to be part of this system.

Using an analogy with railways, we will have potentially six subsystems that could be subject to liabilities:

1. Manufactured vehicle;
2. Onboard signaling system, which will include the five or six elements (the drivers interface might be made redundant) that were described in the V2V operational modes;
3. Car-to-wayside communication, including V2V, V2C, and V2I;
4. Control center, which will host network information;
5. Road infrastructure provider, which will take responsibility for infrastructure signaling as well as road physical conditions; and
6. Car operator or owner.

2.10.6 Vehicle Manufacturer Potential Liabilities

For the sake of this exercise, we will suppose that manufacturers (i.e., Mercedes Benz, General Motor, and Toyota) won't be integrating the onboard signaling system. We believe that as long as these vehicle manufacturers don't supply the onboard signaling system, they won't be associated with taking over the car operation from the driver. Thus, car manufacturers' overall liability will remain more or less the same, as it is the case for train manufacturers of unmanned system. However, there is a caveat to this, which is that train manufacturers must build some intrinsically safe equipment. An example in the railway industry is that whenever a pressure lost in the hydraulic braking system happens, the brake inherently will be applied, as the pressure maintains the brake apart from the wheel. This also most likely means that redundancy might have to be added to critical-related safety equipment. It also means that in case one of these intrinsically safe equipments fails, the car will need to stop in a safe mode.

2.10.7 Onboard Signaling System Provider Potential Liabilities

In most unmanned railway systems, the onboard, wayside, and control center system are integrated by the same supplier. With initiatives such as ERTMS level 3, the EU has pushed for train interoperability between different railway systems, showing the path for decoupling between trains and the railway infrastructure. This shows that system integration of very complex systems can be achieved as long as Governments and the major players of an industry elaborate interoperable standards. This needs to be the model followed by the automotive industry.

Let's assume for a moment that Google decides to position itself as a signaling system provider. This would mean that any car manufacturer (i.e., Chery from China) could install on one of its car model Google's signaling systems composed of the five or six V2V operational elements we've already identified, plus probably the sensors, as the signaling system wouldn't be able to work without them. As this system would replace the human factor from driving, it would in fact be responsible for all wrong decisions made, as the driver is today. Currently, in most countries, the consumer protection laws would impose obligations on the signaling supplier to compensate any person suffering injuries or damages, as a result of such wrong decisions. Although responsibility coming from software problems discovered after a product is sold is less clear cut than for a faulty product, there are legal precedents, especially with cars, as recent quality recall events have shown.

2.10.8 Telecom Provider Potential Liabilities

As we've seen, a critical element of self-driving cars is the possibility to communicate with other cars and the infrastructure equipment, as well as getting useful road condition information from a hosting data center.

> In signaling systems, when the system aren't dedicated to the railway network, telecom providers (i.e.: for GSM-R, a railway GSM standard) are rarely the cause of accidents. The worst case scenario is usually that the signaling system cannot perform according to requirements and the trains must stop. Though not good for the train operator's image, these events are usually manageable.

What would happen if a complete mesh network in one of the most crowded arteries of a megacity would fail? Could the other technologies take over without resulting on an oversaturation of the entire telecom network? Would telecom companies be liable for such massive traffic jams? In our view, probably not, as long as they could reestablish the networks within reasonable delays. Furthermore, as there will be a blend of telecommunication technologies, there will probably be a fallback telecom technology that could be used.

2.10.9 V2C Hosting Centers

The hosting servers would need to store real-time information coming from cars and road operators and send it back on a geo-localized basis. What would happen in the case that they store the wrong information, are hacked (as it recently happened), or worst aren't able to send an emergency geo-localized message? On the same principle of trust, they would be most likely liable for any accidents or damages resulting from such error or omission.

2.10.10 Road Infrastructure Provider

Today, road operators are very rarely liable for accidents. Accidents resulting from road problems, such as potholes or bad signaling, are usually blamed on the driver or bad weather conditions. With unmanned system, this will change.

In railway environments a derailment caused by a broken rail or a faulty fastener cannot be invoked to avoid litigation. Even bad weather conditions cannot provide legal protection. In fact in unmanned system, weather condition monitoring will be mandatory.

Whenever a metrology equipment detects dangerous conditions, it will require that cars automatically reduced speed or stop operation altogether.

2.10.11 Operator or Car Owner

We must make the distinction between a car owner and a fleet operator of unmanned cars. The car owner will still be responsible to ensure that the car is in perfect working order. Thus, the owner will need to ensure that the car goes to the garage. This means that a mechanic could still be responsible for a crashed car, if he failed to inspect and repair all faulty parts. In cases of parking or traffic tickets, the owner of the car would most likely be held responsible for paying the ticket, even if the car and not the owner broke the law.

Would driverless car fleet operators' legal status be similar to the taxi owners' status? We aren't sure that they would only be responsible for ensuring that the provided cars are in working orders.

Railway operators are usually involved in accidents inquiries. Whenever blame cannot be pinpoint to faulty manufacturer's equipment, signaling or infrastructure providers or to an act of God, they can be convicted.

2.10.12 Suggestions to Minimize Legal Barrier to Adoption

Even though driverless cars will be much safer than conventional cars, there will be a time when a faulty electrical contact, a broken piece of equipment, an erroneous piece of software, or an integration problem will generate an accident. In such case, a lawsuit will follow or even class actions could take place if several accidents can be associated with one of the players. Although driverless technology stakeholders have the same basic obligations to offer safe products and may have the same set of legal exposures if they fail to do so, it won't be a simple problem to identify the guilty party. How do you apportion blame between different integrated subsystems, if there is no real system integrator?

In the article[16] on liabilities published in April 23, 2014 "Who Is at Fault When a Driverless Car Gets in an Accident?," Mr. John Villasenor explains that current product liability law is the fruit of decades of precedent that established responsibilities whenever making or selling products. Plaintiffs in products liability lawsuits can choose from various "theories" of liability when seeking to recover damages.

- **Negligence**: it occurs when manufacturers fail to design safe products;
- **Design defects**: it occurs when a characteristic of a product create hazardous situation, due to a flaw in its product design; and
- **Breach of warranty**: Because marketing and selling products create explicit and implicit warranties, products liability also involves contract law.

However, the main difference in our view is that any individual product can work flawlessly but still cause an accident when interfacing with other technologies.

This is why in railway systems, an independent entity reviews the system's safety case and will give its approval to operate, usually after a period of testing. This is doable in confined environments, but much more complex in an application the size of a country. Will safety agencies need to guarantee the safety case of all urban and rural country roads? It is unlikely.

[16]Who Is at Fault When a Driverless Car Gets in an Accident? Author: John Villasenor (Published on April 23, 2014 in the Atlantic).http://m.theatlantic.com/john-villasenor/.

In our view, all subsystem suppliers will need to follow standards and that in the end no entity will give a safe to operate label for the system, but will for each individual equipment or software of this system.

As robots cannot be charged with crime, the trial lawyers will probably in the end go after the richest company, creating a real risk that innovating companies will shy away from developing unmanned technologies. There are solutions to avoid such situation. One of the easiest ways to reduce barrier to adoption is to maintain legal status quo. This means passing on risks to car owners. Manufacturers, signaling, and infrastructure providers might lobby consumers and their insurers to take on risks associated with their driverless vehicles as a purchase condition.

Would consumers agree to take on the risk? For material damages, there would be a good case for consumers to accept such liability, especially if insurance car companies decide to implement a no-fault policy and reduce their policy costs proportionally. As the number of accidents would dramatically decrease and more importantly their severity, it would probably make financial sense also for insurance companies to promote such a program. However, the issue about criminal liabilities would remain as it is unlikely that consumers would accept such risk.

Governments should introduce laws promoting self-driving cars as it is in their best interest to reduce significantly their citizens' fatalities and injuries. A law imposing V2V to all new cars, like the US Government is contemplating, is a move in the right direction. After a transitional period, it could even make sense for governments to outlaw all conventional cars. Although this would create a legal foundation for the unmanned industry, it wouldn't be sufficient to avoid litigation risks. In fact, some entity would still need to cover potential litigation costs. In our view, what would make the most sense would be for Governments to take over the liability for injuries and damages. In fact in many countries, they already do. Injuries in many countries are paid by the public health insurance. Governments could establish a public fund to compensate the victims of accidents involving autonomous cars. In our view, the funds would quickly be compensated by the $billion reduction in health care costs. It would also be in the interest of insurance companies to participate in such funds. Car accidents would still happen and thus people would still require car insurances, but on the other hand, insurance companies would see the bills for heavily injured people reduce drastically, as the severity of accidents would diminish tremendously.

2.10.13 Technical Suggestions to Minimize Potential Litigation

The best suggestion to minimize the impact of litigation is still to reduce risks of potential accidents.

Semi-autonomous cars As seen, we are expecting car manufacturers to propose a progressive approach to self-driving cars. Cars that still require a driver behind the steering wheel are easier to build and potentially cheaper. However, according to some experts, this semi-autonomy may be the worst of all worlds. Drivers pay less attention to road conditions even though cars aren't really built to take control. Drivers would be likely to take over the controls only in emergency situations. How would he or she be able to assess a critical situation in the blink of the eye? In most cases, reflexes would be slow or not to say inexistent if the passenger was taking a nap. This is why Google in May 2014 announced that they will get rid of the human factor altogether. The new car that Google will build in around 100 units will have no steering wheel, pedals, or controls, just a stop and go button (Fig. 2.20).

Brick wall safety concept In our view, an unmanned system cannot allow for uncertainty to threaten passengers' life. Unmanned train system calculates at all time a safety time/distance that is required to ensure complete stop before hitting the train in front of it. In order to avoid class actions (at least at the beginning of this technology implementation), cars would need to follow this same principle. By applying this brick wall safety concept described previously, cars will always be in a position to avoid bumper-to-bumper accidents. To reduce the need for much longer car interspacing, the signaling system should be in a position to shorten the

Fig. 2.20 Picture of Google unmanned car, which Google intends to manufacture. *Source* Picture of the Google car; Copyright of Google Inc., which allows for the unaltered use of its content

necessary additional time required for this safety concept, by eliminating the human perception and reaction time.

Safer than human being concept In order to protect entities against mass litigation, unmanned cars should be at least safer than vehicles with drivers. This will be the case. For instance, accidents resulting from a change of lane should be eliminated as cars will constantly monitor the presence of other vehicle on the right or left side, either through V2V communication or sensor detection. However, we all know that on a non-segregated lane, some unfortunate events cannot be prevented. For instance, someone falling on the street or deciding to jump in front of the car will not be avoided by self-driving cars. A deer deciding to run in front of a car or a dog running to the cars will most likely be hit. But the point is that a "driven" car wouldn't be better at avoiding these crashes. Most likely, humans would be worst, as they always need longer perception and reaction time than robot cars. This should in our view limit mass litigation, though wouldn't avoid criminal enquiries in case of fatalities.

2.10.14 When Will It Happened?

The Nay-Sayers probably believe that it is too complicated or people love too much driving for it to happen. Our belief is that, as for unmanned train systems, self-driving cars will be a common feature of our life pretty soon. It won't be something that will come over night and as we've seen there will be a progression in the class of automated cars.

As we don't have a crystal ball, let's check what the car manufacturers are planning to sell and show what industry experts are expecting.

- BMW in its upcoming i3 electric car model, is planning to introduce traffic-jam features, which will let their car accelerate, decelerate, and steer by itself at speeds of up to 40 km/h. It will require though that the driver leaves at least one hand on the steering wheel.
- Mercedes S-Class cars are already equipped with a system that can drive autonomously through city traffic at the same speed. This speed of 40 km/h is important because it allows for less regulation. Mercedes also claims that some of its technology can detect if a driver is getting drowsy, though it only uses sound alarms to warn the driver. In its S-Class, Mercedes is also planning to include several sensors. A 3-D camera could be positioned on the windshield. Short- and long-range radars would also be installed on front, rear and side of the car. Furthermore, twelve ultrasonic sensors will enable close object detection.

- In 2011, Volvo already declared that by 2020, no one would be killed or injured. It has recently introduced cars, which can take preemptive action, such as tightening the seat belts, charging the brakes for maximum traction, and, even in extreme circumstances stopping the car. It is planning to launch in 2015 technology it refers to as "driver assist", intended for the highway. It will utilize radar and cameras to enable the driver to sit back and enjoy the trip. This equipment recognizes front and behind spacing, limiting the possibility for other vehicle to get near it, as well as detecting lane marking.

Year	Prediction
2014	Volvo will feature Adaptive Cruise Control with steer assist which will automatically follow the vehicle ahead in queues
2015	Audi plans to market vehicles that can autonomously steer, accelerate, and brake at lower speeds, such as in traffic jams [88]
	Cadillac plans vehicles with "super cruise": autonomous steering, braking, and lane guidance. This technology will likely spread to other GM models in following years
	Nissan expects to sell vehicles with autonomous steering, braking, lane guidance, throttle, gear shifting, and, as permitted by law, unoccupied self-parking after passengers exit
2016	Toyota plans to roll out near-autonomous vehicles dubbed Automated Highway Driving Assist with Lane Trace Control and Cooperative-adaptive Cruise Control
2018	Google expects to release their autonomous car technology
2020	Volvo envisages having cars in which passengers would be immune from injuries
	Mercedes Benz, Audi, Nissan and BMW all expect to sell autonomous cars
2025	Daimler and Ford expect autonomous vehicles on the market

Market forecasts of research institutes vary also enormously. We've picked the results of five different institutes, so that readers can decide by themselves what the future looks like.

- In 2014, HIS Automotive a research company reported that by 2025, self-driving cars sale would account for only 230,000 units. This first group of autonomous cars would most likely have Level 3 capability (limited self-driving functionalities).The market shares of this technology would raise to 9.2 % in 2035, as completely unmanned vehicles would become available. 7 million of those 11.8 million self-driving vehicles sold that year would rely on a mix of driver input and autonomous control, with the remaining 4.8 million vehicles completely unmanned. IHS expects that by 2050 e-drive cars will outnumber conventional cars on the

road. By then, the majority of cars sold will be unmanned, with conventional car sales becoming increasingly rare.
- The company Navigant Research is more bullish about sales of autonomous vehicles. According to their 2014 report, self-driving vehicles will grow from fewer than 8000 units sold in 2020 to 95.4 million in 2035, representing 75 % of worldwide car sales. By that time, North America is forecasted to account for 29 % of worldwide sales of level 4 and 5 self-driving cars. China will represent 24 %, while Western Europe will account for 20 %.
- Strategy Analytics research company, meanwhile, expects autonomous cars that are highly automated (but not exactly self-driving) to have a market share of around 15–20 % in 2025–2030.
- Expert members of the Institute of Electrical and Electronics Engineers (IEEE) estimated that up to 75 % of all cars will be autonomous by 2040.
- ABI Research forecasts that self-driving cars would become a reality by 2020 and that 10 million such cars would be sold in the USA by 2032.

2.10.15 Self-driving Market

Predictions on how quickly unmanned vehicle technology will be adopted is complicated and very different from one expert to the other. Estimating how much this market will be worth is even more complex. To do so, we first need to separate the cost of the unmanned technology per se (what we called the signaling technology) from the car itself to assess correctly the self-driving market opportunities. We need also to add infrastructure opportunities, be it on the wayside or in the cloud.

Signaling price As we've already seen for batteries and other technologies such as computers, incremental decreases in cost are mostly dependent on the rate of adoption of that technology. The automotive industry being worth so much, any significant system sales will immediately impact costs. Thus and according to industry experts, the price premium for this electronics technology, which should add around \$10,000 to a car's price in 2025, will fall drastically. This amount, which would be initially installed on premium cars, would then drop to around \$5000 by 2030 and about \$3000 in 2035 when no driving feature would be required anymore.

Two ownership models for self-driving cars Such predictions rely on the principle that unmanned cars will continue to be bought by the passengers as they are today by drivers. This is most likely to happen in the early development of this technology. Self-driving cars are likely to be an important selling feature in the early years, which will affect sales cars positively. The question is if the number of unmanned cars will grow initially and then peak, due to new business models enabling passengers not to own their car but to use it whenever necessary, as a taxi.

Some industry experts are saying that sales of cars will be affected tremendously by such models. For instance, Columbia University's The Earth Institute forecasts the reduction of United State's fleet of vehicles by a factor of 10. Price water house Coopers forecasts a collapse in the United States of the number of vehicles from around 245 million to just 2.4 million.

However, and as we will see in Chap. 7, younger generations tend to be less interested in car ownership, and these estimations remain very optimistic or pessimistic, depending on one's appreciation of the car role in modern society. In our view, people will continue buying cars as this is part of the social status. People don't need a Ferrari nor do they need a watch anymore, but they still buy both luxury items. Furthermore, passengers will most likely still want to feel "at home" in their car and not share it with others. Mercedes Benz President Dieter Zetsche in fact presented his unmanned car at the CES stating that (Fig. 2.21):

> *the car is growing beyond its role as a mere means of transport and will ultimately become a mobile living space.*

So how much money are we talking about for this market? If we do the math for 2035, a point in time when the market will be already reasonably mature, we would have 95 million cars x \$3000 = \$285 billion. Needless to say that is probably one of

Fig. 2.21 In January 2015, Mercedes presented its version of a driverless car at the CES show in Las Vegas (Picture of the F015 driverless concept car, Copyright of Mercedes Benz and kindly lent for this book)

the biggest opportunities for any given market. Furthermore, this excludes the infrastructure market, which would benefit to telecom companies such as Erickson, Cisco, and Huawei, as they stand to gain from the million of wireless routers and wireless equipment that will need to be installed. In fact, Machina a consulting firm[17] suggests that this market will be worth $422 billion by 2022, a significant portion coming from new connected services, which don't even exist today.

2.10.16 Testing the Driverless Application

There are already some driverless car systems being tested on both sides of the Atlantic.

From 2014, driverless taxis will be carrying passengers during demonstration projects in five European cities, during six to eight months. The EU project CityMobil2 brings automated vehicles to designated roads inside the city centre.

In the United States, the Michigan University in partnership with several car manufacturers (Toyota, GM, Ford, Bosch, Xerox, Econolite) has started the construction of a new city dedicated to testing unmanned vehicles on a 12 ha field. This area called Mobility Transformation Center, will integrate four lanes, signaling, junctions, lighting systems, buildings, pedestrians and building areas. In fact most real life conditions will be possible to emulate.

2.11 Security

Security issues are very different from a private or public transport perspective. In private transportation, the usual risks are having your car stolen or being high jacked. In the future when cars will be driverless, new security threats will happen such as hackers getting access to your car's computer and creating malfunctions on the vehicle's equipment, causing a crash. In public transport, on top of the risk for the passenger of being robbed, mobbed or even assaulted by a gang of angry hooligans, there is the risk of people vandalizing the infrastructure as well as terrorists putting a conventional or dirty bomb, a lethal virus or a poisonous gas in the station or on board a train driving at 350 km/h. The answers to such threats can vary tremendously from one risk scenario to the other but will always need to contemplate three main factors:

[17]Report written by analysts Machina Research, "Connected Car Industry 2013" and commissioned by Telefónica SA.

- **Human factors**: how do staff, police, guards, dog patrols, and even passengers react in a threatening situation;
- **Technology**: there are as many technologies as threats that can be thought of. Most technologies come from other areas with few adaptations to the transportation sector. Lately, some technologies have been adapted to the specific needs of Public transport; and
- **Procedures**: what are the reactions that need to be taken in function of different threat scenarios.

Experience has shown that it is the association of these three factors that can achieve the best results. These factors can be applied to the main principles that we've regrouped under four areas:

- **Reassure and Deter:** Through police patrolling or random checks, police may be able to deter potential thieves to organize and assault passengers. By being visible often, law enforcement officers can deter but also reassure passengers. Surveys on security technologies such as CCTV have indeed shown that they can play a reassuring role with the public and can deter people from committing crimes, since they know that their acts can be recorded;
- **Detect, record, and alert:** When a crime is committed it must be detected
- **Assess; confirm**; and
- **Investigate; review; identify; show evidence during trials**.

2.11.1 E-Mobility Security Solution

Defense or security industry specialists might argue that the security solutions are just an extension of existing technologies applied in buildings, casinos, airports, or even on the battlefield and shouldn't be included within these e-mobility technologies. We believe on the contrary that the very nature of public transport networks —characterized by many access points, frequent stops, large geographical areas, many passengers and large fleets of moving vehicles—requires specific security solutions. In the public transport environment and unlike in airports, security Personnel cannot search passengers systematically at specific checkpoints. Furthermore, unlike street, building, or large infrastructure surveillance, public transport security systems must be designed with stringent electromechanical and anti-vibration standards. Additionally, the integration of security systems, such as CCTV on board the train, with the wayside communications infrastructure and potentially into a command and control center, adds incremental complexity not found in other systems. This complexity requires new software and telecommunication technologies, which are part of this e-mobility revolution.

This section is dedicated to the new IP technologies that are unique to public transport and will not describe, access control, biometrics, baggage screening, and other security technologies that are either applied without any changes or rarely used in such environment. Thus, it will mainly focus on CCTV systems and highlight how they will evolve to better tackle the central challenge of public transport security, which is to balance passenger security concerns with accessibility, convenience, and affordability. It will argue that the integration of fixed and mobile assets can better manage security threats. Finally, it will give an insight into how bundling digital CCTV images with other public transport system technologies will improve both the operational performance of the public transport system and increase the situational awareness of security decision-makers. The concepts and ideas coming from the following section were inspired by a report[18] on the benefits of CCTV within Public Transport, written by the Security and ITSI committees of the UITP in 2010, in which the Author had the honor to participate.

2.11.2 End-to-End Security Solutions

Each public transport operator has a unique set of risks and vulnerabilities, and security solutions for metros, commuter lines, light rail vehicles, buses, or intercity operators will vary substantially. That said, the level of sophistication that an operator wants for its solution will depend on its experience to date with security systems already installed, and the likelihood of future disturbances in its system.

Although some public transport operators are still looking for simple security solutions to implement in their existing networks, more and more operators are looking for end-to-end solutions where all subsystems, fixed or mobile CCTV, access control, employee/visitor systems, GPS, passenger information system, etc. are integrated into the operational control center. This fundamental trend can be explained by the development of the e-mobility technologies as well as by the need to align with professional security strategies being implemented to address these various public transport risks and vulnerabilities.

2.11.3 Technological Trends in Security

As seen for safety risks, security threats are event driven. In other words, a perceived risk is always linked to a location (within a bus, on a station platform, etc.) at a given moment (when there is a crowd, after a football game, etc.), involves

[18]CCTV: a Tool to support Public Transport Security; Factors to consider before installing or upgrading; UITP 2010, co-authors M. Babington, L. Barr, D. Bernard, K. Clark, G. Dunmore, B. Hart, N. Koide, T. Kritzer, G. Lucisano, J.C. Pinero, A. Silva Neves, K. Takemoto, J.P. Van Keymeulen, S. Van Themsche.

someone (criminal, bomber, suicidal person, etc.), and something harmful (bomb, gun, toxic gas, fire, knife, etc.).

There are three fundamental security strategies that can be implemented to manage these events:

- **Proactive (Prevention):** Stop the event before it occurs;
- **Reactive (Preparedness & Response):** Act to limit the impact of the event, if the previous strategy failed; and
- **Forensic (Recovery):** Get information to put the system back in operation immediately.

Obviously, from a security point of view, the proactive measures are much better. Stopping terrorists before they explode their bomb is better than finding who did it afterward. From a security perspective, this is the challenge technology providers are facing: going from current forensic approach toward a reactive and ideally proactive phase. Today's reactive and proactive approaches are better addressed by human beings. They can more quickly understand a situation and with good training and adequate policies define the right action to be taken. However, as more and more information is gathered (i.e., through CCTV, access cards, biometrics), human beings have difficulty coping with the amount of data and need more and more automation software for audiovisual and metadata data processing.

Railway operators are becoming aware of these human limitations and are selecting e-mobility technologies based on their track record and cost:

- **IP networks**: Use of Internet Protocol networks onboard vehicles and in fixed environment networks enables the use of standardized software and hardware technology. This quickens the passage from analogue to digital technologies, which as a consequence will push the convergence of images, audio and data;
- **Open software platform**: Open based software, such as SOA solutions, delivers a universal mechanism to interconnect all applications from different systems. This distributes real-time information from disparate data sources and provides powerful new means of unifying different databases across networks;
- **Increasing processing power**: This enables the possibility to perform demanding processing tasks such as video and audio analytics closer to the network's edge, such as within the train or at the camera level;
- **Video analytics (intelligent/smart systems)**: Emergence of software using algorithms enabling patterns detection. By enabling a computer to search for events, the images generated by CCTV or sound detection system can become useful input data for these algorithms and detect events;

- **Broadband communication**: The increasing ability of wireless broadband communications to support transmission of images with sufficient bandwidth; and
- **Bundling of CCTV** with other operational requirements: Another trend is for security features to be bundled with safety, control, maintenance, and other operational features in the system.

2.11.4 Limitations of Analog Security Systems

Security Personnel in major transport networks is not physically able to get access to critical information quickly enough. Indeed, within a medium-size metro network environment, there are more than 1000 fixed cameras, with views centralized at one or several operational control centers. It is thus extremely difficult to detect an event before it occurs (especially during rush hours), or react quickly upon it when it is identified. Good training of security Personnel and scorecards of highest risk locations usually helps detecting hazardous situations, but this is expensive and becoming more and more complex with each new camera added to the system.

In the onboard environment, the situation is even worse where the absence of physical connection impedes security Personnel from viewing live images. Videos are simply stored on a hard disk caddy to be retrieved manually if needed and reviewed usually in the days following the incident. This forensic role is furthermore supported by watermarking technology, which ensures that video sequence can be used as evidence in court.

Having said that we will now see that IP technologies are shaping the security solutions which are being implemented in the railway environment:

- Digital cameras and Video Recorders;
- Audio, video, and metadata integration;
- Increased processing power;
- Increased storage capacity; and
- Emergence of new standards, especially compression technology.

2.11.5 IP Cameras

Analog cameras, which are quickly fading from the market, used digital system to transform light into bits of information and then encoded it back into an analog signal for coax transmission. However, this process created limitations, from which IP cameras are free.

Furthermore, analog cameras at high resolution (4CIF) had a significant problem with interlacing. Images would become blurry whenever there were lots of movements, which obviously is the norm in metro environment.

IP cameras (also called network camera) employ progressive scan technology that better suits depicting moving objects clearly. This more advanced image capture technology means that the whole image is captured at one time, thus providing crystal clear images.

Analog cameras couldn't either provide resolution above television standards (NTSC/PAL specifications), which corresponds to 0.4 megapixels at 4CIF. Operators are now requiring much higher resolution. An IP camera's higher resolution provides more detail and can cover larger areas. Furthermore, in the IP camera system, images are digitized once and stay so with no unnecessary conversions and image degradation.

Network camera technology enables PTZ control over the same network that transports the video. With a Network Dome camera, the PTZ commands are being sent over the IP network, resulting in major cost savings and greater flexibility. What's more, network cameras can integrate input and output signals such as alarms and controlling locks.

2.11.6 *Integrated Audio*

For many public transport applications, audio has become increasingly important. With an analog system, audio is not possible unless running separate audio lines to the DVR. A network camera can solve this by capturing audio at the camera, synchronizing it with the video or even integrating it into the same video stream, and then sending it back for monitoring and recording over the network. The audio can also be fully bidirectional to allow communication over speakers.

2.11.7 *Compression Technology*

Compression technology is based on the assumption that a video frame contains a large amount of redundant information that can be eliminated without a great loss in perceived picture quality. Compression methods are effective up to a certain point, beyond which the image quality quickly degrades.

The technology for compressing video pictures has evolved over time, from its origin as a storage system of still photographs on computers (JPEG) where the compression ratio was only 8 to 1, to the current MPEG-4 standard, where compression technology using wavelet can ensure a compression ratio of 100 to 1. Thanks to such compression technology one day of recording requires only 14 GB, a hard disk capacity, which could easily be found in a hardware store.

It is foreseen that in the near future, MPEG-4 technologies will maintain their leading position in the public transport environment, extending their reach by integrating functions such as H.264 (or MPEG-par 10). Other compression technologies are emerging from the mass media market but at this point in time, it is difficult to say what will replace the MPEG-4 standard in the public transport environment.

2.11.8 Wayside IP CCTV Solutions

There are several benefits to installing an IP CCTV system. The most important is that it shifts the security infrastructure responsibility from security experts to the IT departments. In other words, the IP CCTV network becomes one subsystem of the entire IT department. This improves operational efficiencies tremendously by leveraging the IT department's technical expertise, vendor relationships, and support processes to reduce the costs of deploying and maintaining a video system. It also helps standardizing equipment (i.e., storage and servers) across departments and vendors, which results in lower maintenance costs.

IP CCTV benefits are:

- **Infrastructure cost reduction**: as it permits the use of inexpensive, standard network cabling to power cameras and transmit video instead of more costly coaxial cables. It also reduces the amount of cable and conduit required by eliminating separate power and signaling cables and aggregating cable runs to network switches located throughout a facility.
- **Ubiquitous access to video**: Whether local or remote, security enforcement personnel can access video from anywhere on the network, with all the access policy controls of the organization's other IT services.
- **Leveraging established network infrastructure**: Public transport operators can leverage the use of established, highly secure network infrastructure, proven network connectivity and health monitoring tools and robust storage systems to provide a high degree of confidence that video is available when needed.

- **Unlimited storage capacity**: Operators can allocate more storage for video with longer retention requirements.
- **Synergies with other departments**: Security department can also share video investment with operations and marketing for non-security applications thereby increasing overall return on investment.

2.11.9 Integrated Security Event Management Systems

New open software platforms will strongly influence IP CCTV network technology development and largely influence the design of future CCTV architectures. SOA and event-driven software platforms will create a powerful environment for service-led event-driven networks, in a wired and wireless world. New SDP technology such as VoIP and SIP, or specifically designed for images such as IMS, will, when integrated within the Next Generation Network, enable security suppliers to provide any new service (i.e., any security application) defining it directly at the service layer without considering the transport layer. In other words, any security service will become fully independent from its environment. High-level application technology such as UPnP (Universal plug and play) will allow public transport operators to easily and quickly integrate technology coming from different suppliers. We will describe in detail these technologies in Chap. 5 Connected cities.

Reactive security strategy All these software technologies when combined enable an integrated security event management system, which provides the shift from a forensic approach toward a reactive security strategy. In this strategy, key decision-makers have a complete awareness of what is happening on their various networks, including their mobile networks comprising each single bus, tram, or train.

Proactive security strategy We've mentioned that the best security strategy was proactive, as it is always better to detect any criminal act before it is perpetrated. Such an integrated security platform, including the operator's rules and policies, will become a tool to inform security Personnel about potential threats. Not only will they be informed about the nature of the potential threat, but also they will be able to have a complete view of the situation through devices such as PDAs or 4G cellular phones. The potential for video analytics in the public transport is only limited by our imagination. The applications will be totally dedicated to the specific reality of the public transport environment.

2.11.10 *Total Integrated Public Transport System*

The convergence of voice, images, and metadata is ineluctable. Telecom and network equipment providers are investing tremendous efforts and resources for this to happen. There are few other market sectors other than public transport where these forces will have such a strong impact on the business fundamentals of the industry. Systems that are seen as independent today, such as security (CCTV, access control, biometrics recognition, bomb sniffing devices, etc.), signaling, passenger information systems, train control, remote control, global positioning, remote maintenance, and more generally other applications involving operations and marketing, will share the "services."

People counting is a good example of how CCTV will affect public transport operators. In the medium term, operators will be able to count how many passengers are in their system by using cameras. Too many passengers in the station will require a flow control where, for instance, train headway could be reduced or passenger flow could be controlled at the gate. In order to do this, the exchange of data (services) will be needed between the security system and the signaling. Passengers on the platform could be automatically informed of the next train arrival schedule or alternative lines. The same application will be able to give valuable information for the operator's marketing department where they will be able to know statistically where people embark and in which station they get off. Operational functions such as train control will be informed automatically of passenger load. According to passenger load thresholds, power would be decreased by switching off air conditioning or auxiliary equipment, resulting in a significant reduction of energy consumption.

As demonstrated through the above example, Operational Control Centers (OCC) will become even more the decision-making center of all the different systems converging probably physically in one master room. Every potential stakeholder involved in a crisis situation will have in the OCC all the elements to take a coordinated decision to avoid criminal or terrorist acts or limit their impact.

2.11.11 *Video Analytics*

Video analytics are basically algorithms that are applied to the bits of information coming from cameras (or microphones in the case of audio analytics). Computer vision, speech, facial, and object recognition are some of the areas where algorithms are being developed. It also makes heavy use of digital geometry and signal processing. Most of these systems work on the principle of pattern recognition.

Pattern recognition It can be defined as the act of analyzing raw data and taking an action based on the category of this data. Pattern recognition aims to classify data based on either a priori knowledge or on statistical information extracted from the patterns. The patterns to be classified are usually groups of measurements or observations, defining points in an appropriate multidimensional space. This is in contrast to pattern matching, where the pattern is rigidly specified.

The classification or description scheme usually uses one of the two following approaches:

- **Statistical pattern**: Recognition is based on statistical characterizations of patterns, assuming that the patterns are generated by a probabilistic system; and
- **Syntactical pattern**: Recognition is based on structural interrelationships of features and not only on simple numerical feature vectors, as used in statistical classification.

Although video analytics are being developed for defense and general security purposes, software engineers are designing analytics to meet the specificities of the railway environment. Furthermore, because of the mobile aspect of train or buses, computational power must be considered a constraint that doesn't really exist in fixed environment.

Indeed, applying these algorithms can require enormous computational power. The amount of memory or computer time required can become astronomical when the issue to identify goes beyond a certain size. To take into account memory capacity and data power, security system designers in railway environments will need to consider where they place the system's intelligence: centrally or in a distributed manner.

2.11.12 Distributed Intelligence

Although some experts believe that the best strategy is centralized intelligence, we think CCTV networks in railway environment will evolve toward distributed intelligence. There will be for each operator a unique design on where to apply intelligence. This means that some video analytics will be positioned within the cameras. Most likely, every intelligent piece of software which can be affected by the quality loss during compression or which will try to reduce noise will be put there. Some intelligent pieces of software in the metro onboard environment will most likely be included within the DVRs. By doing this, the operator will be able to both register the audiovisual information (for legal reasons) and tag information linked to alarms. It will also be able to prioritize this information and make sure that it is sent to the security personnel by interfacing with the train control system.

Like in any distributed architecture, there might even be some pieces of video analytics that will be split between different devices. This strategy (that exists in other areas such as automation) improves response time by giving the order to process information only if a state is found in both elements of the CCTV system.

2.11.13 Video Analytics Limitations

Video analytics' objectives are to help security Personnel detect a risky situation or take a set of actions based on the detected abnormal patterns judged risky. As for any good security Personnel, detecting the real risky situation and not a false alarms will define the quality of the analytics. In other words, the operator, whenever assessing the quality of a piece of video analytics, will need to define his acceptable level of false positives and false negatives (as in all statistical tests, there will be a trade-off).

- **False positive rate**: It is the proportion of negative instances that were erroneously reported as being positive; and
- **False negative rate**: It is the proportion of positive instances that were erroneously reported as negative.

Threshold values within the video analytics solutions can be varied to make the test more restrictive or more sensitive; with the more restrictive tests increasing the risk of rejecting true positives, and the more sensitive tests increasing the risk of accepting false positives. With these limitations in mind, new video algorithms are being developed and deployed in public transport. On the wayside, a certain quantity of software applications coming from the retailing and defense industries are being deployed with an acceptable level of accuracy. In the mobile environment, suppliers are trying to cope with issues such as vibration and extreme changes in lighting conditions.

Fig. 2.22 Through a mix of solutions such as sound analytics and anti-intrusion detection, public transport authorities should be able to reduce graffiti and vandalism. *Source* Author

2.11.14 *Video Analytics Technologies*

A few simple video analytics are now starting to be deployed onboard trains and buses, which we've described hereafter. However, the possibilities of creating analytics that can be applied to the public transport environment are only limited by our imagination. For instance, gunshot detection, abnormal behavior detection, and arm detection are a few of these technologies that could emerge in the midterm. They will most likely be developed in the military or law enforcement environments first and then be modified to the specificities of the public transport, we've largely described previously (Fig. 2.22).

- **Camera obstruction detection**: This is a major issue for railway operators, where vandals or muggers paint or break the cameras before committing their crime. In order to limit false positive and negative levels, some manufacturers are merging two functionalities within the DVR:

  - Watchdog functionality, which monitors constantly the state of the cameras and makes sure that power is being absorbed by the cameras (and thus make sure that the problem is not linked to camera electrical malfunction); and
  - Video analytics function that monitors variation in the quantity of light being recorded.

- **Area obstruction detection**: This piece of software detects objects that may interfere with the use of fire exits or other user defined areas intended to be kept free and clear for health, safety or other operational reasons. This capability allows operators to define both the area to be monitored as well as the length of time an obstruction remains in place before triggering an alarm.
- **Anti-intrusion detection**: Cameras using infrared technology detect any people passing by or objects moving within the set perimeter, especially during low-light situations such as night time. New video analytics using conventional cameras are also being developed, basically creating virtual limits, within which any presence is automatically detected. This type of technology is being implemented to monitor restricted areas such as tunnels and depots.
- **Empty vehicle detection**: One operational issue is to detect passengers within a train (especially in driverless applications) before going to the depot. Very often people hide in the train in order to graffiti or vandalize the assets. Video analytics that can detect people within the trains are being developed to cope with this problem.
- **Unattended baggage detection**: This video analytics detects any unattended object located in a specific area. An alarm is triggered according to a set period of time.
- **Automatic target acquisition**: This piece of software enables zoom-in on suspicious persons or vehicles, and tracks it across the full scene in a separate view, as if there were an additional PTZ camera trained on the scene.
- **Automatic car plate recognition**: This piece of software positioned aboard a bus or a tram can detects plate numbers and automatically processes cars that are running or parked irregularly, for example in a bus lane. It uses Optical Character Recognition (OCR) in which the pixels on the digital image of a license plate are transformed into ASCII text.
- **Facial recognition**: The purpose of this video analytics is to recognize people by comparing selected facial features from the image and a facial database. A newly emerging trend in facial recognition is three-dimensional face recognition.
- **Profiling**: The purpose of this video analytics is to recognize a category of population based on criteria, such as gender, race, color of the skin or hair, etc. It can use the same type of algorithms as facial recognition but rather than comparing it to a database of faces (personnel, convicted criminals, missing persons, etc.), it basically compares fixed patterns (color, height, shape, etc.).
- **People counting**: Designed originally for the retail sector, people counting video analytic capabilities are now providing marketing and operations management with another level of storing traffic intelligence

and reporting. Dedicated cameras mounted above entrances, exits or other areas of the station count passengers as they enter and leave an area. The feature provides aggregate counts from multiple points of entry. In the onboard environment, this function is provided by infra red sensing technology positioned on top of the door entrance. Manufacturers are working on video analytics that can substitute infra-red sensors.

- **Loitering detection**: these analytics provide an alarm when a person or group remains in a controlled area for a prolonged period of time. Alert times can be adjustable according to different scenarios, such as squatting, trespassing and soliciting.

2.11.15 Security for Cars

As we indicated before, security is mainly a public transport issue. This isn't to say that carjacking or car thefts aren't important issues. It has more to do with the fact that, besides panic buttons or hidden balise in the car emitting a geo-positioning signal, there wasn't much exciting security technology brought to the market. However, with e-mobility technology, things will change.

As the e-mobility revolution will rely more and more on computers to run, they will get hacked. In fact, some people are already able to get into car systems using Bluetooth or other limited wireless technology. However, when V2V and later on unmanned technologies will be in every car, a new potential security risk will emerge: cyber terrorism. Hackers could not only take possession of someone's car if they were able to pass through fire walls and other IT security systems, but also they could actually send wrong signals from their own cars. Indeed, malicious drivers could send fake signals such as wrong speed, interspacing, and braking information.

The point is that any disruptive technological revolution will bring benefits and new risks. With e-mobility, cyber terrorism is unfortunately coming to the headlines.

Acknowledgments and Disclaimer The use of pictures or references made to studies or companies and their brand does not in any way suggest that the Authors of such studies or the mentioned companies endorse in any way this book or its content. The author endeavors in respecting the copyright subsisting any of the graphics, and texts that he uses, to use graphics and texts he has himself created or to use graphics and texts not covered by copyright. All trademarks and brand names quoted in the book including those protected by copyright of third parties are subject unreservedly to the provisions of current copyright law and the rights of ownership of the registered copyright holders.

Company or Brand Names Stated in the Chapter

- UITP (Union International des Transport Publics)
- Matra (technology later purchased by Siemens)
- VAL now part of Siemens' portfolio of product
- Westinghouse Electric Corporation
- Volvo Group/AB Volvo
- Audi AG
- BMW
- Mercedes Benz is a Trademark of Daimler AG
- Google Inc.
- Ultra Global PRT
- Uber Inc.
- Axa S.A.
- Ford Motor Company
- OnStar™ system: trade mark of General Motors
- Street Views a trade mark of Google
- General Motors Company
- Toyota Motor Corporation
- Chery Automobile Co. Ltd
- Model i3 brand from BMW
- Model S-Class: Trademark of Mercedes Benz
- HIS Automotive
- Navigant Consulting
- Strategy Analytics
- ABI Research
- PricewaterhouseCoopers
- Ferrari S.p.A.
- Erickson
- Cisco System Inc.
- Huawei Technologies Co. Ltd.
- Robert Bosch GmbH
- Xerox Corporation
- Econolite

Chapter 3
Environmentally Conscious Society

Acronyms

ABS	Anti-lock braking system
EPA	Energy Protection Agency
IEA	International Energy Agency
IGBT	Insulated gate bipolar transistor
NHTSA	National Highway Traffic Safety Administration
RBS	Regenerating braking system

There is a paradox. Society as a whole is becoming more environmentally conscious but at the same time, every year the world uses more and more nonrenewable energy. According to the International Energy Agency (IEA), the average energy used per person increased from 1990 to 2008 by 10 %, while world population grew by 27 %. As world population and consumption per capita increased, the overall worldwide consumption growth was of 39 %! Furthermore, developing countries consumption has increased tremendously, not quiet yet catching up the Rich World but bridging the gap constantly. For instance, these are regional energy consumption growth over that period (Table 3.1).

To make matters worse for the environment, this spectacular growth in energy consumption was mainly CO_2 driven with oil and coal (but also to a lesser degree natural gas). In 2008, it corresponded to as much as 81 % of the overall energy produced (Table 3.2).

Needless to say, that alarmists predicting the end of cheap and abundant petrol have been proven wrong. With unconventional energy-schist gas or petrol, and Canadian bituminous tar—kicking in, worldwide proven reserves have increased drastically.

The reality is that even when petroleum reserve will end, the oil and gas industry will be able to find new sources of unconventional energy, such as liquefied coal or even methane recuperated from the permafrost (Japanese and Canadian initiatives are already under way). So, why this paradox? It goes to the principle that if given a choice people will not reduce their lifestyle.

S. Van Themsche, *The Advent of Unmanned Electric Vehicles*,
DOI 10.1007/978-3-319-20666-0_3

Table 3.1 Consumption growth (1990–2008)

Region	Middle east	China	India	Africa	Latin America	USA	EU-27 block
Growth (%)	170	146	91	70	66	20	27

Adapted by Author from IEA data

Table 3.2 Energy by power source (2008)

	TWh	%
Oil	48,204	33.5
Coal	38,497	26.8
Gas	30,134	20.9
Nuclear	8283	5.8
Hydro	3208	2.2
Other renewables[b]	15,284	10.6
Others	241	0.2
Total	**143,851**	**100**

Source IEA
[b]Solar, wind, geothermal and biofuels

As seen, the Oil industry and OPEC leaders are giving them this choice, firstly by increasing the supply of diesel and gasoline. Secondly by allowing the price at the pump to stay lower than 0.80 euro (the rest is only taxes added by the various Government levels), a price attractive enough to continue fueling demand both for gasoline and cars.

Why is energy consumption such an important issue for transportation? The reality is that transportation is one of the main drivers of energy consumption. The following table from the IEA shows that transportation means together accounted for 36 % of the total energy, a number that is in constant growth (Table 3.3)!

Even more concerning for the environment, the automotive industry is likely to experience a significant growth in fleet size.

Car fleet growth All experts predict exponential growth in the number of cars running throughout the world. The worldwide global fleet size of commercial and passenger cars was estimated in 2013 at around 1.1 billion cars. Looking ahead 20 years from now, it is likely to have more than doubled by then. BP estimated in its 2014 forecast[1] that the global vehicle fleet will hit the 2.3 billion car number by 2035.

[1]BP energy outlook 2035, January 2014.

Table 3.3 World energy consumption per sector (end use of energy)

Sector	2000	2008	2000	2008
	TWh		%	
Industry	21,733	27,273	26.5	27.8
Transport	22,563	26,742	27.5	27.3
Residential and service	30,555	35,319	37.3	36.0
Nonenergy use	7,119	8,688	8.7	8.9
Total	**81,97**	**98,022**	**100**	**100**

Source IEA 2010

Not surprisingly, most of this growth is expected to be coming from the developing world (86 %). BP also estimates that non-OECD's fleet will more than triple from 0.4 to 1.5 billion over the outlook period and will overtake the OECD fleet in 2022. Between 2012 and 2035, vehicle density per 1000 population will grow from by 8.4 % p.a. in India and 6.9 % p.a. in China.

The IEA is a little less bullish about the automotive growth, but still estimates that by 2035, there will be around 1.7 billion cars worldwide. This would still increase tremendously the number of vehicles currently on the road. Their car fleet growth estimate is based on vehicles per capita (for comparison, the USA currently has 660 vehicles per 1000 inhabitant):

- 2000: 4 vehicles per 1000 people.
- 2010: 40 per 1000 people.
- 2035: 310 per 1000 people

Luckily for the environment, car efficiency improvements are likely to limit growth in demand for petrol. Thus, non-OECD fuel demand will "only" rise by 82 % despite having three times more cars. In the OECD, this demand is likely to fall by 15 % due to efficiency gains outweighing slowly the growth of vehicle fleet (22 %). These projections, don't actually take into consideration the unmanned car factor or car electrification, which could reduce further oil consumption in the rich countries.

3.1 Governmental Environmentally Friendly Initiatives

Public Authorities and Opinion are trying to influence the transportation choices by imposing new measures to curb pollutants and particles emission with greenhouse impact. Increase in taxes at the pump is one coercitive measure governments are using. Differences in price with or without taxes can reach a staggering 200 % in countries like Turkey and Norway, one of the main fossil fuel producers. Most governments are using various tactics to influence this choice. Some initiatives are coercitives—taxes and bills forcing the industry to be more environmentally efficient—and others are incentive-based: bonuses or price reduction of environmentally friendlier technologies.

3.1.1 Tax on Combustible

In a society that is politically correct, fuel tax is seen as a fair tax that brings in lots of money. The bigger the car and its consumption, the more one pays. By modulating tax levels between fossil fuel types, governments have influenced the combustible choice.

For instance, in all 28 EU Member States the overall tax rate on diesel is lower than on gasoline. This gap in taxation has slightly narrowed in recent years but still remained in 2012, 27 % lower. Therefore and despite being more expensive than gasoline to produce, diesel is cheaper at the pump than gasoline in all EU Member States (except the UK, Bulgaria, Cyprus, Hungary and Romania). This taxation strategy was followed despite evidence to suggest that the external costs of diesel cars are on average higher than those of petrol vehicles (such as pollution and health consequence).

The impact of such tax policy has been extremely successful, as diesel accounted in 2012 for 71 % of fuel consumption by weight, rising from a very low percentage 20 years ago. To understand how successful has been this strategy, we can look to the USA. Excise taxes on diesel have been historically higher there and less than 2 % of the installed American car park is diesel-based. The following chart gives the excise taxes on retail gasoline and diesel fuel in the USA (Table 3.4):

As we can see on the previous table, US excise taxes on gasoline are quite low in regards to other countries, which explains why price (based on 02/2012) at the pump are much lower in North America (USA: \$0.93/L and Canada: \$1.39/L) than in Europe (Italy, \$2.30/L; France: \$2.10/L; Germany: \$2.04/L; Norway: \$2.58/L). On the other end of the spectrum, many countries subsidize gasoline consumption, especially oil producing countries, such as Venezuela, Saudi Arabia, Kuwait, Angola, UAE and Russia are selling gasoline or diesel at a lower price at the pump than what they could sell on the international market. For instance, the price at the pump for a liter of Gasoline in Venezuela was only 2 cents in 2010, according to the World Bank!

Table 3.4 Data from the US Energy Information Administration (January 1, 2013)

In US$ cents	Per US gallon		Per liter	
	Gasoline	Diesel	Gasoline	Diesel
Federal	18.40	24.40	4.86	6.45
Average all US states	23.47	24.00	6.20	6.34

Table 3.5 Price breakdown of gasoline and diesel in France (September 2012)

Prices at the pump	Diesel		Gasoline (95 %)	
In euro	Euro/L	%/L	Euro/L	%/L
1 L of Brent	0.59	41	0.59	36
Refinery	0.11	7	0.09	6
Distribution	0.09	6	0.09	5
Taxes	0.67	46	0.88	53
Total	**1.45**	**100**	**1.64**	**100**

Source Author

Table 3.6 Excise tax at the pump

Country	USA	Canada	Germany	France	Japan
%	14	31	55	54	44

Source Author

The price at the pump is composed of several elements. The following price breakdown is based on French 2012 prices (Table 3.5).

The following table gives an estimation of tax proportion of the gasoline price at the pump, for that same year (Table 3.6).

With excise taxes corresponding already to around 50 % of the pump price, it is easy to understand why European Governments don't pursue more actively further tax increases. The reality is that prices at the pump are already at a level where people think twice about using their car. The price elasticity of demand for gasoline (a measure of variation in demand of a good linked to a change in price) is far from negligible, especially in the long term. Even if people would need their car to go to work, see friends and enjoy life, they would look for cheaper alternative transport means.

An old research,[2] which analyzed several American studies on price elasticity, concluded that:

[2]Explaining the variation in elasticity estimates of gasoline demand in the USA: A Meta-Analysis; Author Molly Espey, publication: Energy Journal (1996).

- On a one-year period, this price elasticity is about −0.26, meaning that a fuel price hike of 10 % will reduce gasoline consumption by 2.6 %; and
- After 1 year, the price elasticity of demand is −0.58; a 10 % hike in gasoline causes quantity demanded to decline by 5.8 % in the long run.

Lately, tax hikes have been rare in the EU as indebted governments need to protect their cash cow. For instance, the total amount generated by oil taxes in France was 34B euro in 2010, which corresponded to about 10 % of all French taxes. In comparison, in the UK, this tax corresponded to around 7 % of all tax sources. Such excise taxes represent a much lower proportion of income sources of the US government.

3.1.2 Carbon Tax

A number of countries have implemented carbon taxes or energy taxes that are related to carbon content. In the Chap. 6 new business model, we describe how the carbon credit works and how it could help public transportation.

3.1.3 New Clean Air Regulation: California Clean Car Law

The California Clean Car's program includes several regulations to reduce the impact of cars on environment. It imposes:

- Stricter fleet average standards for 2015–2025 models to further reduce nitrogen oxide and hydrocarbon emissions;
- Increase in engine durability requirements from 120,000 to 150,000 miles, and impose new particulate emission standards on gasoline-powered cars;
- New greenhouse gas emission limits for 2017–2025 model year cars; and
- Emission limit of 166 grams of carbon dioxide-equivalent per mile by 2025, comparable to the federal standards set by the EPA.

The rules rely on off-the-shelf technologies, including variable valve controls, direct injection, turbochargers, cylinder deactivation, engine stop–start, low-emitting refrigerants for air-conditioning systems, and improvements in transmissions. California's zero emission vehicle program is also intended to commercialize electric, plug-in hybrid, and fuel cell vehicles. The state has set a goal of those technologies reaching 15 % of all new vehicle sales in California by 2025.

3.1.4 Internalization of External Costs

Allowing drivers or passengers to pay less on the highway or on the railway network according to the hour of the day, is a way of optimizing resources and addressing congestion problems. Rail and aviation operators have been using techniques such as yield management to increase their revenue. Few highway or mass transit operators have differentiated their charges according to the time of day and/or week. The reason for this might be that this measure, which would address congestion issues, was difficult to monitor and apply. However, with new e-mobility technology, this can be easily achieved. Urban parking charges that vary with time of day and/or proximity to central business districts are also a good way to address congestion.

3.1.5 Incentive Measures: Tax Credit or Penalties

In order to motivate people to switch to electrical means, several governments (France, USA, etc.) have reduced the acquisition price of cleaner cars by giving rebates. Others have penalized consumers looking for guzzling cars by adding extra taxes on the car's acquisition price. France's "bonus-malus" scheme is an example of a passenger car taxation policy which was proven successful since its implementation in 2008. The scheme aims to accelerate the market penetration of lower emission vehicles by providing financial incentives for people buying lower emission cars (bonus) whilst at the same time penalizing the purchase of vehicles with greater CO_2 emissions (malus).

In the EU, 15 governments provide tax incentives for electrically chargeable vehicles. Examples of policies include tax reductions and exemptions, as well as bonus payments for electric car purchasers.

Historically, the U.S. and state legislature have been kind to hybrid cars. In fact, till 2010, anyone buying a hybrid car in the U.S. was entitled to a maximum of US$3400 in federal tax rebates under The Energy Policy Act of 2005. Although that particular policy along with other incentives like dedicated lane access for hybrids in California, have ended, they helped encourage the adoption of hybrids and motivated manufacturers to build more hybrid car models. Because many European countries lacked similar incentives, hybrid car prices have remained high, making many consumers buy cheaper diesel-engine cars instead.

3.1.6 Congestion Charges

The congestion charges scheme makes use of purpose-built automatic number plate recognition (ANPR) cameras, to record vehicles entering and exiting the zone. Cameras can record number plates with a 90 % accuracy rate. Congestion charges are applied during specific hours and often energy efficient cars are exempted from such charges.

London: In February 2003, London introduced an internet based congestion-charging scheme to improve the accessibility of the city center, air quality and quality of life. The congestion charge is still applied on weekdays from 07:00 a.m. to 18:00 p.m. The daily charge is around $15 with hefty penalties (almost $300) applied to offenders. Vehicles fuelled by alternative sources of energy (registered cars which emit 75 g/km or less of carbon dioxide and meet the Euro 5 standard) are exempted from such charge.
Singapore: Introduced in 1975, the original congestion charge scheme started as a manual system of color-coded paper licenses that drivers had to buy and display at police checkpoints. In 1995 it went electronic, with onboard smart cards. The entry to the restricted zones typically costed between $1 and $2.
Norway: In its two largest cities, Oslo and Bergen, the transport authorities introduced tolls for ring roads in the 1980s, explicitly as a money-raising exercise.
Sweden: Runs in the Swedish capital in 2006 had great results, cutting traffic by 20–30 %, and in 2007 a referendum approved making the scheme permanent. On weekdays, drivers pay around $1.60–$3.00 depending on times of day to enter or leave the capital. Evenings and weekends are free.
Italy: In 2012, Milan implemented a congestion charging, which runs from 7:30 a.m. to 19:30 p.m. each weekday and requires a driver to purchase a ticket for 5 Euro to enter the central city zones.

3.1.7 Public Transport Subsidies

For many decades, governments have subsidized mass transit and bus transports, which is a way to reduce energy consumption.

3.2 Energy Consumption Comparison Between Car Technologies

With the emergence of new electrical cars, one should ask himself if it makes sense from an environmentally friendly perspective to encourage the switch from diesel or gasoline to electric cars. Furthermore, the day all cars are electric is there any advantage for society in favoring public over private transport? In the next paragraphs, we will make a comparison between diesel, gasoline, and electric car consumption as well as between public and private transports. We will then try to compare the same topics, but from a pollution perspective.

First, let's see what the impact of the combustion technology on energy consumption is. To answer this question, we need to understand that the choice of the motor engine technology will be affected only by one of the two factors influencing energy consumption: the car efficiency, which is itself function of engine speed and torque. The other factor—traction energy—is mostly independent from the motor technology as it is directly function of the size (especially frontal size) and weight of the car, as well as on the driver's style (speed and acceleration). As we will see when we compare private and public transports, the traction energy, which is function of aerodynamic resistance (i.e., resistance to air), rolling resistance (i.e., tire friction and type of road pavement), acceleration, and slope resistance will have a significant impact when we compare cars with trains (bigger air drag but much lower rolling resistance). The first questions we will need to answer are: what are we consuming? and how can we compare liquids with electricity?

3.2.1 *Diesel, Gasoline, or Electric Cars*

Conventional cars have an easy to identify source of energy: diesel or gasoline. Understanding what the source of energy for electric car is can be trickier, as it depends on the electrical grid characteristics of the region or country in which such electricity is generated. This is called the electric generation matrix and considers exclusively the energy consumed by a country or region in its power plants to generate electricity. For instance, it doesn't consider energy used by the industry or the gas oil used by people to heat their home. As might be expected, this matrix varies tremendously from one state or country to another.

3.2.2 *Comparable Measuring Units*

In order to compare petro-based combustion technologies with electrical cars, we first need to define a comparable measuring unit. Though the main measuring unit for energy is usually the Joule, we will mainly focus on the kWh (kilo Watt/hour), a widely used metric measure in most parts of the world (the exception being mainly the USA), and convert other units. For convenience, other units also used such as, calorie (cal), British Thermal Unit (BTU), ton of oil equivalent (toe), barrel of Petrol, will be converted using the following formulas:

- 1 kWh = 3412 BTU
- 1 kWh = 860,420 cal
- 1 kWh = 3.6×10^6 J
- 1 kWh = 85.98×10^{-6} toe
- 1 toe = 7.33 barils of petrol

We will also use the Miles Per Gallon Equivalent (MPGe) measure widely used in the USA (see next paragraph).

- 1 MPGe = 0.0470 km/kWh

In order to compare apples with apples, we need also to understand what the characteristics for diesel, gasoline, and electric cars are, at two different locations of energy consumption:

- At the point of combustion, which is on the road when driving; and
- At the point of generation, which means at the power plant site or well.

In both cases, we will try to indicate what are the inefficiencies and losses happening in the system.

3.2.3 Comparison at the Point of Energy Consumption

This comparison requires using the measuring unit MPGe which was defined in 2010 by the American authorities NHTSA (National Highway Traffic Safety Administration) and EPA (U.S Environmental Protection Agency). EPA calculates that 33.7 kilowatt hours of electricity is equivalent to one US gallon of gasoline (1 US gallon = 3.785 L) and the direct conversion measure based on the quantity of energy contained within a liter of gasoline is thus 9.85 kWh.

The MPGe can be applied to many other sources of native fuel than gasoline and is determined by computer modeling according to the following formula:

$$\text{(total miles driven)} \times \text{(energy of 1 gal of Gasoline)}/\text{Total energy of all fuels consumed}$$

The following table gives examples of other native units and values which can be found also in transportation (Table 3.7):

EPA calls this measure at the point of consumption tank-to-wheel for petroleum-based engines and wall-to-wheel energy consumption for electric cars. In other words, EPA measures the energy for which the car owner usually pays. Therefore for electrical vehicles, EPA integrates the energy losses resulting from

Table 3.7 Fuel type and energy equivalent in BTU and kWh per unit

Fuel	Unit	BTU/unit	kWh/unit
Gasoline	Gallon	116.09	34.02
Diesel	Gallon	129.49	37.95
Biodiesel	Gallon	119.55	35.04
Ethanol	Gallon	76.33	22.37
E85	Gallon	82.00	24.03
CNG	100 SCF	98.30	28.81
H_2-Gas	100 SCF	28.90	8.47
H_2-Liq	Gallon	30.50	8.94
LPG	Gallon	84.95	24.9
Methanol	Gallon	57.25	16.78

Source Wikipedia (Fuel type and energy equivalent in BTU and KWH per unit. http://en.wikipedia.org/wiki/Miles_per_gallon_gasoline_equivalent)

AC conversion to charge the DC battery. The EPA MPGe ratings do not account for upstream energy consumption, which includes the energy or fuel required to generate the electricity or to extract and produce the liquid fuel, the energy losses due to power transmission, or the energy consumed for the transportation of the fuel from the well to the station.

3.2.4 Electrical Car Consumption Study

In 2012, the French magazine *Challenge* published the ranking of electrical cars based on the EPA values (but transformed into kWh/100 km) using also EPA's five standard driving cycle tests to calculate the energy consumption of each vehicle. The five driving tests are there to simulate different environments, giving a better representation of real driving conditions. The table hereafter shows the results of this study. We can see that most electrical cars consume between 10 and 14 kWh/100 km. To give an order of magnitude, a small electric car consuming 13 kWh/100 km, traveling in an entire year 16,000 km would consume the equivalent of a standard electric water heater over the same period of time. The same test included also the plug-in hybrid category (Table 3.8).

It should be noted that this study was only based on manufacturers' data and did not come from independent tests. This ranking didn't either consider battery size, car weight (a reduction in 5 % in weight of a given car saves around 3 % of gasoline) and power, which obviously would have made the 18 kW Mia™ Electric car from Mitsubishi not exactly comparable to the 225 kW Tesla's Roadster™!

Table 3.8 Electrical car consumption August 2012

2012	Electric consumption		Autonomy
Car model	kWh/100 km	Equivalent L/100 km	km
Mia Electric	10	1.02	80
Lumeneo Neoma	10.14	1.03	140
Renault Zoe	10.5	1.07	210
Mitsubishi i-MiEV	10.7	1.08	150
Renault Fluence ZE	11.9	1.21	185
Smart Fortwo E.D	12.1	1.23	145
Nissan Leaf	13.7	1.39	175
Ford Focus EV	14.4	1.42	160
Tesla Roadster	15.1	1.55	350
Tesla S	19.9	2.03	426
Toyota Prius	12.32	1.25	25
Chevrolet Volt	14	1.42	80

Source French magazine Challenge (Electrical car consumption: the top 13 in kWh/100 km; 11 August 2012, French magazine Challenge; La vérité sur la consommation des voitures électriques; Author Nicolas Meunier)

To compare these performances with gasoline-based cars, we have found figures per type of conventional gasoline cars, taking into consideration the best and worst stated fuel economy in the New Zealand market (Table 3.9).

Obviously, we cannot compare a large SUV with a MIA™ Mitsubishi electric car, but the 14 kWh/100 km of the Ford Focus™ could easily be compared to the 50.3 kWh/100 km of the light cars. However, before saying that an electric car is at least 3.5 times more efficient than a gasoline-based car, let's analyze the inefficiencies of both technologies to better understand where these inefficiencies originate from.

Table 3.9 New Zealand fuel economy market figures (October 2009)

Type of new car	Best stated fuel economy (L/100 km)	Worst stated fuel economy (L/100 km)	Average fuel economy (L/100 km)	Average fuel economy (kWh/100 km)
Light car	3.70	7.60	5.65	50.31
Small car	3.80	10.50	7.15	63.66
Medium car	4.60	13.40	9.00	80.13
Large car	5.70	14.90	10.30	91.71
Compact SUV	5.80	12.50	9.15	81.47
Medium SUV	5.80	13.20	9.50	84.58
Large SUV	10.80	13.50	12.15	108.18

3.2.5 Engine Efficiency

The engine efficiency is the relationship between the total energy contained in fuel and the amount of energy used to propel cars. Efficiency of each motor is unique and can be calculated by the following formula:

$$\text{Thermal efficiency} = \text{Work done/Heat absorbed}$$

Gasoline engine About 70 % of the work done by a gasoline engine is lost in heat. About half of this heat loss goes through the exhaust and the other half through the cooling radiator. An extra 5 % variation can be explained by the engine's compression ratio.

This is due, in part, from the engine's ability to convert the heat from ignition process into work producing energy. When all this is factored in, the maximum thermal efficiency of a gasoline engine is only around 25 %. It is true though that in the past 4 years, high end cars with Gasoline Direct Injection have increased engine efficiency to around 30 %.

Diesel engine They are usually more efficient, although this diesel cycle in itself is less efficient at equal compression ratios.

Since diesel engines use much higher compression ratios than gasoline's, that higher ratio more than compensates for lower intrinsic cycle efficiency, and allows diesel engine to be more efficient. The most efficient type, direct injection diesels, are able to reach an efficiency of about 40 % in the engine speed range from idle to about 1800 rpm. Beyond this speed, efficiency begins to decline due to air pumping losses within the engine. However, on average the diesel engine efficiency is in the range of 30 %.

For the calculation purposes of this book, we will consider 25 and 30 % efficiency, respectively, for gasoline and diesel engines.

Electric engine have a much higher efficiency. Electric motor's efficiency is around 95 % but inverters that transform the DC power output from the battery into AC are themselves around 95 % efficient.

In electric cars there are no major transmission system and driveshaft components. On some models, you don't even have axles or differentials. The modern petroleum based cars have on the other hand, much more complex drivetrain that eats up about 5 % of the energy, against almost nothing for the electric cars. To be fair, we will add 2 % losses due to power conversion from AC plugs on the wall to DC power in the plug to the electric cars even though DC plugging systems are now available. We also need to consider leakage. When a user charges the battery, not all of the power ends up stored, some of it is used up pushing the electrons through the battery, and we could consider a loss of 1 %.

So, an estimate of the total round-trip tank-to-wheel efficiency is:

- Electric car:
  - 0.95 (motor and drivetrain) × 0.95 (inverter) × 0.98 (battery) × 0.99 (charger) = **88** %
  - This value is in line with what Tesla, an American manufacturer of electric cars advertises.
- Gasoline-propelled car:
  - 25 % (motor) × 95 % (drivetrain) = **23.7** %
- Diesel-propelled car:
  - 30 % (motors) × 95 % (drivetrain) = **28.5** %

To all these numbers, one should subtract energy consumption for moving auxiliary equipment (wipers, lighting, GPS, radio, air conditioning, heating system), which would be in the range of 5 % for all 3 type of cars (being lower for electricity as there is no need to burn fuel and convert it in AC current to power these equipments). It should be noted that all these numbers above might be considered on the high side. Literature mentions values in the range of 16 and 22 % for petrol and diesel efficiency and 80 % for electric cars with lithium battery.

3.2.6 Braking Energy Recuperation

However, our comparison cannot stop there. Indeed one of the main differences between petroleum-based and electric cars is the possibility to recuperate and store the energy generated by the braking system, called kinetic energy. This energy results from transforming braking energy back into electricity, instead of heat or noise.

In a regenerative braking system, the electric motor that is responsible for all or part of an electric or hybrid's propulsion also does most of the braking. When a driver steps on the brake pedal, instead of activating a conventional friction-based braking process, it sends an electronic signal to the electric motor, directing it to run in reverse mode, which creates resistance to slow the application through a process that is analogous to downshifting a standard transmission vehicle. The electric motor running backward acts then as an electric energy generator that can convert the kinetic energy of motion into electrical energy that can be stored for future use.

Everyone is familiar with the small dynamo that bikers put on their bicycle. Well, the electric cars, equipped with regenerating braking system (RBS) and an anti-lock braking system (ABS) can basically recycle this energy which is stored back in the battery for a new usage. As an added bonus, regenerative braking with an electric motor takes most of the load off mechanical brakes, reducing brake maintenance and replacement expenses.

Theoretical recuperation: What is the maximum theoretical recuperation? To calculate such value, we need to use the kinetic energy formula for a rectilinear motion:

$$E = \frac{1}{2} \mathrm{m}\, \mathrm{v}^2 \text{ where energy is equal to mass } (\mathrm{m}) \times \text{ squared Speed } (\mathrm{v}^2).$$

Using a 1125 kg Mitsubishi MiEV™ car, running at 120 km/h, the energy would be 0.174 kWh. This 0.174 kWh would be available if we made abstraction of train drive friction, air drag and slope impact. An experiment conducted in Australian's Engineering School by Guido Wager[3] showed that the MiEV™ Regenerative Braking System was able to generate 0.105 kWh of electricity. This is 63 % of the available kinetic energy at 120 km/h.

Driving the Mitsubishi under different drive cycles and RBS settings, this study showed a wide range of RBS performances. The RBS from the MiEV™ improved the energy consumption between 3 % (US Federal Highway cycle in B-Mode) and 22 % (Federal Test Procedure 75 in D-Mode). Such a large range shows how significant the RBS performance depends on driving patterns and RBS configurations (Table 3.10).

As shown in the table above, the study indicated that on average the regenerative braking system working together with ABS brakes could make an average saving of 18 % on non highway roads. The study showed that there were very large variations between the different driving environments.

[3]Efficiency and performance testing of electric vehicles and the potential energy recovery of their electrical regenerative braking system; Thesis of Australian's School of Engineering; Author: Guido Wager.

As one could expect, electrical cars didn't generate huge savings on highways with RBS, though at 4 % it was far from negligible. The reality is that though driving on highways doesn't require many stop and go (maybe except in some region of Los Angeles), the braking energy recuperated is much higher (speed at squared power). Obviously the savings made by RBS on city roads were the highest at 22 %.

If we include regenerative braking to the overall efficiency of the electric cars, which was calculated at 88 % and using a value of 18 %, we can see that all inefficiencies from wall-to-wheel are more than fully compensated on urban journeys. Don't get us wrong, this doesn't mean that electric cars generate energy, but just that this energy which otherwise would have been lost through heat at the pad brake level, were stored and reinjected in the car.

If we integrate the theoretical recuperation based on the study, we can now identify the inefficiencies at point of consumption (without considering auxiliary equipment consumption): for gasoline 23.7 %; diesel: 28.5 %; and electric cars 106 %. When this is factored in and comparing light cars with a Ford Focus™ for instance, we can come back to the comparison that electric cars are 3–4 times more efficient (Figs. 3.1 and 3.2).

Table 3.10 Comparing RBS performance and energy consumption[a]

Driving NEDC (New European Drive Cycle) drive cycle under different settings			
Driving cycle according to NDEC Lotus Elise	Mode C	Mode D	Mode B
Wh/km without RBS	136	138	142
Wh/km with RBS	120	119	121
Improvement (%)	12	14	15
Driving a FTP 75 (Federal test procedure) drive cycle under different settings			
Driving cycle according to FTP 75 Lotus Elise	Mode C	Mode D	Mode B
Wh/km without RBS	81	79	93
Wh/km with RBS	70	62	77
Improvement (%)	13	22	18
Driving a US Federal HWY drive cycle in D- and B-mode			
Driving cycle according to FTP 75 Lotus Elise		Mode D	Mode B
Wh/km without RBS		143	117
Wh/km with RBS		138	114
Improvement (%)		4	3

Source Guido Wager
[a]See Footenote 3

Fig. 3.1 Internal view of the B-class electric driven car from Mercedes Benz. Copyright of Mercedes Benz (Picture: Internal view of the B-Class electric drive car, Copyright of Mercedes Benz and kindly lent for this book)

Fig. 3.2 The B-class electric drive from Mercedes Benz was rated by the EPA with a range of 140 km and an energy consumption of 25 kWh/100 kms for combined city/highway driving. Copyright of Mercedes Benz (Picture External view of the B-Class electric drive car, Copyright of Mercedes Benz and kindly lent for this book)

3.2.7 A Comparison Done at the Point of Energy Generation

As we've seen, EPA MPGe ratings do not account for upstream energy consumption, which includes:

- For electrical cars, fuel required to generate electricity or the energy losses due to power transmission; and
- For oil-based cars, energy required for extracting and producing the liquid fuel, or the energy consumed for fuel transportation from the well to the gas station.

Though this can seem academic, it is important to really understand how much more efficient electric cars are vis-a-vis fuel-based cars.

3.2.8 Electric Power Generation and Distribution Efficiency

Power generation efficiency In order to compare apples with apples, we will now identify current thermal plant efficiency. Combined cycle power generation (recuperation of electricity and heat) efficiency level is about 50 %, whereas single cycle reaches only 33 %. More modern plant can reach efficiencies of 60 % (i.e., General Electric H System™, Siemens SGT5-8000H™ gas turbine, Mitsubishi J class™). However, when you factor in things like throttling and load following, these numbers go down and most modern plants have an efficiency level closer to 50 %. The other types of power plants have different efficiency level. The following table shows the theoretical efficiency of converting energy from various types of fuels (Table 3.11).

Electric grid efficiency These numbers need also to take into consideration the inefficiency of the electric grid. The following table gives these losses in transmission between sources of supply and points of distribution, as well as in the distribution to consumers, including pilferage (Table 3.12).

So that means the real well-to-wheel comparison for electric cars in the USA including a 6 % loss for its electrical network is for three different classical plants:

- 50 % (Cogeneration plant) × 94 % (line losses) = 47 %
- 90 % (hydroelectric plant) × 94 % (line losses) = 84.6 %
- 35 % (nuclear plant) × 94 % (line losses) = 33.6 %

Drilling and extracting inefficiencies In order to have the full picture, we now need to add the energy required for drilling and extracting the fuel of the cogeneration plant (the same would be necessary for uranium).

Table 3.11 Generation efficiency

Power plant type	Generation efficiency (%)
Hydro electric plant	90–95
Tidal power	90
Coal fired plant	45
Oil fired plant	45
Gas turbine	40
Nuclear fission	35
Wind turbine	30
Biomasse	30
Solar thermal	20
Solar photovoltaic	18
Geothermal	18

Source Eurelectric (Efficiency in electricity generation; Preservation of Resources; Working Group's "Upstream" Subgroup in collaboration with VGB (July 2003) Report drafted by: EURELECTRIC)

Table 3.12 Source World Bank[a]: Transmission and distribution losses (% of output)

Country	Losses (%)
Qatar, Slovak Republic	2
Korea, Israel	3
Germany, Netherlands	4
France, Canada	5
USA, China	6
UAE, Switzerland	7
UK, South Africa	8
Brazil	16
India	21

[a]Electric power transmission and distribution losses Source World Bank. http://data.worldbank.org/indicator/EG.ELC.LOSS.ZS/countries?display=default

For comparison sake, and because of the phenomenal successes of schist gas, let's use the Natural Gas power plants. The drilling and extraction requires energy equivalent to about 9 % of the fuel, and shipping it in a pipeline is extremely efficient, accounting for about 1.5 % of the energy. So that means the total cycle end-to-end is:

– 91 % (extraction) × 98.5 % (shipping) = 89 %

For gasoline, the DOE gives a value of 83 % but this estimation doesn't fully account for the total energy inputs of generation and distribution. Thus, we could probably lower this value to 75–80 %.

3.2.9 Petroleum-Equivalency Factor (PEF)

Is the measure established to calculate the well-to-wheel gasoline-equivalent energy content of electricity (Eg) a fair measure? First, it was established in 2000 by the US Department of Energy. Second, this methodology, which is used by carmakers to estimate credits into their overall Corporate Average Fuel Economy (CAFE) for manufacturing electric drive vehicles, considers an average fossil fuel electricity generation efficiency of only 32.8 %. This is obviously inconsistent with the numbers that we've just seen for more modern power plants.

For the appreciation of the reader, we've calculated the overall efficiency numbers using the US DOE numbers as well as the values for the different power plant technologies. We've used the equation for determining petroleum equivalent fuel economy of electric vehicles, which is the following:

$$\mathrm{PEF} = E_g * 1/0.15 * \mathrm{AF} * \mathrm{DPF}.$$

For those interested in the calculation, the following table describes the variables and the values used (Table 3.13).

Based on an estimation of the DOE values and calculation of the different power generation technologies, the equivalent economy of fuel resulting from switching to electric cars would be between 83 kWh/gal (22 kWh/L) according to the DOE and 229 kWh/Gal (61 kWh/L) for an hydroelectric plant analyzed separately.

3.2.10 Well-to-Wheel Energy Comparison

Now we have a number that we can really use to compare the overall efficiency of petroleum and electric-based cars. So how do the results from well-to-wheel conventional car compare to that of an electric car? If using a cogeneration based on burning petrol, the electric car efficiency is about double the conventional gasoline car (even more if based on natural gas). The most favorable number for electric cars

Table 3.13 Petroleum-equivalent fuel economy of electric vehicles

		Unit	DOE	Co-gen	Hydroelec	Nuclear
E_g = gasoline-equivalent energy content of electricity = $(T_g * T_t * C)/T_p$		kWh/gal	33.7	33.7	33.7	33.7
		kWh/L	8.9	8.9	8.9	8.9
T_g	US average fossil fuel electricity generation efficiency	%	33	50	90	35
T_t	US average electricity transmission efficiency	%	94	94	94	94
T_p	Petroleum refining and distribution efficiency	%	83	83	83	83
C	Watt-hours of energy per gallon of gasoline conversion factor	Wh/gal	33.705	33.705	33.705	33.705
E_g	$(T_g * T_t * C)/T_p$	Wh/gal	12.520	19.086	34.355	13.360
PEF = Petroleum-equivalent fuel						

economy = E_g * 1/0.15 * AF * DPFWh/gal *AF83.469127.240229.03289.068All electric vehicle (AF = 1)kWh/gal8312722989kWh/L22346124Electric cars with petroleum powered accessories (AF = 90 %)kWh/gal75.1114.5206.180.2kWh/L20305421Ratio (%) 248378680264*Source* Author

is achieved with hydroelectric plants, where electric cars would be almost seven times more efficient.

The following table sums up the difference in performance weel-to-wheel (wall-to-wheel for electric cars or tank-to-wheel for fuel cars) and well-to-wheel integrating the different types of power plants (Table 3.14).

We've used values on the high end for wheel-to-wheel efficiency and thus these previous values are also high. Literature will mention values of 15 and 18 % for

Table 3.14 Performance comparison when integrating inefficiencies (not including powering up of auxiliary equipment)

Efficiency: weel (or wall) to wheel	Unit	DOE	Co-gen	Hydroelec	Nuclear
Gasoline cars	%	23.7	23.7	23.7	23.7
Diesel cars	%	28.5	28.5	28.5	28.5
Electric cars	%	106	106	106	106
Efficiency: well-to-wheel					
Gasoline cars PEF	%	20	20	20	20
Diesel cars PEF	%	24	24	24	24
Refining and distribution efficiency	%	83	83	83	83
Electric cars PEF	%	27	41	90	31
Refining and distribution efficiency	%	83	83	100	90

Source Author

gasoline and diesel cars versus our 20 and 24 %. However, as there is no region that only uses one source of energy to generate electricity we will now look at a few electric energy matrixes.

3.2.11 Energy Efficiency According to the Energy Matrix

An energy matrix defines the breakdown of the country power output according to the different types of energy. The following graph from the US Energy Information Administration (EIA) gives a good overview of the overall energy matrix of that country (Fig. 3.3).

It shows that 93, 3, and 4 % of the transportation fuel source was petrol, natural gas, and renewable, respectively. The overall electric power generation matrix included thermal power plant using 1 % of petrol, 24 % of Natural gas, and 41 % of coal (down from 48 % in 2009), with 12 % coming from renewable energy and 21 % from nuclear power plants.

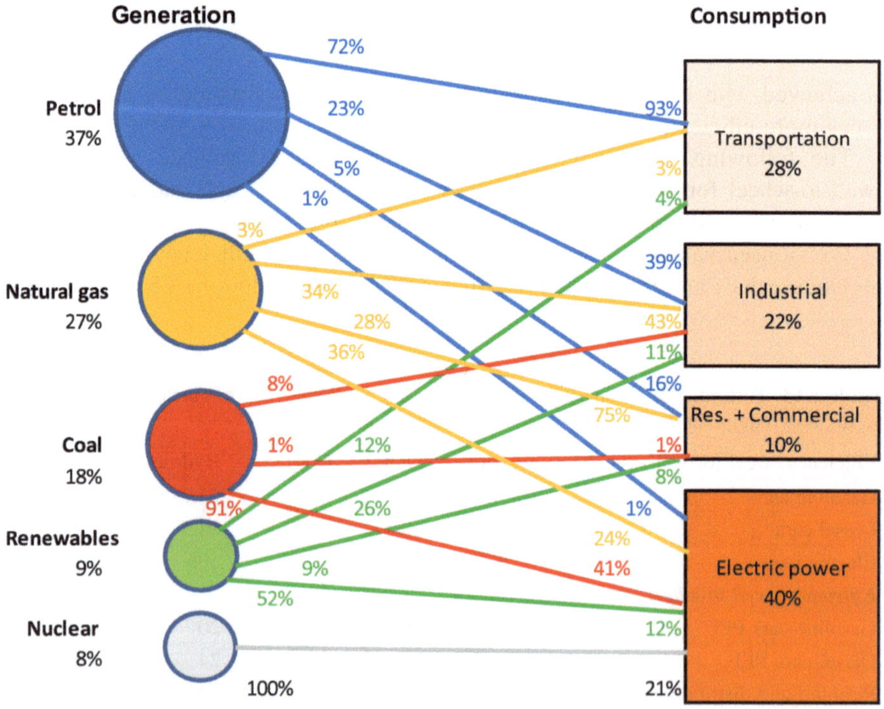

Fig. 3.3 Primary energy consumption by source and sector, 2012 (US Energy Information Administration; Primary Energy Consumption by Source and Sector, 2012)

Table 3.15 Energy matrix of the USA; 2012 electrical production

Power source	Plants	Power capacity (GW)	Capacity factor (%)	Annual energy (billion kWh)	% of annual production
Coal	557	336.3	55.8	1514.04	37.0
Nat Gas	1758	488.2	33.3	1237.79	30.2
Nuclear	66	107.9	86.2	769.33	18.8
Hydro	1426	78.2	40.0	276.24	6.8
Other renewables	1956	80.5	32.3	218.33	5.3
Petroleum	1129	53.8	5.6	23.19	0.6
Misc	64	2	92.6	13.79	0.3
Storage	41	20.9	−2.5	−4.95	−0.1
Imp-Exp				47.26	1.2
Total	**6997**	**1168**	**44.0**	**4095**	**100**

Source Wikipedia (Electric power transmission and distribution losses Source World Bank. http://data.worldbank.org/indicator/EG.ELC.LOSS.ZS/countries?display=default)

The following table gives a more detailed picture of the 2012 US electrical production (Table 3.15).

Based on this table, we can calculate the US electric matrix overall efficiency, to which we've added drilling and shipping inefficiencies (Table 3.16).

Thus in the USA, comparison between the different technologies is:

- Electric car: 106 % × 36.5 % = 38.6 % versus 21 % and 25 % for gasoline and diesel cars.

We've also done the calculation for two other countries with very different electrical energy matrices—France and Brazil—, which shows higher efficiencies.

- France: Electric cars: 106 % × 38.9 % = 41.2 %
- Brazil: Electric cars: 106 % × 78.6 % = 83.3 % (Table 3.17)

These numbers must also be compared to the 21 and 25 % for the gasoline and diesel cars. Thus and to sum up, electric cars are almost twice as efficient as gasoline cars in the States and France, but as much as four times more efficient in Brazil.

Table 3.16 Electric generation matrix of the United States

Energy produced	Breakdown (%)	Efficiency (%)	Dril + ship (%)
Coal	36.97	44	83
Nat gas	30.23	35	89
Nuclear	18.79	35	90
Hydro	6.75	90	100
Other renewables	5.33	25	100
Petroleum	0.57	44	83
			36.5

Source Author

Table 3.17 Electric generation matrix of France and Brazil

Energy produced	France			Brazil		
	Breakdown (%)	Efficiency (%)	Dril + ship (%)	Breakdown (%)	Efficiency (%)	Dril + ship (%)
Fossil fuel	9.6	44.0	87.0	17.1	44.0	87.0
Nuclear	76.4	35.0	90.0	2.6	35.0	90.0
Hydroelectricity	11.9	90.0	100.0	78.8	90.0	100.0
Other renewables	2.3	25.0	80.0	1.5	25.0	80.0
		Total	**38.9**		**Total**	**78.6**

Source Author

3.2.12 National Energy Savings Resulting from an All Electric Fleet

What would be the savings at the level of a country, if it were to switch totally to an electric car fleet? For this analysis, let's use the French consumption of gasoline for private cars.

First, we need to introduce a new conversion measure: the MToe (or Million tons of oil equivalent). International conventions fixed the value of 1 Toe to 11,630 MWh. This means that 1 ton of petrol has as much energy as $11{,}630 \times 10^3$ kWh or 0.086 MToe equals 1 TWh ($\times 10^6$ kWh). In France the 2013 total electricity production was 550.9 TWh, which corresponds to 47 Mtoe. The same year, the overall French petroleum consumption was 43 Mtoe of which 69 % went for private car usage, which is the equivalent of 29.5 Mtoe. The price of import in 2012, for petroleum products reached the value of 54.5b euro.

Supposing that a magician would transform from one day to the other the entire current conventional French car fleet to electric, we would contemplate higher efficiencies in the order of magnitude of 73 %, when including the diesel fleet proportion of 70 % versus 30 % for gasoline and integrating the higher efficiency of electric cars.

- 41.2 %/(70 % × 25 % + 30 % × 21 %) = 73 %

The total savings would thus be in the range of 12.4 Mtoe! Using the 2012 prices, this would correspond to a saving of 15.6b euro ($19.6b) and could finance around three new nuclear power plants each year or thousands of renewable power plants.

The American EIA released the 2012 fuel consumption numbers of finished motor gasoline of 133,462,854,000 gal (492 billion liters). Assuming that most cars are gasoline-based (diesel isn't popular in the USA and represented in 2012 less than 2 %), the overall savings would be, when we factor in the relative efficiencies well–to-wheel:

- 38.6 %/21 % = 83 % higher efficiency

Cars and light trucks accounted for 63 % of the US transportation petroleum use in 2012 (according to Transportation Energy Data Book) and thus the total US savings would be 141 billion liters of gasoline. Using the Brent cost of around 0.75 US$/L (0.60 euro), this would result in savings of $106 billion.

3.3 Evolution of the Electric Vehicle Market

We've seen that from a country's economy perspective, going electric can bring huge savings. Most trains are already electric and in Europe, electrifying the last diesel lines is done whenever economically viable. In the USA, because freight is the dominant market, few lines are being electrified. However, all new commuter and metro lines being added are generally electrified. Many cities are considering electrifying their bus fleet or changing their combustion engines to natural gas. As for electrical cars, the rate of adoption will depend on several conditions being fulfilled. Probably, the most important is the impact of going electric on the citizens' wallet.

3.3.1 Difference in Price at the Pump

The following picture gives a good overview of the 2012 average US price of a regular gallon of gasoline and the equivalent "electrical gallon". According to the American Department of Energy (DOE) the gasoline price is more than three times the price of the equivalent electric recharging price! (Fig. 3.4)

In Europe, prices of electricity are also much cheaper than the equivalent diesel liter. For instance, in France around 2 Euros are necessary for a journey of 100 km with an electric car, and 6 Euros for the equivalent distance with a diesel car consuming around 4 L/100 km. Furthermore, it is possible to benefit from price variation according to the hour of the day. Vehicles can be charged at night when time-of-use rates can be way lower than during the day and during peak time. In some places, prices go down by 50 %!

It is true that if suddenly all drivers were to switch to electric cars, the governments would change their tax policies and apply heftier excise taxes on electricity at the pump.

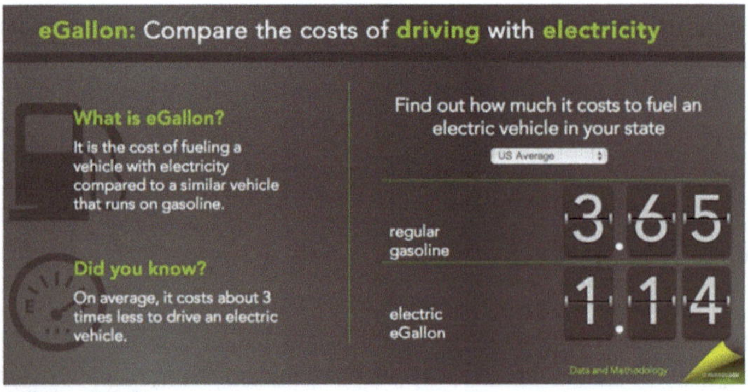

Public Domain DOE

Fig. 3.4 DOE comparison between electric and regular gasoline gallon of energy (2012)

3.3.2 Total Cost of Ownership

Not all is perfect moneywise with electric cars. Acquisition, battery replacement, and overall electric car maintenance costs are still an issue that can reduce financial attractiveness of this technology. These costs, which are part of the total cost of ownership, are still unfavorable, but under massive investments of the car industry and new technology breakthroughs, are falling quickly. But let's look at the battery cost and its likely evolution to apprehend the rate of adoption of electrical cars.

Battery costs Tesla, an electric car manufacturer in cooperation with Panasonic of Japan and other partners are planning to invest $5 billion in a US plant that should at term produce 500,000 batteries per year. With this production capacity, Tesla believes that its battery pack including its power management unit and cooling system could reduce by 30 % its current price of $300 per kWh of storage capacity. According to Tesla, two-third of this saving would come from scale alone. It is true that with 500,000 units, that would double the worldwide production of lithium-ion battery output.

Other areas of technological improvements that could lead to further price reduction, such as better design of the electronics that manage power, as well as overall vehicle weight reduction will help sustain the continuous drop in electric car price.

New technologies: Further ahead, scientists are looking at ways of increasing the lithium battery content without starting fire, improving the anode itself. For instance, producing safe batteries with a lithium anode and a sulfur cathode (two very high energy storing capacity materials) would allow to hold around five times more energy than the current batteries, weight-per-weight. Students from Stanford University have in 2015 created the buzz with an Aluminum Ion battery (a negatively charged anode made of aluminum and a positively charged cathode made of graphite).

Battery evolution law: For the last decade, lithium-ion battery capacity has been increasing every year by around two third. In other words and to quote the Moore's law (predicting successfully for almost 50 years the doubling of transistors on integrated circuit every two years), batteries' capacity have doubled every three years and the trend is likely to continue, especially with the building of this mega-factory. With that trend, it is anticipated that by 2025, the equivalent of a Nissan Leaf™ battery pack will cost less than $1800, making its propulsion system (motor and battery) cheaper than a comparable gasoline motor. Assuming even modest increases in storage capacity, electric cars will rank better on initial cost, range, performance, ongoing maintenance and fuel costs. By that time, it is anticipated that electric cars will likely overtake gasoline-powered ones.

Total cost of ownership evolution The consulting company McKinsey has elaborated a bottom-up cost model, which estimated by how much such battery pack prices could fall from its 2013 price tag of around 500$ per kWh (Tesla's price is said to be nearer $300). The study[4] suggested that by 2025, this price could fall to around 160$. According to McKinsey at those prices, the electric car would be competitive from a total cost of ownership perspective. If this still to build mega-factory fulfills its promises and brings down price to $200 per kWh, electric cars will already be at par costwise with conventional cars. The study concluded that three factors could accelerate the passage from petroleum-based to electric cars:

- Scale effect and manufacturing productivity;
- Lower component prices; and
- Battery capacity boosting technologies, such as advances in anodes, cathodes, and electrolytes (with potential identified technologies), which could increase the battery capacity by as much as 80–110 %, by 2020–2025.

If electric cars can generate so much savings for society and are becoming increasingly attractive from a financial perspective, does it mean that suddenly everyone will start buying them. To answer yes means, on top of reducing this total cost of ownership gap with conventional cars, addressing four specific issues: battery capacity, battery efficiency, recharging time, and charging infrastructure.

3.3.3 Battery Capacity

Car manufacturers are looking at various battery technologies with the right mixture of anode, cathode, and electrolyte properties that can bring the following characteristics:

- Store as much energy as possible in the smallest possible volume;
- Have the lowest weight possible to reduce the energy consumption;
- Maintain the lowest self-discharge rate, causing charge to diminish over time;
- Perform well through many charge/discharge cycles over the battery's life;
- Recharge rapidly and easily; and
- Have the highest number of times the batteries can undergo a deep discharge and still accept charge.

The following table gives information on the power and energy comparison according to the type of technology (Table 3.18).

---

[4]Battery technology charges ahead; Mckinsey Quaterly (July 2012); Author: Russel Hensley, John Newman and Matt Rogers.

Table 3.18 Power and energy output according to battery type Web page: All about batteries

Battery type	Energy/weight Watt-h/kg	Energy/volume Watt-h/L	Power/weight Watt/kg	Energy/US$ Watt-h/$
Lead–acid	30–40	60–75	180	4–10
Nickel–zinc	60–70	170	900	2–3
Lithium-ion	160	270	1800	3–5
Lithium polymer	130–200	300	up to 2800	3–5

3.3.4 Battery Efficiency

Car and battery manufacturers are in a race to find the most efficient material combination. In searching for the best batteries to store energy, car makers want to store as many kWh of energy in the smallest, lightest, and least costly package. The minimum energy to be stored by a battery is around 35 kWh, a level of energy sufficient to drive a small car for around 150 km. Lithium-ion batteries have been chosen by most automakers. The reason why they have done so is that such technology can provide up to twice the power density of a NiMH battery (Fig. 3.5).

Furthermore and unlike nickel–cadmium batteries, lithium-ion cells have no memory effect that, following repeated use, diminishes their storage capacity. They also don't drain themselves when the car is parked for extended periods.

Each electric car battery pack consists of multiple cells, in the size of 160 × 225 × 7.6 mm. For instance, the Leaf™ from Nissan has 192 and the Volt™ from General Motors has 288 cells. The individual cells are wired together to form modules. All lithium-ion cells function similarly. Inside each

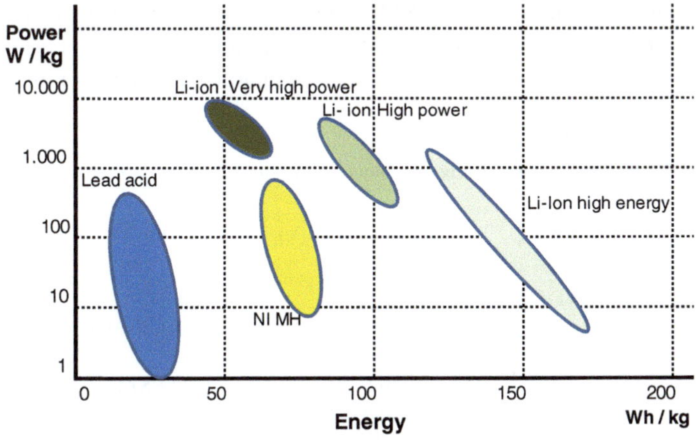

Fig. 3.5 Battery efficiency comparing energy versus power output per kg; various versions of lithium ion are also illustrated. Illustration made by Author

cell, there are multiple cathodes and anodes, sandwiched with separator material and a conducting electrolyte gel consisting oflithium salt in an organic solvent. In a fully charged cell, lithium ions are stored temporarily in the anode. When the battery delivers energy to the propulsion motor, the ions pass via the electrolyte through the separator to the cathode. That process is reversed during charging.

New development on lithium sulfur batteries could change the car manufacturers' opinion about lithium-ion battery. Indeed, sulfur has many oxidation states, more than many other metals. This property combined with sulfur's lightness and relative low cost to produce means that such a battery could theoretically store up to five times more energy per gram than conventional lithium-ion battery at a very competitive price. A few companies have already announced that production of such batteries should start in 2015 (e.g., Nissan).

3.3.5 Energy Charging Time

How much time does it take to recharge a battery? The correct answer is it depends on the type of batteries, the type of outlet being used to charge, and the current that is provided. The following table gives a view of recharging time approximation for the Nissan Leaf™ (Table 3.19):

Most recharging stations require a dedicated plug, such as the J1772 or the Japanese plug called CHAdeMO. In the USA to get more juice, an electric car owner can build a level-2 charging station that will allow 240 VAC with 30 amps current (Europe is already in 240VAC) (Fig. 3.6).

If the owner adds an optional 6.6 kW onboard charger, the charging time may be reduced to only 4 h. Finally, the fastest charging option is a DC charging station, which takes the Leaf™ from 0 to 80 % of its charge in about 30 min.

Table 3.19 Charging time according to voltage and current, possible due to specific charging technologies

Empty car recharging	Level 3	Level 2 with option	Level 2	Level 1
Voltage and current type	480VDC/125 Amps	220 VAC/40 Amps	220 VAC/40 Amps	120 VCA/15–20 Amps
Equipment		6.6 kW onboard charger	J1772 SAE plug	J1772 SAE plug
Time in hour	0.5	4	8	15

Source Author

Fig. 3.6 Picture of European home charging equipment. *Source* Author

These charging options are more high-tech than a simple wall outlet. The plugs include pins for data, letting the charger communicate with the car. This communication channel lets the car tell the charger how much electricity it wants at any given time during the charging cycle, helping preserve battery life and prevent overheating. For most automakers, J1772 has become an accepted standard for Level 2 charging. Level 3 charging may be a harder standard to create, as the CHAdeMO Japanese group has become somewhat entrenched while other automakers seem pretty keen on J1772 Combo. This J1772 combo plug works with a single car connector for both DC and AC charging, and is fully compatible with the J1772 plug. The car connector adds two pins at the bottom for the DC fast charge system (Fig. 3.7).

Electric vehicle owners can charge off AC or DC power at home as well as ultrafast DC charging at public facilities. This fast DC should get the car to 80 % charge in about 20 min. Tesla has developed its own plug standard. It wishes to achieve a five-minute charge. This will require not only further improving in the

Fig. 3.7 Charging sockets on a 2012 Nissan Leaf™. The socket on the left is a CHAdeMO connector for fast DC charging, and the socket on the right is an SAE J1772 connector for normal AC charging (Picture from Musashi1600 (Own work) [CC BY 3.0 us) (http://creativecommons.org/licenses/by/3.0/us/deed.en), Licensed under [CC BY 3.0 us); via Wikimedia Commons)

charging system, but also improving the interface with the electrical grid to sustain a 120 kW charging station (Chademo standard enables 50 kW charging and normal plugs can only produce 10 kW). Drawing large amounts of power from the grid also incurs demand charges from the utility, increasing the cost of the system.

3.3.6 Charging Infrastructure

Car owners need to be able to charge their car whenever necessary. Obviously, they can do it at home or at their work place if the recharging infrastructures are available, but may have also to do so at any other time elsewhere, if they haven't planned well their trip. Thus, a critical factor for successful electric car adoption is the installation of charging infrastructure in commercial locations. Widespread deployment of charging infrastructure at commercial locations must either:

- Be subsidized by public authorities or even by businesses to attract clients; or
- Generate sufficient income to provide a decent return on the investment made by the infrastructure investor.

Access fees provide one mechanism to provide this return on investment. Three payment models can be found:

Fig. 3.8 CHAdeMO high-speed charger at Vacaville, California Park and Ride lot (Picture from C-CarTom (Own work) [CC BY-SA 3.0). http://creativecommons.org/licenses/by-sa/3.0) or GFDL)

- Amount of time connected to the recharging unit, factoring in an average energy transfer price;
- The driver pays according to a set fee per kWh, the quantity of recharged energy; and
- By means of a subscription wherein all in-network charging is included in a monthly fee (Fig. 3.8).

3.3.7 Trolleybus

Electric buses are nothing new. They could be found all over the world under a different name: the trolleybus. Trolleybuses draw electrical power through a pantograph connected to an overhead catenary, like most trains do. This technology has been around for many years but has seemed to never really pick up. There are several reasons for this:

Fig. 3.9 Two section trolleybus in Genoa Italy. *Source* Author

- Most bus manufacturers are as well car manufacturers, that had no real interest in promoting electrical vehicles;
- Catenaries are expensive to install and maintain and create visual pollution;
- The grid to which these catenary are connected to, are usually the same as for the residential area. They aren't really ruggedized for public transport and thus are prone to frequent failures. In mixed traffic, this means that trolleybuses aren't really reliable and can block important arteries; and
- Trolleybuses don't have the same flexibility as conventional buses, as they need to be connected to the catenaries. In mixed traffic they have problem overpassing a stalled car (Fig. 3.9).

3.3.8 Catenary-Free Buses

The number of city buses (almost all diesel-powered) sold worldwide in 2011 was 64,000. The figure is expected to more than double to 135,000 by 2020. For environmental reasons, many cities are contemplating replacing their diesel-based technology, by greener solutions, such as electric, compressed natural gas, hybrid technology (a combination or diesel and battery power), and hydrogen fuel cells.

Improvements made to car batteries can be applied to buses. However, the advantage of electric buses in regard to cars is that space and esthetic restrictions are much less important. A set of batteries can be integrated and positioned on the top of the roof or under the passenger seats somewhere. However, buses have a different problem. During acceleration, the power required to move a full bus is extremely high and batteries are usually not good at delivering high source of power during a small period of time. Supracapacitors (see next section) on the contrary deliver high surge of power for a small period of time. The ideal solution would be to have a mixed system using both technologies, but this hasn't been done yet and could be pretty expensive in regard to using a single solution.

That might not be necessary if a new technology called free-catenary operation can be introduced to the market. As we will see in the next section, most of the energy consumed during trips on a segregated line happens during acceleration. This happens at the station and during the following 80 m. During the remaining time, energy is consumed to fight mainly road tire friction, before buses start decelerating and recuperating braking energy.

Technologies developed for the tramway market are being applied to buses. The first technology is based on magnetic induction. To simplify, it uses the same principle as for an electric toothbrush. Electric cable curved to make loops create a magnetic flow which is controlled by power electronic. Whenever an energy pickup based under the bus passes over the "charging slabs" embedded in the road pavement, they capture the magnetic flow, which is then converted back into electrical current onboard the bus. The slabs connected to the electricity grid can power up to 200 kilowatts. Batteries are charged as long as the bus is stopped at the station, typically around 20 s.

A second technology based on a pantograph which is lifted at the station to get energy from the catenary is also being tested. This more conventional connecting technology uses also batteries to store energy.

3.4 Energy Consumption Comparison Between Private and Public Transport Means

One of the first questions we asked ourselves was if all cars were electric and without a driver, does it still make sense for governments to favor public over private transport? Identifying how much energy each transport mode consumes and doing a comparison first by type of vehicle and then afterwards per passenger transported, will help answer this question. In the previous section, we wrote that the traction energy required for moving a vehicle was function of aerodynamic resistance, rolling resistance, acceleration, and slop resistance. To be more precise, the traction energy is given by the following formula:

- $P = F \cdot v$ or
- $P = (½ \cdot r \cdot C_x \cdot A \cdot v^2 + f \cdot m \cdot g + k \cdot m \cdot a + m \cdot g \cdot \sin J) \cdot v$
- P = [(Aerodynamic resistance) + (Rolling resistance) + (acceleration) + (slope resistance)] × speed;

where:

P instantaneous power in Kw
r air density: (around $\rho = 1.2$ kg/m^3)
C_x aerodynamic coefficient (N) (related to the "streamline" of the vehicle) (The Toyota Prius™ has a C_x of 0.25)
A frontal surface area (m^2) (around 2.5 m^2 for a small car such as a Yaris™)
v vehicle speed (km /h)
f rolling resistance coefficient (around 0.012 for tires running on asphalt or concrete)
m vehicle mass (kg)
g gravitational deceleration (9.81 m/s^2)
k correction coefficient on the mass to take into account rotating parts
a vehicle acceleration (m/s^2)
J slope angle

Air drag is minimal at very low speeds but rises rapidly with increasing speed. With good design of rolling bearings, rolling friction does not increase as rapidly with speed as does air drag, so air drag begins to exceed rolling friction at speeds above 60 km/h and dominates energy requirements at speeds above about 150 km/h (Fig. 3.10).

The previous graph for a 4-car train set Electrical multiple Unit (EMU) illustrates these conditions. On a flat road and for a car accelerating quickly from 0 to around 100 km/h, about 15–20 % of the energy consumed by a car is used to fight this air drag force. The road to tire resistance is much smaller and can reach no more than 5–10 %. The remaining 70–80 % is therefore due to acceleration. However, once the acceleration is finished, the traction effort on a flat road is mainly used for fighting resistance to air and ground.

If we look at the main formula's elements which are affected by transport type, we can see that the difference in energy consumption between transportation means will be generated mainly by three factors:

- Difference in mass between the three different vehicle types;
- Friction coefficient difference between a steel wheel and rubber tire; and
- Change in frontal surface between car, bus, and train, with its impact on air drag.

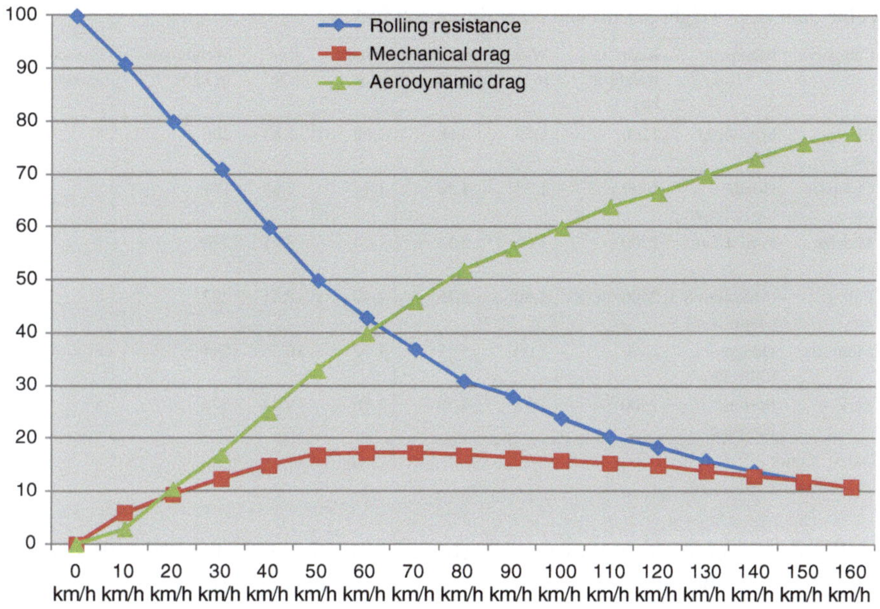

Fig. 3.10 Percentage of total resistance according to speed for of a 4-car EMU. *Source* Author

3.4.1 Weight Comparison

What is there in common between a 2000 kg car, a 15,000 kg bus, and a 30,000 kg train car? Actually a lot, if we transform everything into the weight per area (in kg/m^2). In order to compare inefficiencies, we will use the total external dimension (excluding mirrors' space) and not the useful area, as is commonly calculated in public transport.

Most cars weigh around 2000 kg. The following graph gives an average kerb weight (also called weight at tare) according to car types. This weight can obviously change slightly in function of options and brands (Table 3.20).

Calculating the weight of a bus isn't as straight forward as a car, as it may have various sections and even two decks. Buses weight can vary by as much as 7000 kg, going from 11,500 to 18,000 kg.

The following graph gives data comparable to the previous table and in particular the weight per area, without any passengers (Table 3.21).

Trains can vary even more as there are several train sections called car-train that might be independent or reunited by gangways. There are also major differences in shape, size, height, and weight according to their use: metro, commuter, tramway, double-decker, regional, and high-speed trains (Fig. 3.11).

The following table gives an overview of the different information required to calculate weight per area (Table 3.22).

Table 3.20 Car weight per m^2 and other dimension features

Category	Brands	Kerb weight in kg	Width in m	Length in m	Height in m	Area in m^2	Weight/area in kg/m^2	Capacity (passenger)
Electric car	Mitsubishi MIeV	1200	1.59	3.68	1.62	5.83	206	4
Compact car	Honda Civic Sedan	1300	1.75	4.50	1.44	7.88	165	5
Midsize car	Ford Taurus	2300	1.66	5.13	1.54	8.51	270	5
Full size car	(Mercedes S Class)	2500	1.60	5.25	1.47	8.41	297	5
Minivans	Dodge Caravan	2500	2.00	5.13	1.72	10.26	244	5
SUV	Porsche Cayenne	2500	1.93	4.79	1.70	9.24	270	5

Source Author

If we now compare the average kerb weight of the three transportation modes, we can see the following data: Cars ~220 kg/m^2, Buses: ~400 kg/m^2, trains: ~620 kg/m^2, which correspond approximately to a ratio of 33, 66, and 100 %. That means that on average a car is one-third and half as light as a train and a bus. This is important because mass impacts the energy required for acceleration, as well as fighting rolling and slope resistances.

3.4.2 Acceleration Force Comparison

Energy consumed during acceleration is extremely important in urban environment, where drivers must stop and go. On highways or on a regional train journey, acceleration forces don't interfere. The acceleration force is function of mass and gravitational deceleration (around 9.81 m/s^2) as well as a small correction coefficient on the mass to take into account rotating parts. The following table gives values for 3 acceleration scenarios (at tare) and the corresponding power in Newton (N) (Table 3.23).

3.4.3 Rolling Friction Force Comparison

Metal-to-metal friction is a source of rail's advantage in transport energy efficiency. Steel wheel on steel rail generates only about 20–30 % of the rolling friction that rubber tires running on pavement do.

Table 3.21 Bus weight per m^2 and other dimension features

Vehicle	Brand	Kerb weight in kg	Width in m	Length in m	Height in m	Area in m^2	Weight/area in kg/m^2	Capacity (passenger)
Single deck	ENV RO 300	8850	2.5	12.0	4.15	30	295	70
Single deck	ENV RO 300	8800	2.5	12.5	4.15	31	282	76
Single deck	Mercedes Citaro	11,240	2.5	12.0	4.15	30	375	75
Double-deck	Scania NUD/Alexander	11,720	2.5	10.5	4.30	26	446	101
Double-deck	Scania NUD/Optare	11,776	2.5	10.5	4.30	26	449	101
Bendy bus	Mercedes Articulated Citaro	17,020	2.5	18.0	4.15	45	378	149
Coach (2 axle)	Scania K124/Van Hool	12,580	2.5	12.0	4.15	30	419	49
Coach (2 axle)	Iveco/Beulas	13,140	2.5	12.0	4.15	30	438	49
Coach (2 axle)	Volvo B10M/Berkhof Axial 50	13,340	2.5	12.0	4.15	30	445	51
Coach (2 axle)	Mercedes Coach	9840	2.5	9.4	4.15	24	419	38
Coach (2 axle)	Mercedes Coach	13,000	2.5	12.0	4.15	30	433	55
Coach (2 axle)	Setra 415 GTHD Coach	13,240	2.5	12.2	4.15	31	434	55
Coach (3 axle)	Scania KEB/Berkhof Axial 70	15,840	2.5	14.2	4.15	36	446	65
Coach (3 axle)	Setra 416 GTHD Coach	15,240	2.5	13.4	4.15	34	455	55

Source Author

Fig. 3.11 Picture of a French regional double-deck train, which car-trains are reunited by gangways. *Source* Author

This is due to the fact that rails are much smoother than pavement or concrete and because steel wheels are much more rigid than rubber tires that tend to deform. The greater rubber tire deformation absorbs energy that is wasted as heat and noise. Furthermore, the lower the tire pressure, the greater is the deformation. The difference of contact surface between steel and rubber wheels is huge at only 3 cm^2 for the trains' wheel compared to 283 cm^2 for the rubber tire, which explains also this difference. Rail wheel flange, which keeps trains on tracks, slightly offsets this advantage.

The following table shows that a metro weighing ten times as much as cars will generate lower friction forces. We can also observe that friction is independent from speed (Table 3.24).

3.4.4 Air Drag Force Comparison

Let's now analyze in more detail the impact of train and bus' higher frontal surface. Using a simplified model (reality is always more complicated, as the train's drag is function of the number of cars, its situation in a tunnel or outdoor, wind forces, etc.), and multiplying the height by the width to get the frontal area and using an average coefficient we would have the following data (Table 3.25):

Not surprisingly, conventional train air drag (not aerodynamic high-speed trains) is 25 times higher than cars.

Table 3.22 Train weight per m^2 and other dimension features

Vehicle	Brand	Kerb weight in kg	Width in m	Length in m	Height in m	Area in m^2	Weight/area in kg/m^2	Capacity (passenger 4/m^2)
Metro car	Vancouver EMU Rotem (2-car train set)	76,000	3.00	41.0	3.60	123	618	334
Metro car (Light)	ART Bombardier vancouver (Mark II) (2 car)	48,000	3.20	34.7	3.81	111	432	290
Metro car	New York R143 Kawasaki (4 car set)	150,000	2.97	73.4	3.70	218	688	972
Metro car	New York R160 Kawasaki/Alstom (4 car set)	154,400	2.97	73.4	3.70	218	708	972
Metro car	VLU London Bombardier (8 cars)	194,400	2.68	133.3	2.88	357	544	1448
Metro car	Paris MP99 CC Alstom Bombardier (6 cars)	144,000	2.45	90.2	3.27	221	652	720
Monorail	Sao Paulo Bombardier 7-car train	93,800	3.14	86.0	4.05	270	347	680
Tramway	Citadis Alstom	44,200	2.30	29.4	3.36	68	654	178
Tramway	Brussel Flexity Bombardier (5 sections)	39,200	2.30	32.0	3.40	74	533	178
Commuter train	M8 Long Island LIRR Kawasaki (2 car)	130,900	3.20	51.8		166	790	
Commuter train	Mi 09 bi-level Alstom RER A Paris (5 cars)	288,000	2.90	112.0	4.32	325	887	1305
Commuter train	X60 Alstom (6 cars)	206,000	3.26	107.2	4.28	349	589	904
Commuter train	DeutschBahn Class 423 (4 sections)	105,000	3.02	67.4	3.78	204	516	544
Commuter train	DeutschBahn Class 425 (4 sections)	114,000	2.84	67.5	3.75	192	595	488
	Desiro Classic (2 cars)	68,000	2.83	41.7		118	576	220

(continued)

Table 3.22 (continued)

Vehicle	Brand	Kerb weight in kg	Width in m	Length in m	Height in m	Area in m^2	Weight/area in kg/m^2	Capacity (passenger $4/m^2$)
Diesel Multiple Unit								
Regional train	FLIRT, SBB (4-car train set)	120,000	2.88	74.0	4.15	213	563	478
Regional train	E351 Series Nippon Sharyo (4 car) S21-S22	154,600	2.84	84.8		241	642	232
Regional train	BR Class 357 Bombardier (4-car train set)	83,000	2.80	158.0	3.78	442	188	406
Regional train	Coradia Duplex Alstom (4 car) Z23500	260,000	2.92	107.5		314	828	350
Regional train	Francilien (SNCF Z 50000), 8 cars	235,000	3.06	112.5	4.28	344	683	922
Regional train	UK HSR (BR 395) Hitachi 6-car train set	265,000	2.81	121.3	3.82	341	777	862
North American HST	Acela Express (8 cars) Bombardier Alstom	565,000	3.16	202.9		641	881	304
HST	Zefiro 380 (8 cars) Bombardier China	462,000	3.40	215.3	4.16	732	631	664
HST bi-level	TGV Duplex (10 cars) Alstom-SNCF	380,000	2.90	200.0		580	655	545
HST	N700 Series Shinkensen japan (16 car set)	715,000	3.36	405.0		1361	525	1323

Source Author

Table 3.23 Power required to run vehicles according to acceleration scenarios

Vehicles	Weight per vehicle	Power for acceleration in N		
		0–30 km/h in 10 s	0–60 km/h in 10 s	0–60 km/h in 20 s
Car	2,000	1,667	3,333	1,667
Bus	9,000	7,500	15,000	7,500
Train	28,000	23,333	46,667	23,333

Source Author

Table 3.24 Rolling resistance to run vehicles

Vehicles	Wheel and environment	Weight per vehicle	Friction coefficient	Friction in *N* at speed of		
				30 km/h	%	60 km/h
Car	Ordinary car tires on concrete	2,000	0.015	294	107	294
Car	Car tires on tar or asphalt	2,000	0.030	589	214	255
Bus (normal)	Ordinary bus tires on concrete	9,000	0.015	1,324	482	1,324
Metro (European)	Steel wheel on steel rail	28,000	0.001	275	100	275

Source Author

Table 3.25 Air drag force according to 30 and 60 km/h

Vehicles	Frontal area	C_x (aerodynamic coefficient)	Drag forces in *N* at		
	In m^2	In N	30 km/h	60 km/h	%
Car	2.8	0.25	29	117	4
Bus	9	0.7	263	1,050	38
Train	11	1.5	688	2,750	100

Source Author

3.4.5 Energy Consumption Comparison at Vehicle Level

Traction power can be calculated in Newton or in kW. We will use once again kW for comparison purposes. The following table integrates all the above forces and allows us to understand their role in energy consumption in function of three acceleration scenarios. They give values with passengers, considering a weight of 80 kg (Table 3.26).

We could conclude from the previous table that cars consume much less power and thus energy. However, this would be an error as we need now to factor in two important elements: the passenger capacity and the occupancy rate. Intuitively,

Table 3.26 Power for acceleration in kW at different speed and acceleration

Vehicles	Weight per vehicle	Power for acceleration in kW			Power at steady speed	Capacity
	At 4 passengers/m^2	0–30 km/h in 10 s	0–60 km/h in 10 s	0–60 km/h in 20 s	At 60 km/h	Passengers
Car	2,400	74	268	148	28	5
Bus	15,000	473	1,695	945	195	75
Train	37,600	1,034	3,947	2,067	187	120

Source Author

we can feel that a car that carries 5 passengers with a surface of 10 m^2 cannot be compared with a train that carries 4 passengers/m^2. Even more, we are using 4 passengers/m^2 as this is the standard in rich countries but for many big cities, metro capacity is calculated at 6 or 8 passengers/m^2 with real capacity reaching in some megacities the amazing 10 passengers/m^2!

3.4.6 Power Comparison at Maximum Capacity

When we factor in the passenger element, the calculated power per passenger transported changes completely (Table 3.27).

From the above table, we can see that buses and trains are much more load efficient than cars at full capacity and maximum speed of 60 km/h. Buses need less energy than trains during acceleration but whenever the train reaches its cruising speed it starts consuming less energy. We've estimated the impact of going to 6 and

Table 3.27 Power for acceleration in kW per passenger transported, at different speed, acceleration and according to variation in train maximum capacity (4, 6, and 8 passengers per m^2)

Vehicles	Power for acceleration in kW			Power at steady speed	Capacity
	0–30 km/h in 10 s	0–60 km/h in 10 s	0–60 km/h in 20 s	At 60 km/h	Passengers
Car	15	54	30	6	5
Bus	6	23	13	2.6	75
Train with 4 passengers/m^2					
Train	9	33	17	1.6	120
Train with 6 passengers/m^2					
Train	6	25	13	1.2	160
Train with 8 passengers/m^2					
Train	5	20	10	0.9	200

Source Author

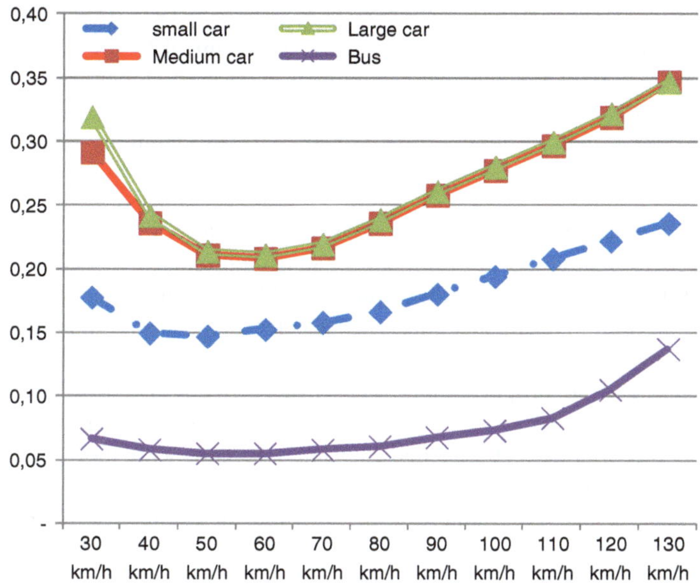

Fig. 3.12 Car and bus fuel consumption at various speeds (in kWh per passenger km) for full cars [adapted from the World Bank HDM Model]

even 8 passengers/m^2 during cruising speed, for a typical train with one-third seating area. The results show an even higher differential, with power required to transport passenger reduced from 1.6 to 1.2 and 0.9 kW, respectively.

The following graph, adapted from a report of the World Bank HDM Model, visualizes the energy consumption at various speeds and at full loading for buses and cars (Fig. 3.12).

3.4.7 Energy Consumption Comparison with Real Occupancy Rate

The previous calculation was done without taking into consideration the occupancy rate. Cars can transport 4–5 passengers. However, during peak hours most lanes are full of cars with no other passengers than the driver. The result is that cars are mostly carrying air.

For instance, in the United States the occupancy rate has hardly budged from 2000 to 2010 at around 1.55 passengers per car. Most Western European countries (UK, Denmark, the Netherlands, Norway, Germany, Austria, Spain and Italy) are with an occupancy rate almost identical to the USA of 1.54 passengers (2007 data). The average number of passengers per car for Eastern European countries was slightly higher at approximately 1.8 passengers per vehicle (i.e., Czech Republic—1.4, Slovakia—2 and Hungary—1.9).

Why is this occupancy rate so important? The point is that any additional passenger has a very limited impact on energy consumption. For instance, the second transported person in a small car weighing 1300 kg will increase the total mass by less than 6 %, which means that during the acceleration the total impact will always be less than 6 % per extra passenger. In fact, the bigger the car, the less impact that second passenger has proportionally. Furthermore at full speed of 100 km/h on a flat road, the impact of an extra passenger is less than 2 %. Thus as a recommendation, city planners should dedicate lanes to cars with more than one passenger and encourage car pooling.

Going back to our US example, if the average car occupancy were to go from 1.55 to 2 passengers (assuming the linear replacement of other cars), the fuel savings for the US economy and environment would be in the range of 40 %. The following table compares various transport mode occupancy rates in the USA. It also calculates the energy in kWh per passenger kilometer (Table 3.28).

These numbers are obviously very specific to the United States, a country which uses little public transport. In comparing the various transport modes, we also need to factor in two important elements: the citizen's culture of public transport and the peak hour capacity. Indeed, in Europe the culture of public transport is much stronger. The following graphs give a vision of a continent with such strong public transport culture (Figs. 3.13, 3.14 and 3.15).

Table 3.28 Energy consumption per passenger kilometer, adapted from the US Transportation Energy Data Book (2009)

Transport mode	Average passengers per vehicle	kWh per passenger-km	Occupancy in %
Cars	1.55	1.67	31
Cars (intercity travel)	1.2	2.00	24
Rail (intercity Amtrak)	20.9	1.15	70
Motorcycles	1.16	1.16	58
Rail (transit light and heavy)	24.5	1.19	20
Rail (commuter)	32.7	1.33	14
Personal trucks	1.84	1.73	37
Buses (transit)	9.2	2.00	12
Taxi	1.55	1.67	31

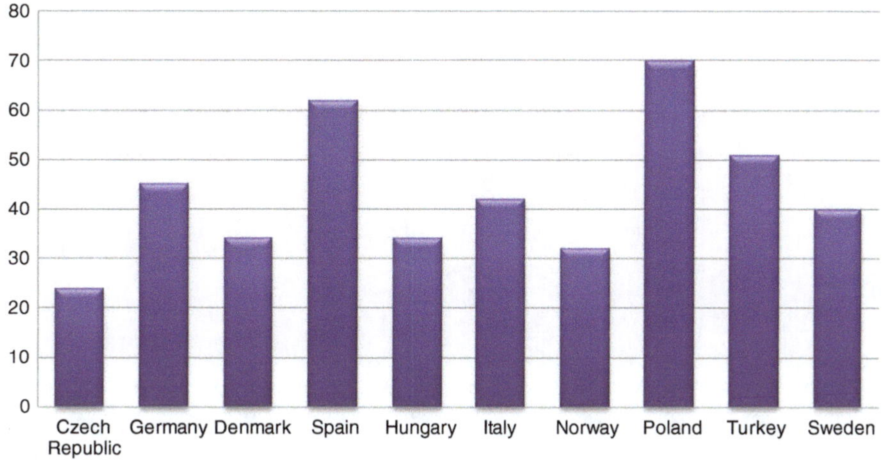

Fig. 3.13 European train occupancy. *Source* European Environment Agency (2008) (European Environment Agency (2008). http://www.eea.europa.eu/data-and-maps/figures/train-occupancy-rates) in %

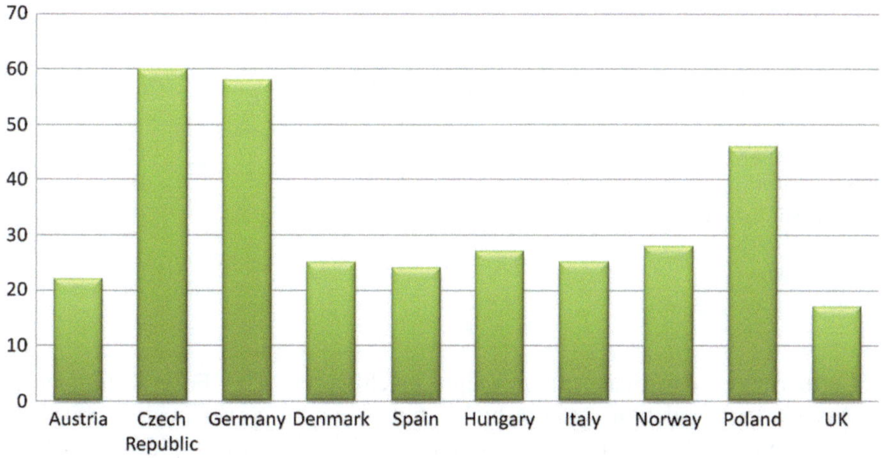

Fig. 3.14 European bus occupancy. *Source* European Environment Agency (2008) (European Environment Agency (2008). http://www.eea.europa.eu/data-and-maps/figures/train-occupancy-rates) in %

The overall occupancy rate in public transport on the old continent is much higher than in the USA, with an average of around 40 % in 2008 for trains and 30 % for buses. With these numbers, the overall European energy savings per train looks much more attractive (Table 3.29).

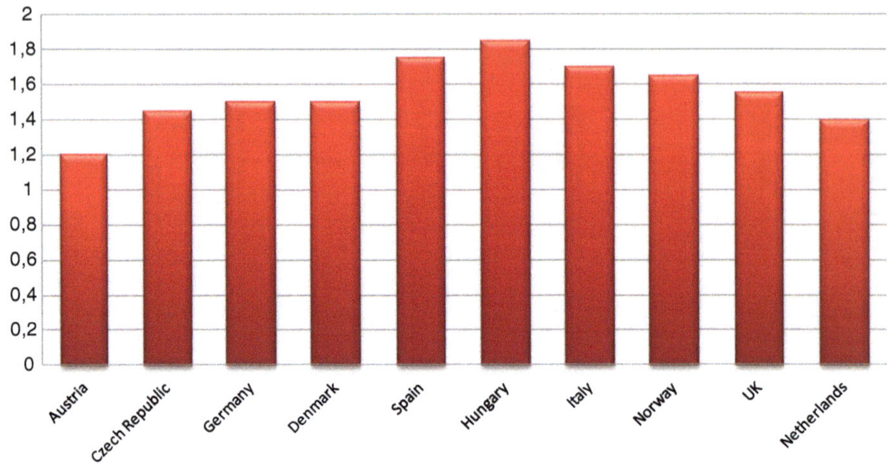

Fig. 3.15 European car occupancy. *Source* European Environment Agency (2008) (European Environment Agency (2008). http://www.eea.europa.eu/data-and-maps/figures/train-occupancy-rates) in number of passengers

Table 3.29 Comparison between transport mode energy consumption, in kWh per passenger kilometer

Transport mode	Occupancy in %	kWh per passenger-km	Comparison in %
Cars (intercity)	24	2.00	100
Buses (transit)	30	0.82	41
Rail (transit light and heavy)	40	0.61	30

Source Author

Furthermore, American trains weigh much more. The Federal Railway Association regulations require higher buff strength, corner posts, collision posts, anti-climb mechanism, and so on. All of this increases significantly the train's weight: around 900 tons buff strength for locomotives and end cars and 350 for coaches, compared to Europe's 200. As a consequence, the North American Acela High speed power cars weigh 90 metric tons versus 68 for the TGV power cars they're derived from. A 25 % decrease in weight would reduce even further the train's energy consumption.

The following graph compares energy efficiencies of tramways, Paris commuter (RATP—SNCF) and metro, buses, as well as motorbikes, diesel, and gasoline cars. These numbers give an idea of how more efficient metro and tramway cars are in the Paris region in relation to cars. The data is given in kWh per kilometer per

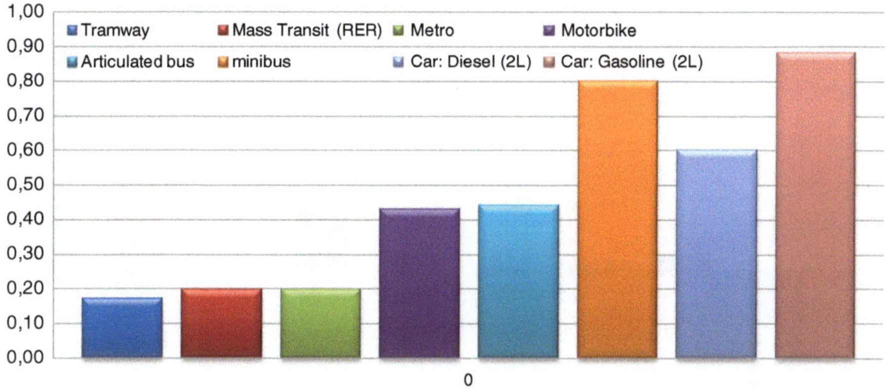

Fig. 3.16 Energy efficiency in kWh, Adaptation from Ademe study (January 2008) (EFFICACITES ENERGETIQUE ET ENVIRONNEMENTALE DES MODES DE TRANSPORT (January 2008); Study done by Deloitte for ADEME)

passenger and thus takes into consideration efficiency, capacity, and occupancy levels in the French Capital (Fig. 3.16).

3.4.8 Train Energy Losses and Recuperation

The values indicated above date already from 2008. The following graph gives the view of the railway energy losses for such older technologies (Fig. 3.17).

Since then, technologies have enabled better energy efficiencies. For instance, manufacturers have worked hard on reducing the energy use of auxiliary equipment such as Heating, Ventilation, and Air-Conditioning (HVAC), lighting, by recuperating some of the heat losses and by employing LED lighting.

3.5 Greener Technology

New technical solutions will further reduce energy losses resulting from the use of converters, brakes, and gear boxes.

3.5.1 Silicon Carbide Inverter

An inverter is an electronic equipment that changes DC to AC current. The inverter can be designed according to the input and output voltages, frequency, and overall

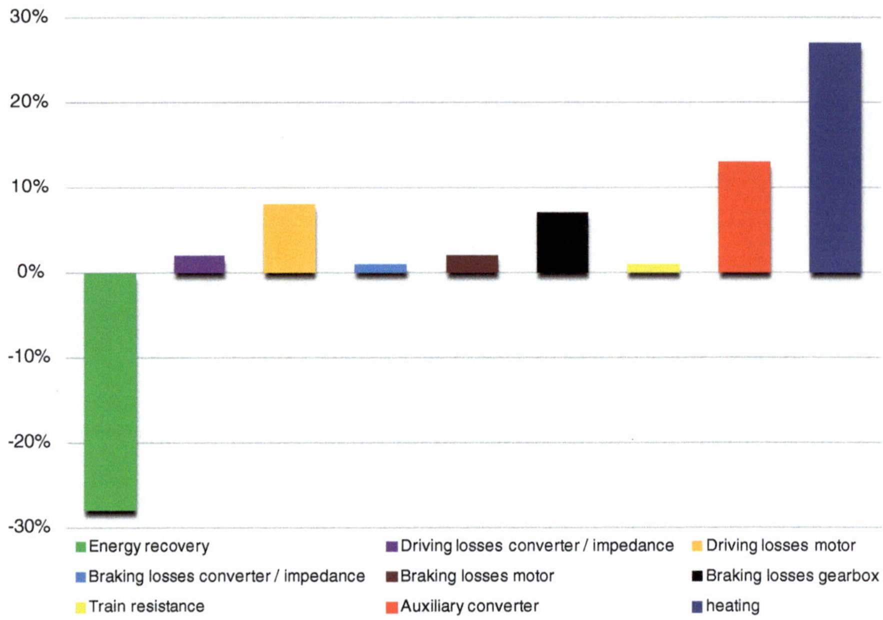

Fig. 3.17 Classical EMU energy recuperation and losses. *Source* Author

power handling. Since the beginning of this century, most inverters have integrated high capacity insulated gate bipolar transistor (IGBT) modules, a power semiconductor used for electronic switch. IGBT combines high efficiency and fast switching. Trains equipped with IGBT inverters can readily use dynamic braking, where motors become generators and feed the resulting current into an onboard resistance (rheostatic braking) or back into the supply system (regenerative braking). Thus depending on train system and overhead catenary receptivity, it can regenerate as much as 30 % of the total energy consumption, as seen on the previous graph.

In December 2013, Mitsubishi Electric Corporation announced the launch of a railcar traction inverter system for 1500 V DC catenaries that incorporate the world's first all-silicon carbide (SiC) power modules made with SiC transistors and SiC diodes.

The all-SiC inverter greatly reduces power loss, size and weight compared to conventional IGBT power modules. Mitsubishi is claiming in their press release, that their SiC traction inverter system's switching loss will be approximately 55 % less than conventional inverter system incorporating IGBT power modules. Size and weight should be reduced by about 65 % compared to conventional inverter systems with IGBT power modules. The new system also increases regenerated energy through the use of regenerative brakes in all speed ranges.

Thanks to these SiC solutions, Mitsubishi claims that total energy consumption of railcar systems, including their motors, should be reduced by about 30 % compared to conventional systems.

3.5.2 Permanent Magnet (PM) Motor

In modern trains, AC asynchronous traction motors have been the norm since they took over DC-powered motors. Under the impulse of more powerful power electronic modules and advances in the area of software control, a new technology has emerged in 2010, allowing for spectacular changes: permanent magnet motors. Unlike for conventional asynchronous motors where the flux is created within the stator windings, the PM rotor creates its own flow through the use of magnets. PM motor consists of a permanent magnet rotor driven by a rotating magnetic field realized through 3-phase AC-fed coils. It is called synchronous, because the rotor will rotate at a constant speed which is synchronous with the rotation of the magnetic field. In order to achieve strong permanent magnetic fields with small light magnets, rare earth magnets are used (e.g., Nd–Fe-B). These motors save energy as their:

- Power/weight ratio that is greater than 1 kW/kg versus 0.8 kW/kg for previous generations of motors;
- Motor losses, which are mainly due to heat when current flow runs through copper wires, are about half of the conventional motors; and
- Greater compactness (25 % smaller) for more convenient installation on the bogies, with simpler ventilation circuits and overall simplified drive train, reduces weight.

Furthermore, permanent magnet motors can control precisely the speed rotation up to 0 RPM. By changing the current flow and making spin the rotor in the other direction under the action of the permanent magnet, the motor can act as a service brake, while regenerating energy. This opens the possibility of getting altogether rid of the service brakes in the future, reducing even further losses as electrical braking could replace mechanical braking, even at low speed.

3.5.3 Direct Drive Mechanism

In a conventional propulsion system the traction motors provide torque through transmission equipment such as gear box. A direct drive mechanism is one that takes the power coming from a motor without such reductions, avoiding losses due to such transmission.

The choice of a direct drive with conventional motors would have required bigger traction and heavier motors, which explains why it never really happened. Permanent magnet synchronous motor is a promising candidate for a wheel-mounted construction, due to high specific torques. In such a wheel-mounted permanent magnet synchronous motor the torque of the motor is directly transmitted to the wheel without using gears and flexible couplings. As the drive sits directly on the rim of the wheel, it provides much higher energy efficiency.

3.5.4 Direct Drive with PM Motors Controlled by SiC Inverters

Combining the three technologies together opens opportunities to reduce significantly drive and brakes losses from the motor and the converter, eliminating altogether losses from the train drive. As we've just mentioned, it should also recuperate energy lost in the form of heat coming from the service brakes, which could be reused in the system. It is difficult to assess how much energy would be saved in the future through this, as there is no application yet. This is our guess:

- **Gearbox losses**: eliminating the 5–8 % energy inefficiency;
- **Motor losses**: a 25 % reduction in driving and braking mode, which could result in a 2–3 % overall higher efficiency;
- **Converter losses**: a 50 % reduction in driving and braking mode, which could result in a 1–2 % overall higher efficiency;
- **Enhanced kinetic energy recuperation** by eliminating mechanical braking for service braking, even at low speed (not emergency braking): 1–3 %

Without being over optimistic, we could imagine that these new combined technologies should reduce energy consumption by around 5–10 % and enable additional energy recuperation in the range of 5–10 %.

3.5.5 *Energy Recuperation and Wayside or Onboard Storage*

We've seen that electric cars can recuperate braking energy. This technology originated in the railway industry, where such energy has been recuperated for years. Unlike cars though, electric trains have several choices for reusing braking energy. Trains can obviously use it for onboard consumption but when a train is braking, it usually doesn't need such energy. In the past, there was no onboard storage equipment and thus, the railway industry started reinjecting such energy back into the network via system overhead wires called catenary or through third rail for metro cars (Fig. 3.18).

The problem with this strategy is that such energy needs to be captured immediately by accelerating nearby trains. Time is of the essence as the line cannot accumulate too much energy without creating a risk of damaging the cables and the equipment. Distance also is important as wire resistivity (resistance to electron, which is function of the metal type and width of the conductor) will increase energy losses in the network.

Fig. 3.18 Picture of third rail in metro operation. *Source* Author

How much energy are we talking about? Trains are heavy and as kinetic energy is function of mass, the recoverable energy is considerable. We've already seen that the kinetic energy formula applied to a running vehicle (rectilinear motion) is:

- $E = ½ mv^2$ Where m is the mass in kg and v the speed in m/s^2

A 6 car train-set such as the X60 from Alstom, which weighs 206 tons at tare, when braking at a speed of 100 km at full capacity (4 passengers/m^2) would generate almost 30 kWh (107 MJ). For a mass transit network, recuperating such energy of an entire fleet is vital. In fact, metro networks are characterized by short distances between stations of 1–1.2 km and by operational needs for steep acceleration to reach full speed and deceleration to come to a standstill as soon as possible. If measured, most acceleration and deceleration forces are applied during the first and last 100 m respectively. During the remaining 800 m metros reach cruising speed, where energy is basically used to fight drag and rolling friction and power up the auxiliaries.
Thus optimizing the system in order to always have a train accelerating while another one is braking, allows for significant energy savings, which we've seen for older train technologies is in the range of 30 %.

Several options have been around for using the recuperated kinetic energy and new technologies enable even greater possibilities.

- *Auxiliary equipment consumption*: Consuming directly this energy in auxiliary equipment such as HVAC, lighting, wipers, etc.
- *Consumption by nearby trains*: Reinjecting into the line this energy to be captured by nearby accelerating metros or trains. However, if other vehicles are not available to use this regenerated power, the voltage on the distribution network rises to the point where the braking vehicles must stop transferring regenerated energy to the network. In such case, the braking train will dissipate the kinetic energy as heat either through onboard or wayside resistors. The energy is anyhow lost in both cases.
- *Wayside storage*: In the last few years, metro operators have become more concerned about energy consumption. With the development of more powerful electronic chips, equipment manufacturers have been able to create energy storage equipment, which are positioned in the stations and can capture the energy which would be lost when no accelerating metro can be found nearby. Two types of electronic wayside technologies are available: supracapacitors and line converters. Another technology called flywheel has also been adapted for wayside storage.

Supercapacitors These power electronic capacitors play two key roles:

- Delivering quick bursts of energy during peak power demands; and
- Capturing excess power that is otherwise lost and quickly store it.

Super-capacitors' main strength is that they discharge and recharge quickly. Additionally, they will tolerate many more charge and discharge cycles than batteries. They bridge the gap between conventional electrolytic capacitors and rechargeable batteries:

- They store as much as 10,000 times more energy per volume than conventional capacitors but deliver less than half as much power per unit time; and
- Their energy density that is approximately 10 % of conventional batteries but their power density is generally 10–100 times greater.

In the railway energy, system integrators have integrated several super capacity units and created a bank of supercapacitors, which storing capacity can be tailored made to the railway network. For instance, one of the energy storage system design is scalable, with an energy storing capacity ranging from 0.25 to 5.0 kWh.

In South Korea, Metro operators in Seoul have claimed to achieve power savings of more than 20 % using super capacitor. Several 48 V multi-cell ultra capacitor modules were installed on 750 and 1500 VDC lines in seven Seoul metro stations. They were also installed in the Korea Train eXpress depot.

Flywheels They play the same role as supercapacitors but instead of using power electronic they use a heavy, high-speed rotating metal disk lying on conventional bearings that builds up kinetic energy as it spins. The faster the flywheel spins the more kinetic energy it stores.

We've used the analogy of the bicycle dynamo to explain kinetic energy. To make an analogy, we could think about the toy cars that use friction motors based on the principle of flywheels. A flywheel is a rotating disk that stores energy as kinetic energy. The flywheel rotates with a connecting rod, known as the shaft. The shaft is where the energy moves in and out of the flywheel. Two factors control the quantity of energy flywheels store: its moment of inertia and rotational speed. Indeed according to Newton's Second Law (applied to rotating objects) torque is equivalent to the moment of inertia multiplied by rotational acceleration.

- **Moment of Inertia**: it is defined by the rotating object's mass, radius, and inertia constant. The inertia constant is dependent on the shape, which usually comes as a ring with spokes (constant = 1.0) or a solid disk (constant = 0.5).
- **Rotational Speed**: The energy a flywheel stores is proportional to the square of the rotational speed. The rotation itself is limited by the material's density and strength. This is why more modern flywheels are made of carbon fiber and employ magnetic bearings allowing for much faster rotational speeds.

In wayside applications, the vehicle's braking energy is transferred to the flywheel through the catenary by applying torque to it, thereby increasing its rotational speed. Conversely, the flywheel releases back to the grid its stored energy by applying torque to the shaft. Modern materials make flywheels smaller, lighter, and able to spin more quickly, thus more capable to store more energy per occupying space. This makes it more and more attractive for onboard applications especially for automotive applications.

Line converters Line converters on the other hand don't store energy. They simply bypass the electric network of the railway system and reinject the recuperated energy back into the main high voltage line of the energy distribution concessionaire or within a different voltage line such as, for instance, the offices of a railway station. In other words, it basically enables the recuperated braking energy to be consumed elsewhere in the network for a completely different purpose.

Onboard storage Supercapacitors have been tested in tramway and metro application as far as 2005. However, so far, onboard supercapacitors success has been extremely limited. One of the main reasons is that they are heavy and bulky, as many 3VDC cells must be put in series together to achieve the typical 24VDC railway requirement. Furthermore, investment made in battery research by the automotive industry is likely to improve much faster battery performance. A combination of supercapacitors and batteries is also possible but in our view it is unlikely that supercapacitor can compete in the future with batteries, as they cannot provide energy for enough time to get rid of catenary wires.

Free catenary operation The advances in battery performance we referred to, have pushed the railway industry to come up with onboard banks of batteries that can store energy for non-catenary operations and by the same token recuperate the kinetic energy. Introduction of catenary-free applications are popping up in places such as France, Qatar, Dubai, and Brazil encouraging the production of even more efficient battery systems. As we've seen, the peak of energy demand is at the station when starting acceleration and during the first 80 m. After that most metro will only consume energy to fight air drag and rolling resistance. If overhead catenaries are available at the station and during this first 80 m and the dwell time is sufficient for restoring energy, then tramways should be able to run without any catenary.

Energy optimization software Modern signaling technology has integrated the need for metro and mass transit operators to reduce energy consumption. As seen, the cheapest way to recuperate the energy of a braking vehicle is to send it back into the grid via the catenary or third rail for other nearby trains to use. This is now possible through train regulation whereby the signaling software controls the arrivals and departures of nearby trains to maximize the energy transfer. In other words, when a metro starts braking the signaling software ensures that nearby metro cars start accelerating.

3.6 Final Energy Consumption Comparison

At full load (4 passengers/m^2) and constant speed (i.e., 60 km/h), buses and metros require around the same power: 190 kW. Cars on the other hand because of their lower weight only consume 28 kW (see Table 3.26). These results calculated on the basis of European technologies, with 80 kg per passenger weight and for an average cruising speed of 60 km/h, are representative of urban travels. However, such values are meaningless unless we consider the transportation's average occupancy rate (as well as energy recuperation when braking).

When we factor in these elements, we get very different results for Europe and the USA, as occupancy is significantly impacted by culture of public versus private transport (i.e., American bus occupancy is only 12 %). As a consequence, buses and cars in the US consume more or less the same energy per kilometer and passenger transported, while in Europe, buses consume half the energy (Table 3.30).

In Europe, conventional cars consume around seven times more energy than rail and twice as much in the USA, wheel–to-wheel. When comparing on a well-to-wheel basis, integrating the countries' generation matrix, rail transit vehicles and buses would still consume four and two times less than conventional cars, respectively. Furthermore, these numbers don't take into consideration for railway mass transit:

- Newer energy efficient technologies which could still improve mass transit results by up to 20 %
- Higher occupancy rates at 6 passengers/m^2 or even 8 passengers/m^2 which could increase even further the railway energy efficiency by around 20 and 40 %, respectively.

We saw also in this chapter that electric cars are far more energy efficient than conventional diesel and gasoline automobiles. The energy efficiency well-to-wheel in the USA and France is, respectively, 39 and 41 % versus gasoline (21 %) and diesel (25 %) energy efficiency.

Table 3.30 Comparison between car, bus, and train of energy consumption, adapted from the Ademe study and the US BTU information

Transport mode	Efficiency			Real kWh per passenger-km			
	Wheel-to-wheel	Well-to-wheel		Wheel-to-wheel		Well-to-wheel	
Vehicle type	Global (%)	Paris (%)	USA (%)	Paris	USA	Paris	USA
Cars petroleum-based	26	21	21	0.78	1.66	0.88	2.00
Buses (transit) diesel	28.5	25	25	0.39	1.66	0.44	2.00
Rail (transit light and heavy)	106	41.2	38.6	0.11	0.71	0.20	1.19

Source Author

When we factor this in, we could say that in the USA, electric cars would be around 80 % more energy efficient than buses and as energy friendly as mass transit. This could come as a surprise, but we shouldn't forget that the US occupancy rate of buses and metro cars is low and most buses run on diesel, which cannot recuperate energy. If we were to consider only big cities like New York with a culture of public transit, these numbers would probably change.

This is why the numbers we've presented for Paris are intuitively less surprising. When considering well-to-wheel energy consumption in Paris between mass transit and private vehicles, buses and metros are much more energy efficient than electric cars. Buses are around 25 % more energy efficient than electric cars. Metros and commuters are a little over three times more energy efficient. This once again doesn't consider the newer greener technologies, the fact that buses will also become electric, or the extra capacity that could be brought by increased occupancy levels in buses and trains.

On the other hand, it doesn't consider also the impact of "unmanned taxi services" as seen in Chap. 2, which could reduce significantly the consumption per passenger km, as it would allow for much higher average occupancy rate. Increasing average occupancy rate to around four passengers would mean that these services would be almost as environmentally friendly as metro operation and more than buses.

3.7 Pollution Comparison Between Car Technology

Public authorities must define different transportation and social policies, which can have a huge impact on the environment. From an environmental perspective, the best strategy is to avoid consumption in the first place.

For instance, policies to incentivize companies to allow people working from home is a good way of reducing pollution. Creating pedestrian areas for people to walk or dedicating bicycle lanes for people to go to work isn't only good for the environment, it is also good for the citizen's health. However, we all know that in many cities, people live too far from their main activities and must use transportation means. Thus, encouraging the spread of industries and commerce throughout the cities is a way of making sure that all citizens can find work or goods in their neighborhood without having to use transportation means. Correlating the deliverance of construction permits to the number of additional jobs is also a way to impose restriction on commuting needs.

The second best solution for public authorities is to insure that policies or investments are made to ensure that the system is used to its full operating capacity.

Policies encouraging companies to provide their employees with flexible working hours are positive because they allow people to avoid morning and evening rushes, thus reducing traffic jams or the necessity for additional public transport capacity. On the other hand, drastic measures such as forbidding people to use their cars according to their plate number will generate immediate benefits. From a society perspective, the policy will have a short term impact on all citizens, but in the long term will affect mainly the less fortunate. Rich people will buy a second or third car, usually second-hand, with less energy efficient technology. From an environmental perspective, the remedy is than worse than the initial disease.

Operating all transportation systems at full capacity is an additional way of being environmentally savvy. Working at 6 or even 8 passengers/m^2 during peak hours makes sense from an environmental perspective rather than duplicating the fleet size and staying with loads of idle equipment for most of the remaining hours.

Obviously it makes much less sense from a passenger's point of view, but finding the right balance between comfort and efficiency is important. Any avoided purchase of an extra car, bus or train will enable material economies and bring additionally the following benefits:

- **Avoided unnecessary maintenance**: What isn't built doesn't need to be maintained or recycled, thus reducing consumables and spare parts.
- **Reduced energy consumption**: 90 % of overall energy consumption is linked to operations, over the vehicles life. However, by right-sizing the fleet, energy used for manufacturing, which can represent for a train

around 4.5 % can be avoided, as well as the 0.5 % needed at the end of life for the equipment's recyclability. The remaining 5 % which are used during maintenance can also be reduced by rightsizing the fleet.

Another selection strategy is to consider recyclability needs. Choosing transportation modes with equipment that last longer, will reduce the recycling needs.

Railway technology is built for 30 years whereas the average bus and car life is 10 year. This means that when comparing transportation means, two extra buses must be ordered and recycled over the same period of 30 years. Furthermore, the railway equipment and components are usually built to last longer. Most modern vehicles can be Designed for Environment (DfE), taking into consideration the most stringent international norms. Using recyclable equipment or being able to de-construct easily the parts for recyclability should be part of the vehicle selection criteria.

The fifth strategy is to reduce the energy to be consumed when driving the vehicles.

We've seen that the physical forces that are running against any moving vehicle are the friction coming from the contact of wheels on the road or tracks, the air drag, the acceleration forces and forces to climb ramps. Reducing tire friction, improving the coefficient of air penetration, reducing weight and speed can reduce significantly the energy consumption of the vehicles. Improving also the efficiency of all the different equipment as we've seen in the previous section are all initiatives that will reduce the environmental footprint of these transportation means.

In the following section, we will calculate the impact of the various transportation means in terms of pollution based on the result from the previous sections. The focus will be on air pollution, as water pollution would be a consequence of pollutants emission reaching lakes or river through rainfalls.

3.7.1 Air Pollution

As for energy consumption, air pollution can be calculated in two different ways: wheel-to-wheel and well-to-wheel. However, the place of pollution isn't neutral for calculation of the impact, as pollution within cities has consequences on the health of the citizens. In other words, where we pollute is almost as important as how much we pollute.

Indeed, fumes cause pulmonary diseases and premature deaths. Children are particularly vulnerable, as poor air quality triggers asthma which is the number one cause of hospitalization among children and has a major impact on healthcare system. When we know that around 7 million people die every year worldwide because of pollution (according to the World Health Organization), we can measure how important city pollution is.

Let's first explain what the factors that produce air contaminants are.

Conventional petrol-based pollution Air pollution happens during the combustion phase. This physical reaction uses the air's oxygen content to combine with the fuel, which is a mixture of several hydrocarbons.

During this phase, the oxygen and the hydrocarbons will interact and recombine, giving a mix of water vapor, carbon dioxide (CO_2), and sometimes carbon monoxide (CO) and partially burned hydrocarbons. In addition, at high temperatures the oxygen tends to combine with nitrogen present in the air, forming oxides of nitrogen (usually referred to as NO_x). This mixture, along with the unused nitrogen and other trace atmospheric elements, is what we see in the exhaust.

Vehicle fuel usage and its tailpipe emissions are the single largest human-made source of carbon dioxide, nitrous oxide, and methane. Additionally, vehicles that are stationary, idle, or traveling at reduced speeds due to congestion pollutes more. Therefore, technologies that reduce fuel consumption and idling could play a significant role in reducing greenhouse gas emissions, particularly in major cities.

3.7.2 *Carbon Dioxide (CO_2)*

During combustion, two molecules of oxygen and one of carbon recombine to form carbon dioxide. The level of CO_2 produced is function of the fuel's origin, as well as the proportion of the different other elements mixed with these liquids.

In the USA bout 2.68 kg of CO_2 are produced from burning a conventional liter of gasoline (for instance, not containing any ethanol) and around 2.4

using gasoline with 10 % ethanol. For diesel, around 2.95 kg of CO_2 are produced from burning a conventional liter and 2.6 kg with 10 % ethanol. In Europe the value that are usually considered are respectively 2.3 kg for gasoline and 2.63 kg for diesel. Adding ethanol helps reduce CO_2 emission but on the other hand reduces energy efficiency. Indeed, ethanol efficiency is only 2/3 of gasoline. Adding 10 % ethanol would reduce mileage by about 3 % during summer and reduce it even further during cold winter days.

The EIA estimated in 2012 that gasoline and diesel burned during transport resulted in the emission of 1511 billion metric tons of CO_2. This total was equivalent to 83 % of total US CO_2 emissions by the transportation sector and 29 % of total US energy-related CO_2 emissions.

CO_2 generated per person Each country has its own environmental footprint. In the UK, calculation gives 3.67 tons of CO_2 emissions per person. Almost all of this is used in internal combustion engines. Adding 0.14 tons of CO_2 for oil extraction and 0.43 tons lost in oil refining gives a total energy wasted in oil exploration and refining of 0.57 tons, which means that only 3.10 tons of the total 3.67 tons is actually used in internal combustion engines (15 % pollution inefficiency). Allowing for an extra 0.43 CO_2 tons arising from fossil fuel used in motor vehicle manufacturing, each UK citizen produces 4.1 tons during transport per year.

Car owners of bigger cars, pollute obviously more than others. Vehicle emission data is available for new car registrations and can help identify by how much. To take into consideration real life conditions (driving style, passengers, luggage, poor maintenance, weather, under-inflated tires, use of air-conditioning) we added an extra 15 % to DEFRA values (Table 3.31).

The figures above show that larger cars pollute one-third more than smaller ones.

Table 3.31 Emission of CO_2 per car type[a]

Size	Engine size (L)	km/L	g of CO_2 per km
Small car	<1.5	15.73	624
Medium car	1.5–2.0	14.03	688
Large car	>2.0	10.20	960

[a]European Environment Agency (2008). http://www.eea.europa.eu/data-and-maps/figures/train-occupancy-rates

3.7.3 Nitrogen Oxide: NO_x

Nitrogen oxide can refer to a compound of oxygen and nitrogen (NO), or a mixture of such compounds (NO_x). NO_x compounds play several negative roles. For instance, production of nitric acid originating from NO_x transformation is one of the main causes of acid rains. A similar transformation process can happen at the level of human mucosa, which irritates the respiratory system. This gas plays also an important role in smog appearance and creates problems at the ozone layer. Their role in PM formation is also important (NO_2 and SO_2 can combine to form PM).

3.7.4 Nitrous Oxide: N_2O

At room temperature, N_2O is a colorless nonflammable gas, with a slightly sweet odor and taste. It is a major greenhouse gas and air pollutant. Considered over a 100-year period, it has almost 300 times more impact per unit mass on global warming than carbon dioxide.

3.7.5 Particulate Matter: PM_{10} and $PM_{2,5}$

PMs are very thin solid or liquid particles that remain in the air, and have a diameter smaller than 10 (PM_{10}) or 2.5 micrometer ($PM_{2.5}$). Some particles are naturally produced (i.e., pollen and fire). However, a great proportion is man-made. In cities, it is estimated that around 20 % of all PMs are due to transport. PMs are bad for the human health. The toxicity of these particles is linked both to their size and their composition. Because of their very small size, these particles can penetrate very deeply into the lungs alveolus or even in the blood vessels.

3.7.6 Volatile Organic Compound (VOC)

These elements are organic matter, which evaporate easily and are present in their gas state at ambient temperatures. These compounds play a fundamental role in the production of smog. Furthermore, some of these compounds (i.e., benzene) induce cancer.

3.7.7 Health Impact of Pollution

The following details about the impact of air pollution on human health were inspired from an article published[5] in January 2014.

Cardiovascular disease Many molecules produced by transport have a direct impact on human health. Of all these molecules, PMs are the ones causing the most concern. Any particle smaller than 10 micrometers, can enter the lungs and for every increase of 10 $\mu g/m^3$ in PM_{10}, lung cancer rate can rise by 22 %. The smaller $PM_{2.5}$ is worst as they pass into the bloodstream and as a result can damage blood vessels.

> The likely explanation is that pollutants may trigger inflammation in the body, activating chemicals that are toxic to the heart and cause people to breathe faster, increasing heart rate and blood pressure. A University of Michigan study[6] found that higher concentrations of $PM_{2.5}$ were linked to faster thickening of the inner two layers of the common carotid artery.

Lung disease Air pollution not only triggers asthma attacks but increases their severity and frequency and may even cause the condition. A 2013 European study[7] concluded that 14 % of chronic childhood asthma is due to exposure to traffic pollution near busy roads. Jo Waters states in the article[8] that:

> Pollution is also thought to worsen symptoms in patients with chronic obstructive pulmonary disease, which includes emphysema, chronic bronchitis and cystic fibrosis. Vehicle exhaust gases, including ozone, nitrous oxide and sulfur dioxide, can also act as irritants, making lung tubes tighten and narrow, causing wheezing and breathlessness.

---

[5]Can car exhaust fumes cause dementia? Asthma. Heart attacks. Cancer. Even diabetes. Why experts fear traffic pollution may be linked to a list of health problems. Author: JO WATERS; PUBLISHED on 27 January 2014. http://www.dailymail.co.uk/health/article-2547008/Can-car-exhaust-fumes-cause-dementia-Asthma-Heart-attacks-Cancer-Even-diabetes-Why-experts-fear-traffic-pollution-linked-list-health-problems.html

[6]A University of Michigan study found that higher Air pollution linked to hardening of the arteries Apr 24, 2013 Laurel Thomas Gnagey. http://www.ns.umich.edu/new/releases/21420-air-pollution-linked-to-hardening-of-the-arteries

[7]Road traffic pollution as serious as passive smoke in the development of childhood asthma (March 21, 2013); Source European Lung Foundation.

[8]See Footnote 5.

Diabetes A study[9] published in 2013 concluded that children's exposure to traffic pollution was associated with an increased risk of developing insulin resistance, which can lead to type-2 diabetes. The study showed that the higher the PM concentration is, the more important the risks are.

Premature birth According to a study[10] published by Californian Universities, women living in polluted urban areas have a 30 % increased risk of giving birth prematurely.

Cancer The International Agency for Research on Cancer (IARC), part of the World Health Organization, has classified outdoor air pollution as an agent causing cancer. In its 2013 warning, it concluded that air pollution causes lung and bladder cancer. Furthermore, not only did it classify air pollution as carcinogen. The PM alone was classified as carcinogen. The IARC reported that out of the 3.2 million people who died worldwide from air pollution in 2010, 223,000 succumbed from lung cancer.

Dementia Air pollution is thought to influence dementia sickness such as Alzheimer. A 2012[11] study of postmortem brain examinations following accidents in the very polluted city of Mexico, builds on a growing body of research suggesting that air pollution exposure can cause changes in the brain similar to those seen in Alzheimer's patients. More than half the brains of urban fatalities showed signs of amyloid- B plaques, against none for rural fatalities. Forty percent of the urban fatalities had pretangle material (material also associated with Alzheimer) compared to close to nil for the rural people studied.

3.7.8 Greenhouse Gas Effect

Pollution emission from petrol-based engines also has a direct impact on the environment.

Climate change Despite compelling evidence of climate change, governments from many countries have avoided taking harsh and unpopular measures to reduce

[9]Air pollution increases risk of insulin resistance in children; Published in 2013 in Diabetologia, the journal of the European Association for the Study of Diabetes (EASD), and is by Elisabeth Thiering and Joachim Heinrich, Helmholtz Zentrum München, Neuherberg, Germany, and colleagues.

[10]Traffic-related air toxics and preterm birth: a population-based case-control study in Los Angeles County; study published by Californian Universities, Author: M.Wilhelm, J.K. Ghosh, J. Su, M. Cockburn, M. Jerrett and B. Ritz.

[11]Alzheimer's-like brain changes seen in young who breathed polluted air. (Jan 30, 2012); Source: Calderon-Garciduenas, L, M Kavanaugh, M Block, AD Angiulli, R Delgado-Chavez, R Torres-Jardon, A Gonzalez-Maciel, R Reynoso-Robles, N Osnaya, R Villarreal-Calderon, R Guo, H Zhaowei, Z Hongtu, G Perry and P Diaz.

greenhouse gas emission. For instance, the US federal government, Canada, China, and other countries have not committed to CO_2 stringent reduction.

How does CO_2 emission impact the environment? Firstly it affects a specific region of the atmosphere called the troposphere. Indeed, the greenhouse effect issue concerns the warming of this lower part of the atmosphere (it is about 10–15 km thick, varying with latitude and season). In this zone, greenhouse gas molecules such as CO_2, methane (CH_4), Nitrous oxide (N_2O), and N_2O are caught. Warming will occur because these greenhouse gases, while they are transparent to incoming solar radiation, absorb infrared (heat) radiation from the Earth that would otherwise escape from the atmosphere into space. The greenhouse gases then reradiate some of this heat back towards the surface of the Earth.

Effect on ozone layer Substances that contribute to ozone depletion usually have high concentrations of chlorine or bromine atoms and include chlorofluorocarbons (CFCs). Vehicle emissions contain few of these substances and therefore have little effect on ozone depletion. Hydrocarbons are recognized by EPA as having no ozone depletion impact.

3.7.9 Wheel-to-Wheel Pollution of Different Transport Modes

The following table shows the different gas emission at the point of energy consumption (in g/L) (Table 3.32).

The EPA provides also the average pollutant emission and fuel (Table 3.33)

The EPA gives also the pollutant emission per type of vehicle in teragram equivalent (Table 3.34)

3.7.10 Well-to-Wheel Pollution of Different Transport Modes

We've seen that polluting outside of cities is critical for human's health, even more so when we acknowledge a worldwide massive shift of population from rural to urban areas. However, we cannot forget that many inhabitants still live in rural areas. Furthermore, pollution has also a huge impact on wildlife and the environment through global warming.

Table 3.32 Web page[a] environment Canada gas emission factors for energy mobile combustion sources, in g/L

Mobile combustion	Emission factor for transportation		
Transport mode	CO_2	CH_4	N_2O
Gasoline vehicles			
Light-duty gasoline vehicles (LDGVs)			
Tier 2	2289	0.14	0.022
Tier 1	2289	0.23	0.47
Tier 0	2289	0.32	0.66
Oxidation catalyst	2289	0.52	0.20
Non-catalytic controlled	2289	0.46	0.028
Diesel vehicles			
Light-duty diesel vehicles (LDDVs)			
Advance control	2663	0.051	0.22
Moderate control	2663	0.068	0.21
Uncontrolled	2663	0.10	0.16
Natural gas vehicles	1.89	9×10^{-3}	6×10^{-5}
Propane vehicles	1510	0.64	0.028
Motorcycles			
Non-catalytic controlled	2289	0.77	0.041
Uncontrolled	2289	2.3	0.048
Railways			
Diesel train	2663	0.15	1.1

[a]Fuel Combustion Environment Canada (Web page); Date Modified: 2013-06-21. http://www.ec.gc.ca/ges-ghg/default.asp?lang=En&n=AC2B7641-1

Table 3.33 EPA average emissions and fuel consumption for passenger cars, based on average mileage of 12.000 and 0.0419 gal per mile (October 2008)[a]

Pollutant/fuel	Emission in g/km driven	Annual emission in kg
VOC	1.668	12
THC	1.737	13
CO	15.161	113
NO_x	1.118	8
PM_{10}	0.0071	0
$PM_{2.5}$	0.0066	0
CO_2	594.194	4,421
Gasoline consumption in L	0.156	1,885

Adapted by Author

[a]Average Annual emission and fuel consumption for gasoline fuel passenger cars and light trucks; Source EPA Office of Transportation and Air quality EPA420-F-08-024

Table 3.34 Total US pollutant emission per type of vehicle and fuel type[a]

2012	In teragram equivalent	CO_2	CH_4	N_2O
Gasoline	Total	1100	1.1	12
	Passenger cars	755.7	0.8	8
	Light trucks	286.6	0.3	3.5
	Buses	0.8	0.1	0.5
	Motorcycle	4.2	+	+
Diesel	Total	435.4	+	0.4
	Passenger cars	4.2	+	+
	Light trucks	13.2	+	+
	Buses	16.3	+	+
	Rail	40.2	+	+

[a]See Footnote 12

Table 3.35 Total pollutant emission per activity and power plant type; EPA 2012[a]

2012	In teragram equivalent	CO_2	CH_4	N_2O
Electricity generation	Total	2042.2	0.5	18.3
	Coal power plant	1512.2		
	Natural gas	492		
	Petroleum	18.8		
Transportation	Petroleum	1735.4	1.7	16.5

[a]North American Power Plant Air Emissions; Source: the Secretariat of the Commission for Environmental Cooperation (CEC) of North America. http://www3.cec.org/islandora/en/item/10236-north-american-power-plant-air-emissions-en.pdf

In order to calculate the impact of transport modes, we must analyze the impact of the different types of power plant on CO_2 emission.

The EPA, provided the pollutant emission in teragram of equivalent CO_2 (Million metric tons) for their electrical generation matrix (Table 3.35).

These numbers per se don't mean much but if we cross the values of this table with the previous one, we can deduct that proportionally, the electric generation matrix in the USA emits:

- More CO_2 than the vehicles (around 17 % more);
- Much less methane (less than one-third); and
- More N_2O (around 10 % more).

These results are obviously linked to the importance of coal in the American generation matrix. The following graph gives additional information on other pollutants. It shows a North American direct comparison between coal, petrol, and natural gas power plants for level of emission of NO_x, SO_x (Sulfur Dioxide), CO_2, PM_{10}, and Hg (mercury) (Fig. 3.19 and Table 3.36).

A newer study from the American National Oceanic and Atmospheric Administration concluded that Natural gas power plants in the USA emit 40 % less CO_2 than coal power plants. Their study performed exclusively at the level of the

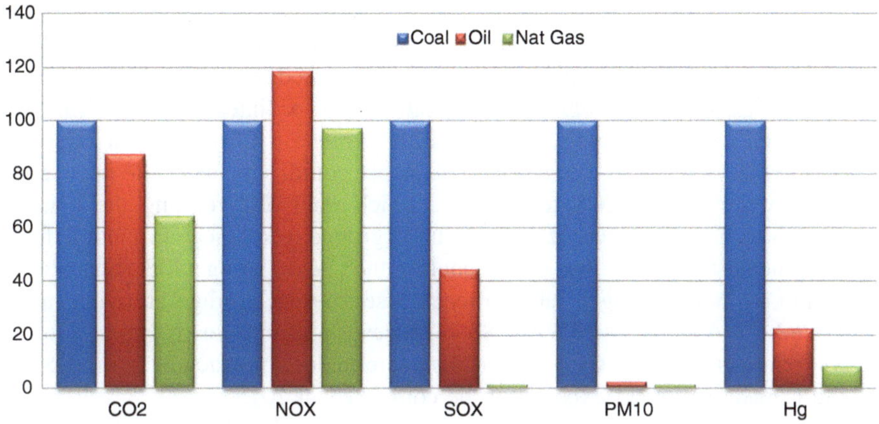

Fig. 3.19 Emission of pollutants in % of Coal (Average Annual emission and fuel consumption for gasoline fuel passenger cars and light trucks; Source EPA Office of Transportation and Air quality EPA420-F-08-024)

Table 3.36 Typical emission of pollutants

Pollutants	Emission rates in kg/MWh		
	Coal	Oil	Nat gas
CO_2	1021.04	759.09	515.29
SO_2	5.9	5.45	0.05
NO_x	2.72	1.82	0.77

[a]North American Power Plant Air Emissions; Source: the Secretariat of the Commission for Environmental Cooperation (CEC) of North America. http://www3.cec.org/islandora/en/item/10236-north-american-power-plant-air-emissions-en.pdf

power plant and based on measurements collected between 1997 and 2012, concluded that the average emission of:

- Coal-based power plants: 915 g of CO_2/kWh of electricity produced;
- Natural gas power plants: 549 g of CO_2/kWh produced; and
- Combined cycle natural gas plants: 436 g of CO_2/kWh produced.

Reduction of CO_2 emission In 2012, the EPA set new standards to reduce CO_2 emission. New large plants (> 100 MW) fueled by natural gas could emit no more than 450 kg of CO_2/MWh of electricity produced. Smaller natural gas plants would have to achieve a less stringent rate of 500 kg/MWh. Coal plants would have two options, either of which would require Carbone Capture and Storage technology (CCS).

CSS technology: New large coal plant would need to meet 500 kg/MWh, which they couldn't through efficiency alone. Indeed, the most efficient type of coal plants, using ultra-supercritical boilers or integrated gasification combined cycle technology, can currently achieve a CO_2 emission rate of around 770 kg/MWh. Thus new coal plants could only meet the standard through the use of CCS technology, which traps CO_2 exiting the plant, transports it, and injects it into an underground geological formation for permanent storage. Newer plants could either begin using CCS soon after startup, or begin using it later to reach a seven-year average emission rate around 500 kg of CO_2/MWh. CSS is however still in its early phase. Only two commercial-scale CCS power plants are under construction in the USA and Canada and were expected to be completed in 2015.

3.7.11 *Electric Generation Matrix*

The following graph (values from 2008) shows the impact of the electrical generation matrix (in gram of CO_2 per passenger × km traveled) for the Paris region. For instance in France, most of the energy used by the metro and commuter lines comes from the French utility EDF, whose primary energy source is nuclear (Fig. 3.20).

This graph shows the impact of not using fossil fuel to produce electricity. If Paris metro and commuter lines were to use the average European 2008 values, the emission of CO_2 per passenger km traveled would still show that electric public

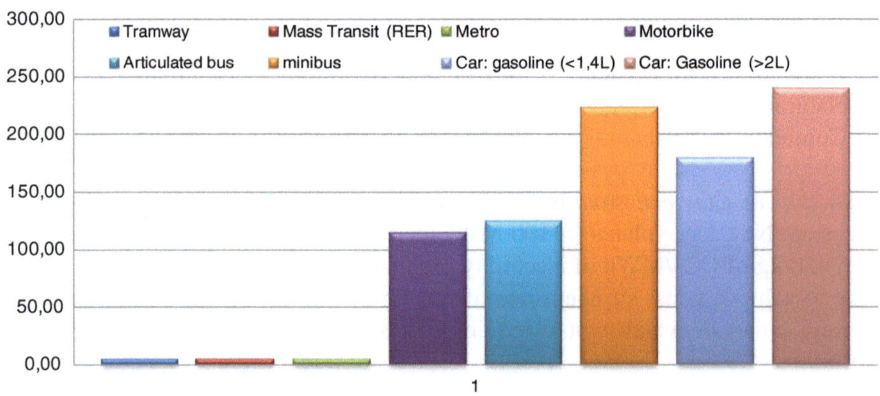

Fig. 3.20 Emission of CO_2 per passenger traveled in the Paris region (Ademe 2008) [Transportation Energy Data Book; (2009)]; expressed in gep (gram of equivalent petrol)

transport means are much cleaner than conventional gasoline technology, but less so (Fig. 3.21).

3.7.12 Emission of CO_2 Per kWh

If we go back to the electrical generation matrix of France, Brazil, and the USA, and use values of 915 g of CO_2/kWh for coal and 549 g of CO_2/kWh for Natural gas per kWh of electricity produced, we find that each kWh generates 70, 130, and 500 g of CO_2, respectively (Table 3.37).

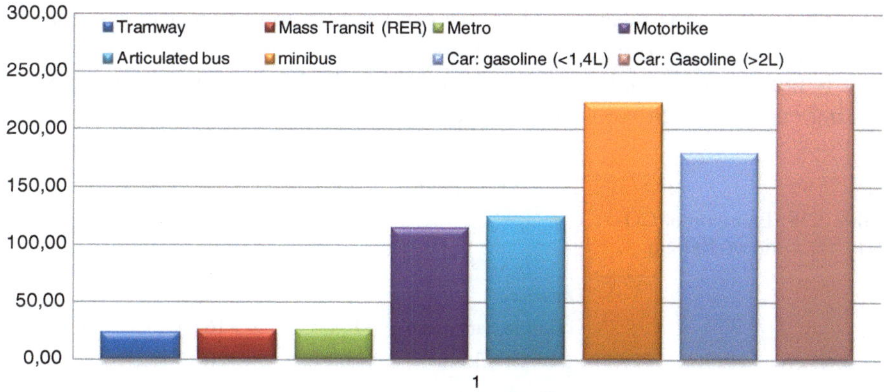

Fig. 3.21 Emission of CO_2 per passenger traveled in the Paris region if using European values (adapted from Ademe 2008) [Transportation Energy Data Book; (2009)] expressed in gep (gram of equivalent petrol)

Table 3.37 Emission of Co_2 according to different generation matrix

Electrical matrix	France	Brazil	USA
Energy produced	in %	in %	in %
Fossil fuel total	9.6	17.1	67.8
Coal	5.9	11.9	37.0
Nat Gas	2.8	3.5	30.2
Petroleum	0.9	0.3	0.6
Nuclear	76.4	2.6	18.8
Hydroelectricity	11.9	78.8	6.8
Other renewables	2.3	1.5	5.3
g of CO_2/kWh	70.0	130.0	500.0

Source Author

When we compare these numbers for instance with EDF announced values of 56 g of CO_2 per kWh of electricity produced, we can say that this approximation is pretty accurate. The EPA calculates that on average the American electricity grid emits 550 g of CO_2 per kWh (1222 lbs per kWh) or 10 % more, which isn't that bad either.

Now if we compare these numbers with the 2289 and 2668 kg/L generated by gasoline and diesel, we need to calculate how many kWh a liter of gasoline and diesel generates. We've seen in previous section that 1 L of gasoline and diesel is equivalent to 9.85 and 10.98 kWh, which translates into 208 and 232 g of CO_2 per kWh, respectively. We've seen in the previous section that an electric car would require about 11 kWh to run 100 km, while a medium gasoline car would consume 80 kWh for the same distance. Thus, we can easily compare electric and conventional cars on a well-to-wheel basis (Table 3.38).

It is important to emphasize that these values don't even consider recuperated braking energy, which would improve electric cars results by around 18 %.

Table 3.38 Emission of CO_2 in kg for a trip of 100 km

Vehicle type	France	Brazil	USA
Electric cars	0.77	1.43	5.5
Diesel cars	18.56	18.56	18.56
Gasoline cars	16.64	16.64	16.64
kg of CO_2/kWh	0.07	0.13	0.50

Source Author

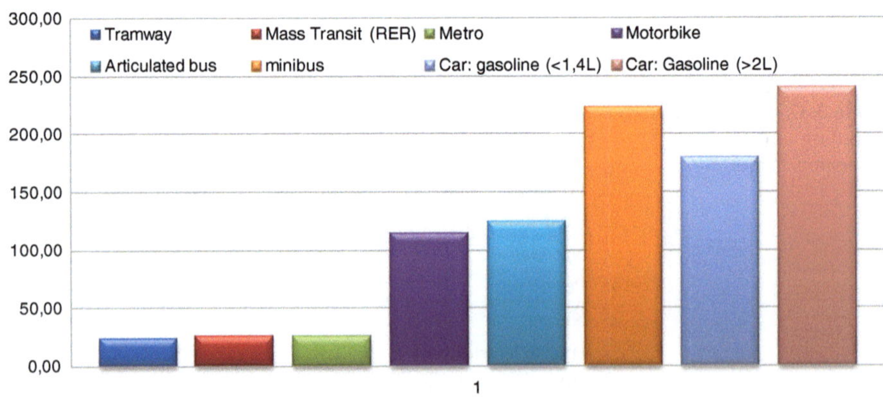

Fig. 3.22 Transport mode energy efficiency expressed in gep (gram equivalent petrol). *Source* Ademe 2008

3.7.13 Emission of CO_2 Per Transportation Means

In order to make an accurate comparison, we would need to take into account the occupancy rates as we did for the energy calculation. We can also just use the values calculated by EPA that give values for passenger per km traveled. In the USA, commuter rail, subway trains, long distance trains (i.e., intercity rail) emit on average 108, 100, and 116 g of CO_2 per passenger km, respectively.[12] According to the same source, on average, a bus trip emits 66 g of CO_2 per passenger km. If we assume the average occupancy rate of 1.5 passengers per car (1.1 at peak hour), a gasoline and diesel car would emit 110 and 123 g of CO_2 per passenger kilometer on average and around 150 and 170 g of CO_2 at peak hour. The emission of CO_2 per passenger km traveled of mass transit systems in Paris is extremely different than the numbers for the USA, as we can see in the following chart. This is caused by the higher occupancy rate as well as to the characteristic as of the French energy matrix that relies almost exclusively on nuclear power plants (Fig. 3.22).

3.7.14 Conclusion About Pollutant Emission

We've analyzed the pollutant emission at the point of energy consumption and generation. Wheel-to-wheel (wall-to-wheel) pollutant emission is exclusively linked to conventional petroleum-based vehicles. Electrical vehicles don't pollute in cities. This is extremely important because that's where the pollution problems are the most acute. Thus, forcing urban citizen to adopt electric transport, be it private or public, is an extremely sound policy, which would reduce significantly health costs.

Health costs We've seen that pollutant emission can generate cardiovascular and lung diseases, diabetes, cancer, premature birth and dementia. According to the World Health Organization in 2012 around 7 million people died—one in eight of total global deaths—as a result of air pollution exposure. According to the US EPA, Office of Air and Radiation, (March 2011), total combustion emissions in the US accounted for about 200,000 premature deaths per year due to changes in particulate matter concentrations. The largest contributors for both pollutant-related mortalities are road transportation, causing a total of approximately 58,000 early deaths per year, and power generation, causing approximately 54,000 premature mortalities per year.

[12]Inventory of U.S. Greenhouse Gas Emissions and Sinks: 1990–2012; Source American EPA; (2014). http://www.epa.gov/climatechange/pdfs/usinventoryreport/US-GHG-Inventory-2015-Chapter-3-Energy.pdf

The Ontario Medical Association calculated that in 2005 the total cost of illness due to air pollution was US$6.43 Billion and that mortality represented 82 % of the cost, while pain and suffering represented 7 %. Health care cost and lost in productivity only accounted for 6.5 and 5 % respectively. Furthermore, in the UK, an old study (1999) found that at that time, the cost of lung disease linked to air pollution was in the range of 6 % of the National Health Hospital System, in line with the Canadian values.

Health cost in % of GDP Now if we want to put a global value to air pollution costs, we must try to calculate the impact on a country's economy. A 2011 study[13] by Muller and Mendelson estimated the cost of air pollution at 0.7 % of the US GDP. We don't have these same values for 2014, but as unfortunately the quality of air in most big cities hasn't improved we could use the same values to assess the worldwide cost of pollution. Based on an estimated negative impact on the economy of between 1 and 4 % of the world countries' GNP, air pollution could account for a total yearly cost for society of over $600 billion.

It pays to invest in air pollution remedies According to a 2012 study[14] since 2001, every increase in $PM_{2.5}$ (10 μg/m^3) in six US observed cities was associated with an adjusted increased risk of all-cause mortality of 14 %, and with 26 % and 37 % increases in cardiovascular and lung cancer mortality, respectively. Another American study found that a decrease of 10 μg/m^3 in the concentration of $PM_{2.5}$ was associated with an increase in mean life expectancy of 0.35 years. According to a report published in the magazine Nature,[15] aggressive cuts in greenhouse gas emissions could prevent 300,000–700,000 premature deaths annually by 2030; 800,000 to 1.8 million by 2050; and between 1.4 million to 3 million by 2100. The authors of this report estimate the value of the health benefits derived from cutting one ton of CO_2 at between $50 and $380, or more than the projected costs of cutting one ton of CO_2 within the next several decades for global warming purposes.

Global warming costs The impact of global warming can already be felt. Temperatures are being more extreme and causing destruction coming from stronger storms and hurricanes, with the impact of wind and flooding. For instance, the typhoon Haiyan that hit the Philippines in 2013, was the strongest ever, with

---

[13]Environmental Accounting for Pollution in the United States; Author Nicholas Z. Muller, Robert Mendelsohn, and William Nordhaus; Economy American Economic Review 101 (August 2011).

[14]Chronic Exposure to Fine Particles and Mortality: An Extended Follow-Up of the Harvard Six Cities Study from 1974 to 2009; Authors Lepeule, Johanna; Laden, Francine; Dockery, Douglas; Schwartz, Joel. Environmental Health Perspectives (March 2012).

[15]Co-benefits of mitigating global greenhouse gas emissions for future air quality and human health; Published in Nature climate change; Authors: J. J. West; S. J. Smith, R.A. Silva, V. Naik, Y. Zhang, Z. Adelman, M. Fry, S. Anenberg, L. W. Horowitz & J.F. Lamarque. http://www.nature.com/nclimate/journal/v3/n10/full/nclimate2009.html

wind reported at 250 km/h. It is estimated that the number of fatalities was over 6000 and the cost of destruction on the Philippine's economy to be somewhere between \$6 and \$15 billion. The IPCC (Intergovernmental Panel on Climate Change) released in 2014 a report of the impact on global warming on the economy. In this report it is stated that between 0.02 and 2 % of worldwide GNP would be reduced by an increase of 2 °C. This report also states that without deep cuts in CO_2 emission, the world is on its way to pass that value. Unfortunately, the world is already on its way to lose that amount and without doing anything, could see further loses.

3.8 Other Environmental Considerations

3.8.1 Battery Recycling

The opponents to electric cars will rightfully say that the batteries which weigh between 150 and 250 kg and last on average 5 years, contains toxic components such as lead, cadmium and lithium. As they are also much heavier than standard lead batteries that weigh only 20 kg, they tend to require more energy to produce and recycle. However, it should be noted that most components in the batteries can be recycled. As the number of batteries will increase, the economics of recycling will also improve from their current level, which are already pretty decent.

- *Lead–acid batteries*: They can be 95 % recycled.
- *Reuse as storage facility*: Studies of electric car batteries have shown that even when a car battery has only 70 % of capacity left—too little to serve in a car—it may still have around 10 years of useful life left as grid storage devices. Thus, modern batteries could be reused for 15 extra years with the right business model.
- *Nickel–metal hydride batteries*: Found in most conventional hybrids, they are very valuable to recycle, because nickel sells for a high price and is easily reusable. A few companies are already running commercial nickel–metal hydride battery recycling centers. These plants achieve a 90 % recycling level.
- *Lithium-ion batteries*: They are more difficult to recycle, in part because car manufacturers use various combinations of chemical components with different recycling values. Furthermore, their recycling is still in its infancy. Lithium–nickel–manganese–cobalt batteries found in many modern plug-in cars, bring high recycling value because nickel and cobalt are costly to produce. It is anticipated that the cheaper chemical elements planned for future models, will still be recyclable. Recycling of lithium-ion batteries is a little tricky. They must be frozen to defuse the lithium before being sheared, shredded and separated into their respective parts to be resold. The US has one lithium battery recycling pilot plant under construction in Ohio and there are active commercial plants in Europe.

3.8.2 *Enabling Renewable Energy Storage*

Turning wind and sun into energy isn't costly, as long as equipment is installed at the right location. The main problem is reliability. In many places during several days of the year, rainfalls or wind stops blowing. Furthermore, nights fall inexorably everywhere, stopping solar panels from generating energy. However this shouldn't impede countries which lack favorable climate for renewable energy from becoming a "green country". For instance, the EU has decided to invest massively in renewable energy, which would see the current 100 GW of production capacity in 2012 double by 2020.

Germany is one of these European countries, which isn't especially favored climatewise, but is committed to using large quantities of renewables. In September 2010 the German government announced the following new ambitious energy targets: Renewable electricity in 2011: 20.5 %; 35 % by 2020; 50 % by 2030; 65 % by 2040; and 80 % by 2050. Though this is a positive step for the environment, this decision is likely to create huge problems for energy utilities in the short term and may actually even generate blackouts in some regions. For the moment the grid is holding even though on some days, renewables output can almost top 60 % of total production. For instance, in that country on October 3, 2013, renewable output peaked at 59 %, providing 36.4 % of the total output of that day. The grid was able to cope but utilities had to stop production of other power stations.

In order to be able to cope with such ambitious objectives, Germany will have to invest massively in two areas, pushing even higher the price of the kWh, already the most expensive in Europe:

- New transmission lines linking all European parts to take advantage of climate diversity (i.e.: wing blowing up North and sun shining down in the South); and
- Investment in storage capacity, probably using a decentralized model, where the energy is stored at the point of consumption rather than of generation.

Electrical cars may help in reducing the cost of investment in storage capacity. Indeed, one of the cheapest potential solutions to solve this issue would be to store energy when it is produced and use it during off-peak hours. The most obvious way to store energy is through the use of batteries. However, technically there is no current battery technology, which could store enough energy to smoothen production peaks. So why not imagine millions of batteries linked to a smart grid and working together? Those parked and plugged-in electric vehicle could sell the electricity from the battery during peak loads and recharge either at night (at home) or during off-peak (at work).

With new reliable vehicle-to-grid technology, each vehicle could turn its 20–50 kWh battery pack into a distributed storage facility, helping balance the network loads, providing emergency source of power, or being used by households during hours when rates are higher. The amount of energy stored by cars isn't small and could represent 2–5 days of average household consumption (typically around 10 kWh). However and in order for this strategy to work, a large disadvantage of current battery technology would need to be improved. Indeed, currently each storage cycle stresses the battery with one complete charge–discharge cycle and this would need to be overcomed.

Going back to our calculation about turning the conventional car fleet into an all electrical one, we've seen that for a country like France, the 2013 private car petrol consumption was 29.5 Mtoe or 343 TWh. Because of the higher efficiency of electric cars, 12.4 Mtoe would be saved, which means that 17.1 Mtoe would still need to be produced in addition or almost 200 TWh. The French energy matrix would need to be increased by 36 % to cope with such growth in demand. Now let's assume a realistic once per day discharge to the grid, as well as once per day consumption of the energy, and thus two daily recharges from the grid.

This means that a complete electrical car fleet could consume and store a little more than 50% of the newly increased energy matrix production. If applied to Germany and all things being equal, this would solve partially that country's problem of renewables till 2030. However, Germany would need to invest even more in production capacity be it in renewables or in conventional thermal plants.

3.8.3 Reduced Land Intake

Land in big cities is scarce and very often expensive. Furthermore in megacities, green parks or just green areas are hard to find, most of the city being covered by dwellings or roads, impermeabilizing soil. As a consequence cities are more prone to flooding, unless huge water retention tanks are built in the most vulnerable areas.

Heavy metro systems are usually built in tunnels, and thus have a very low land intake. Only stations occupy land but are often integrated under shopping malls or tall buildings and thus land intake becomes negligible. Monorails using elevated structure are well suited for being built between two sides of an avenue. Their land intake is minimal: about two times 1 m^2, every 30 m of guideway.

Tramways or buses running in mix traffic will occupy no more space than the cars' own land intake. This space isn't free but in most cities it already exists, as people require a car access to their home. On the other hand, buses or tramways running on segregated lines require vast amounts of land. As a matter of fact, two different types of land intake are necessary for BRTs such as the Transmilenio. As we will see in Chap. 5 on megacities, this very successful BRT occupies extremely large areas of land (width of 16 m) that need to be asphalted and in some cases overpasses need to be built to access the stations located in the middle of the lanes.

BRT operations in cities like Sao Paulo can occupy less space as they only use one dedicated lane. On the other hand, tramways on segregated lines can occupy areas where the soil is actually grass or pebble stones, enabling the soil to absorb rain water.

3.8.4 City Integration

Visual impact Heavy metros have very low visual impact. Only metro stations can be visible but are often integrated under shopping malls or tall buildings. Buses running at grade have a limited visual impact, as they are already integrated within our city landscape. Tramways on the other hand, with their catenary system have a significant impact on the look of a city. Catenary-free operations are now being pursued by several manufacturers to eliminate altogether the negative visual impact or limit it to areas where it isn't as damaging.
Elevated structures built for light rail or BRTs need to be large (and heavy) by nature and can have a huge visual impact. Monorail on the other hand, with their sleek design and very light beam (690 mm wide) can easily be integrated within the city. With very small turning radius required (46 m) and high grade capacity (up to 6 %), it can easily follow the existing building and infrastructure, integrating itself well in the city landscape. Furthermore, monorails can actually become a city landmark like the monorail is in Sidney, Las Vegas, or Dubai, attracting tourists to the city (Fig. 3.23).

City separation Once again heavy metros operating in tunnels don't create any physical or psychological barriers within a city, as intercity highways and segregated lines for buses or tramways do. These crossings become bottle necks for road traffic. Monorails with their light elevated structure create very limited city separation.

City revitalization Public transport can play a crucial role in revitalizing urban centers. Heavy metros can reduce the number of parking required for cars or buses circulating in an important city area. Tramways can also play a crucial role in revitalizing an area, especially when it is constructed with space for pedestrian and bicycles. As a heavy transport mean, monorails can also play an important revitalization role, reducing parking and congestion. Using an elevated guideway it is, however, more difficult to integrate in the urban environment than a tramway. Buses are usually not seen as part of the city revitalization solution but part of the problem.

Fig. 3.23 Monorail at Dusseldorf airport Germany. *Source* Author

Traffic flow disruption At grade transit technology can disrupt traffic flow. Segregated lines can seriously limit the car flow to a few over or underpasses. Furthermore, in megacities buses running in shared lanes can seriously hinder traffic, as they need to jump from one lane to the other.

3.8.5 Noise Pollution

Noise emission can have various origins. For instance for buses, air impact on the carbody, motor noise and exhausts as well as running tires all have impacts. These different sources of noise will vary according to the speed of the rubber tire vehicle. For speed lower than 50 km/h, the noise is mainly due to tires running on road surface. For a speed between 50 km/h and 70 km/h the running noise is equivalent to the motor and the exhausts. After 70 km/h the motor and exhaust noise are predominant. Rubber tires running on a road surface will emit more or less the same noise than steel wheel technology. For instance, buses and tramways have a noise emission of around 80 dB, while metros and monorails emit around 85 dB.

However, the big difference is that heavy metro or monorails, for not running at grade, have no (for metro) or very low noise impact (for monorails), when tramway and buses have an immediate impact on environment, pedestrians and neighbors.

3.8.6 *Vibration*

Heavy metro can create vibrations. New rail fixing technologies can improve this nuisance but not eliminate it altogether. Elevated structure doesn't create this type of problem to neighboring buildings. Buses on the other hand, can create some vibration to the neighboring residences.

Acknowledgments and disclaimer The use of pictures, or references made to studies or companies and their brand does not in any way suggest that the authors of such studies or the mentioned companies endorse in any way this book or its content. The author endeavors in respecting the copyright subsisting any of the graphics, and texts that he uses, to use graphics and texts he has himself created or to use graphics and texts not covered by copyright. All trademarks and brand names quoted in the book including those protected by copyright of third parties are subject unreservedly to the provisions of current copyright law and the rights of ownership of the registered copyright holders.

Companies and Brands Stated in the Chapter

- BP Plc
- Mia is a trademark of the Mitsubishi Group
- Roadster is a trademark of Tesla Motors Inc.
- Focus is a trademark of Ford Motor Company
- H System is a trademark of General Electric Company
- SGT5-8000H is a trademark of Siemens AG
- J class is a trademark of the Mitsubishi Group
- Mercedes Benz is a Trademark of Daimler AG
- Tesla Motors Inc.
- Panasonic Corporation
- Mckinsey & Company
- Leaf is the trademark of Nissan Motor Company Ltd.
- Volt is the trademark from General Motors Company
- Prius is the trademark from Toyota Motor Corporation
- Yaris is the trademark from Toyota Motor Corporation
- Mitsubishi Electric Corporation
- X60 is the trademark from Alstom SA
- EDF is a trademark of Electricite de France

Chapter 4
Avoiding Megacities' Standstill

Acronyms

BRT	Bus rapid transit
CAPEX	Capital expenditure
LOS	Level of services
OPEX	Operational expenses
PPHPD	Passenger per hour and per direction
PRT	Personal rapid transit
V2I	Vehicle to infrastructure
V2V	Vehicle to vehicle

The United Nation Habitat introduced the term megacity in the 1970s to describe a metropolitan area with a total population in excess of 10 million people. This term describes municipalities or agglomerations composed of several cities. In 1950, the world had just one megacity: New York. By 1975, there were five and today there are around 28 megacities on different continents. By 2025, around 630 million inhabitants or 8 % of the world's population will live in 37 megacities. By 2050, this figure is projected to grow to a staggering 1.2 billion!

By itself, the urban densification process doesn't create a problem. Actually having more people living in more densely populated area is more environmentally friendly than people living in suburbs, commuting everyday to work. The issue about megacities is that due to such high concentration of people, enormous land consumption, as well as problems of air pollution, water scarcity, poverty, and traffic jams are amplified. Furthermore, when population doubles every 10 years, proposing urban policies and planning strategies to manage the sustainable development is almost impossible. As a result, transportation problems in megacities increase exponentially and private transport there cannot escape being on the verge of coming to a standstill. Two major urban expansion processes are at the heart of this problem:

- Horizontal growth due to urban sprawl; and
- Vertical expansion caused by higher housing densification.

To utilize a useful image, under the combination of these two forces, megacities aren't really exploding, but rather imploding. As a consequence building new roads

S. Van Themsche, *The Advent of Unmanned Electric Vehicles*,
DOI 10.1007/978-3-319-20666-0_4

to cope with additional inhabitants is almost impossible, as it would require expropriating thousands of inhabitants from several buildings. To improve the situation, city developers are only left with two alternatives:

- Build underground infrastructure, which is extremely expensive; and
- Appropriate existing roads to build new public transport, which creates other problems, such as new traffic jams, separation of cities along these corridors, increased noise and pollution.

Furthermore, this takes time, which politicians usually don't have. This is why Mayors tend to revert to drastic measures, such as restricting car use in city areas with the worst traffic jams. Gasoline price hikes, which is also an effective way of reducing traffic jams cannot be applied solely at the city level, but rather at the country or state level. We will focus our analysis on restrictive measures that are available to Mayors.

4.1 Private Transport Restriction Measures

In the rich megacities of this word, Mayors tend to use financial coercive measures, such as municipal VAT taxes, tolls, congestion charges, and high parking costs. In the poor countries, they tend to favor car use restriction.

4.1.1 Congestion Charges

Using high congestion charges to restrict movement of citizens can be perceived as socially unfair, as poor people have less discretionary money. However, it works well, especially during the first years, reducing traffic and pollution.

For instance in London city, the 2003 congestion charges reduced traffic significantly (i.e.: by 30 % after one year). As a result, travel patterns changed with a huge transfer to public transport, particularly buses. Some motorists who would otherwise drive through the inner city during peak periods shifted their route, travel time or destination. Others changed their transport mode to taxis, motorcycles, bicycles or walking. Average traffic speeds within the zone during charging days increased 37 %, from 13 km/h before to 17 km/h after congestion charges were introduced. Peak period congestion delays declined about 30 %, and bus congestion delays declined 50 %. Bus ridership increased 14 % and subway ridership about 1 %.

Table 4.1 Net projected costs and revenues of London Congestion charges; 2011 study (see footnote 1)

Costs and revenues	NPV in M$	NPV in million pounds
Start up costs	280	180
Operating costs	500	320
Total cost	780	500
Charges revenue	1080	690
Penalty revenues	170	110
Total revenues	1250	800
Profits	470	300

To add to this virtuous cycle, congestion charges usually bring extra profits. The following figures extracted from a 2011 UK study[1] shows the benefits of this measure (Table 4.1).

Furthermore, more than ten years after its introduction, reductions in the number of cars are still significant. Other price restriction measures have also a positive impact on reducing the need to build new road infrastructure. Initiatives usually implemented in smaller cities (i.e., electronic road pricing and time-of-day-based tolls) are also efficient at reducing traffic jams during peak hours.

4.1.2 Private Car Restriction

In 1989, Mexico City mayor city introduced the program “Hoy No Circula” (“today it doesn’t circulate”). The program bans most drivers from using their cars one day per week, based on the last digit of the vehicle’s license plate. Other South American cities such as Bogota, Santiago, and Sao Paolo have followed suit and copied the Mexican program. These programs were aimed at reducing traffic jams and as consequence air pollution, but have largely failed. In fact, research by Lucas Davis (2008)[2] showed that despite the program’s high costs, this program failed to produce any improvement in Mexico City’s air quality and traffic. What happened is that many Mexico City residents, who could afford it, bought a second car with a license plate that didn’t match the same day or used taxis. Worst they usually bought a second-hand car which polluted more, making air quality worst. In fact, such measures ended up being worst for poor citizens.

[1]London Congestion Pricing Implications for Other Cities; Author Todd Litman from the Victoria Transport Policy Institute, (24/11/2014).

[2]The effect of driving restrictions on air quality in Mexico City. Author: Lucas W. Davis (University of Michigan 2008).

Fig. 4.1 Example of vehicle capacity: monorail and bus. *Source* Author

4.1.3 Promoting Car Pooling and Financial Restrictions

Many studies have shown that higher gasoline prices and electronic pricing of road usage have a much better track record on reducing traffic jams and lowering pollutant emission than driving restriction. One other measure that can be taken by mayors to postpone the investment in new transport infrastructure is to promote car pooling. As we will see, new social media and e-mobility technologies can increase car pooling's attractiveness and efficiency. As a consequence, dedicating an express lane for cars with four passengers can basically quadruple the existing road capacity, at almost no extra cost. However, in megacities all these measures cannot postpone for long the need for massive increase in transport capacity (Fig. 4.1).

4.2 System Capacity

4.2.1 Holistic Approach to System Capacity

In big cities, residents are asking for more public transport and want their city officials to dedicate more resources to it. Unfortunately, many of these megacities can be found in the Third World or Developing Countries, which have few existing public transport infrastructures on which they could rely as the backbone of their future transport network. Furthermore, transport authorities are confronted to a shortage of investment money for such big infrastructure projects. When the money is there, city planners still need to convince angry neighborhoods hit by the not in my backyard syndrome, fight righteous green lobbyists who tend to be against any tree cutting project even if society and the environment as a whole can benefit from such initiative, and deal with the various transportation industries, which will always push for their own technical solution.

As we've seen, one of our premises is that all transportation means are in competition between them. Thus, it is extremely important to understand two fundamental aspects: the capacity and real cost of each transportation means. As we are working with cities on the verge of breaking down because of traffic jams, we will calculate the maximum capacity of the different motorized systems with and without newer driverless technology. We will exclude from this analysis the impact of creating bicycle lanes and pedestrian areas, which can bring many benefits but are outside of the boundary of this book's analysis.

When we consider the maximum capacity, we need obviously to consider the transportation means capacity per se but not only. Other factors must also be added to avoid penalizing metro and monorail technologies that occupy almost no land and thus don't block the existing road capacity. In other words, a tramway or a BRT that occupies land will obviously retire many cars, but by occupying the roads, these systems will by the same token, block traffic at junctions creating further reduction in road capacity. Don't get us wrong, it could be politically and socially sound to favor public over private transport, but we cannot forget that land is scarce and expensive in megacities. The concept of appropriation that we are defending isn't usually considered by city planners. Instead of building a BRT or a tramway line, houses or even better, parks with bicycle lanes could be envisioned. Instead of replacing roads by BRTs or tramways, monorails, and metros could be running over or under a private express road lane with wireless prepayment cards financing the same project.

4.2.2 Increasing Capacity of Existing Infrastructure

To avoid fundamental changes to existing roads or train networks or having to invest in new systems all together, transport authorities have a few cards in their sleeve: they can improve the vehicle and increase the network throughput. For instance, larger and longer vehicles, running faster and closer to each other can make better use of existing infrastructure. Thus, they can play with four variables: the capacity of the vehicles, the number of vehicles running on the system, their average speed, and the headway. Though train, bus, and car obey to these same four variables, the way they are impacted can differ.

4.2.3 Vehicle Capacity

Vehicle capacity is a function of the number of passengers that can be carried by the vehicle. This is also defined by four elements: physical characteristics, safety standard, comfort level, and occupancy rate.

Physical characteristics We need to consider two types of physical characteristics of the vehicles:

- **Envelop**, which basically considers the vehicle's height, width and length; and
- **Useful area**, which considers strictly the area passengers can use.

To solve capacity issues, vehicles can be bought larger, longer or even with two storeys, as long as their envelop can fit within the networks' own physical characteristics. For railways and metros, there are many network physical limitations such as station clearance and length, tunnel envelop calculated with jerking, type of electric grid connection, etc. The environment of cars and buses is much more standardized. Roads always have a minimum width that doesn't really impact manufacturers. Buses' maximum length is affected mainly by the number of sections, which can be restricted by tight curves. Cities envisioning running double-deck bus fleets must also check overpass height constraints.

Capacity can also be increased by improving the useful area of each vehicle. Manufacturers can improve their vehicle's design by for instance, putting equipment on the roof or adding gangways between metro cars. Transport Authorities might as well decide to change the train or bus comfort level, by reducing the number of seats or inner space dedicated to seating area. They can also take a more drastic attitude, which is to buy trains or buses that can withstand higher volume of standees per area, affecting directly the level of passenger comfort.

4.2.4 Maximum Number of Vehicles

There are a maximum number of vehicles that can run on a track or lane at any given time. However, till this number is reached there is always a possibility to add extra capacity. Once this number is reached, there is no point in adding more vehicles, as the extra vehicles would only stay idle or would need to slow down to maintain the same safety spacing.

4.2.5 Average Speed

Line or lane speed is a function of the vehicle's inherent characteristics, the safety and comfort of passengers, as well as the limiting factors of the network operation. For instance, vehicle's acceleration, maximum speed, and deceleration are key factors influencing the overall system speed. In most metro applications, because distances between two stations are short, trains are almost always accelerating and decelerating. Metro car motorization is adapted for those specific conditions. However, metro operators must also take into consideration, the passengers' safety and comfort when accelerating and decelerating. After all they don't want when braking, their clients to fly through the windshield. Another limiting factor to speed is the network operation. The duration for opening doors and getting people out and in, called dwell time, is an important speed limitation. Other factors, such as bus or train injection on the line or crossings will also impact time. Average speed clearly changes according to railway application type. For instance, metro average speed will be in the range of 30 km/h, while intercity trains may reach 250–380 km/h depending on technology.

Though car acceleration or speed possibilities are usually more a function of the driver's wealth and his willingness to show it off, in most countries even the cheapest automobiles aren't limited by their inherent motorization characteristics (Germany's autobahn is obviously an exception). Infrastructure specificities, weather conditions, or legal speed limitation are the main speed restricting factors. Slope steepness, curves, state of the asphalt, rainy, or icy conditions are some of these factors that will affect car speed. Traffic light and potentially any people, cars, or objects blocking the road access will also influence speed.

Bus speed will be influenced by a mixture of all these, which explains why in general buses running in urban areas have problem going over 20 km/h.

4.2.6 Headway

The distance between two vehicles moving in the same direction is important for network capacity because it limits the number of vehicles you can add to this system without reducing the overall speed.

4.3 Road Capacity with Drived Cars

There is a lot of literature on road capacity concept, as this is both something everyone is familiar with and at the same time complicated to calculate. Many models exist based on various mathematical formulas. We will focus on two essential data to explain how to calculate simply a road's capacity at a cross section: headway and traffic volumes. Additional information such as traffic flow conditions, density, intensity, and mean speed will also come in handy. For ease of comprehension we will avoid using the word capacity to express the number of cars per hour, as many specialists do (i.e., pipeline capacity, road capacity), and instead use the word intensity or flow rate. Capacity will be dedicated to passengers and expressed in PPHPD.

4.3.1 One-Lane Highway Intensity

Headway is defined by the time between successive cars measured by rear-to-rear bumpers. As we've seen in the section on PRTs, space between two vehicles running on the guideway, influences directly the system's capacity. The spacing influenced by the stopping distance, is the sum of the distance a car will run due to the driver's perception and reaction time and the distance necessary for stopping the car (for more precise information, see formula in that section) (Fig. 4.2).

In the PRT section, we've calculated the capacity taking into consideration the "safety brick wall" principle. However, in the day-to-day life, the relationship between spacing and speed on regular highways is set by human driving habits. Studies have shown that drivers running on modern and good road infrastructure will tend to maintain a minimum following distance, which can be transformed into a roughly constant average time gap between 1.5 and 2 s. This exact time gap will vary depending on the driver's behavior and experience.

Highway traffic studies have measured flow rates as high as 2000 vehicles per lane per hour. Doing the math, this comes to 1.8 s between cars (including car length). However, observing 1.8 s between cars is rare. Indeed drivers tend to calculate their safety distance from bumper to bumper, not considering their car length. This obviously doesn't change anything at high speed, but when cars are almost at a stop, it can double space and time. So if we use a 2 s gap, the overall flow rate will be 1800 vehicles per hour. This is usually considered by transport authority as the bare minimum and educational campaigns have been pushing for 3-s headway in many countries. The lines that anyone can see on autoroutes are there to remind drivers of this safe headway requirement. Though we won't measure maximum headway taking into consideration weather conditions, it should be noted that bad weather will reduce flow rates by around 15 %, as driver need longer braking distance and increased perception time under heavy rain.

There are various other models to define road intensity but we will focus on flow rates measured over a period, which gives a realistic value. Using such model and

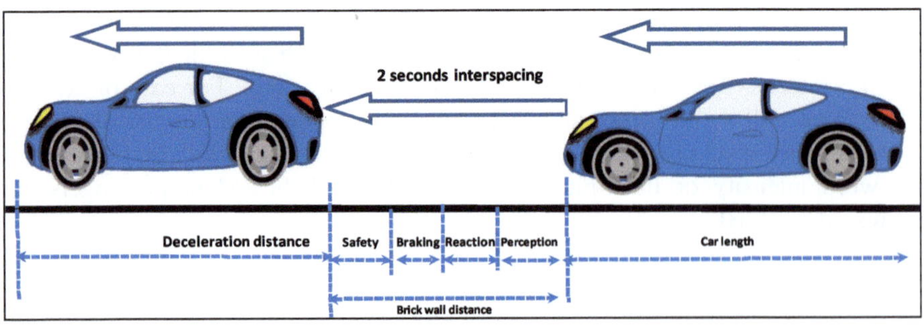

Fig. 4.2 Factors integrated in the stopping distance calculation. *Source* Author

under normal traffic conditions, the intensity is function of speed and density of vehicles on the road.

$$I = N\mathrm{v} * X * V$$

where

- **Intensity or service volume**: the number of vehicles passing a road section (Ncs) in a unit of time (*T*); I = Ncs/T (i.e., 1 vehicle/s)
- **Density**: the number of vehicles (*N*v) present on a unit of road length (*X* i.e., km) at a given moment; *D* = *N*v/*X* (i.e., 100 vehicles/km)
- **And mean speed**: *V* (i.e., km/h)

4.3.2 Level of Service (LOS)

The following table gives traffic flow figures and LOS-based of American highways (Table 4.2).

A LOS means that less than 20 cars/km can run at more than 100 km/h, which allows for 700 vehicles to be counted on the highway during 1 h. Roads will experience over the day several LOS levels. For example, a highway might be at LOS D for the morning and evening peak hours, but have traffic consistent with LOS C outside peak hours, and come to a halt once every Friday evening when people leave for the countryside.

The LOS in North America, as per Highway Capacity Manual are:

(a) **Free flow**: Motorists have complete mobility between lanes. The average spacing between vehicles is about 170 m;

(b) **Reasonably free flow**: LOS A speed is maintained and maneuverability within the traffic stream is slightly restricted. The lowest average vehicle spacing is about 100 m.

Table 4.2 LOS based on American highway with a 70 km/h design and 3.5 m lane width

LOS	Flow conditions	v/c limit	Service volume	Speed	Density
			Veh/h/lane	km/h	Veh/km
A	Free	0.35	700	>100	<20
B	Stable	0.54	1100	>95	<33
C	Stable	0.77	1550	>90	<50
D	High density	0.93	1850	>75	=67
E	Near capacity	1	2000	>50	=112
F	Breakdown	Unstable		<50	>112

Source Adaptation by author of information from Highway Capacity Manual (2000) [Published by Transport Research Board; National Research Council (HCM 2000)]

(c) **Stable flow, at or near free flow**: Ability to maneuver through lanes is noticeably restricted and lane changes require more driver awareness. Minimum vehicle spacing is about 67 m.
(d) **High density**: Speeds slightly decrease as traffic volume increases. Freedom to maneuver within the traffic stream is much more limited and driver comfort levels decrease. Vehicles are spaced at 50 m. Minor incidents are expected to create delays.
(e) **Near capacity**: Flow becomes irregular and speed varies rapidly because there are virtually no usable gaps to maneuver in the traffic stream and speeds rarely reach the posted limit. Vehicle spacing is around 30 m, but speed is still at around 80 km/h. Any incident will create serious delays.
(f) **Breakdown**: Every vehicle moves in lockstep with the vehicle in front of it, with frequent slowing required. Travel time cannot be predicted, with generally more demand than capacity.

At low lane intensity, drivers can go as fast as the upper legal speed limit allows them to. As intensity increases, so does the flow rate. However, speed starts decreasing as drivers try to maintain their following distance. At a certain point, when intensity becomes high enough, the speed falls to a point where drivers are unable to maintain their minimum following distance, and then the flow rate decreases drastically.

The following graph shows a simplified correlation between flow rate and intensity (Fig. 4.3):

It shows that the maximum highway traffic throughput is achieved around 20–25 cars/km.

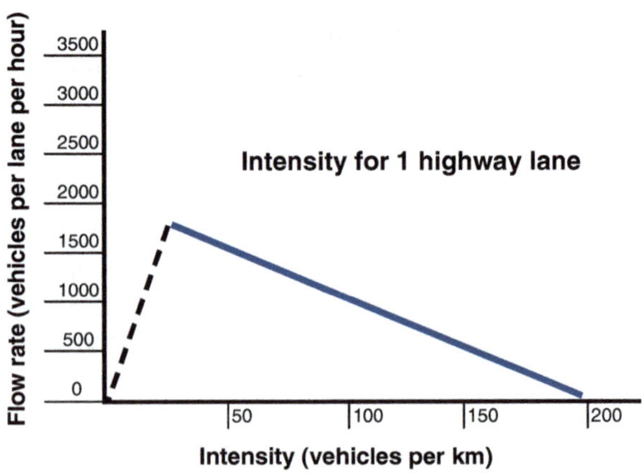

Fig. 4.3 Relation between flow rate and intensity in number of cars. *Source* Author

Table 4.3 Impact of lane width on US highway intensity; 2004 study (see footnote 3)

Lane width	2.6 m	3.0 m	3.5 m	4 m
Fw heavy vehicle	0.88	0.94	0.99	1.03
Fw cars	0.93	1	1.02	1.03

Table 4.4 Impact of heavy vehicle on US highway intensity; 2004 study (see footnote 3)

Percentage of heavy vehicle (%)	0	2	4	6	8	10	15	20	25
FW heavy vehicle	1	0.99	0.98	0.97	0.96	0.95	0.93	0.91	0.8

Table 4.5 Bus stop adjustment factor table: 2004 study (see footnote 3)

Number of lanes	Number of stopping buses per hour				
	0	10	20	30	40
1	1.00	0.96	0.92	0.88	0.83
2	1.00	0.98	0.96	0.94	0.92
3	1.00	0.99	0.97	0.96	0.94

4.3.3 Highway Intensity

The total highway capacity is influenced by many factors:

- $I = I_o \mathrm{NL}\ f(w + HV + p)$; where I is the intensity function
- I_o = optimal lane intensity (as seen above);
- NL = number of lanes
- F (w + Hv + p) = function of lane width and lateral clearance factor, heavy vehicle factor, and driver population factor;

The relation between number of lanes and intensity isn't linear. Other factors such as lane width and lateral clearance will reduce the overall intensity. Furthermore, heavier vehicles such as buses and trucks have a much higher safety braking distance to respect, reducing the highway intensity. Finally, the human factor will also impact this overall intensity, by as much as 15 % (i.e., if drivers don't know the road or are older).

The following tables show the impact of these factors for US highways taken from a 2004 road capacity study by John Van Rijn[3] (Tables 4.3 and 4.4):

In urban environments, there are also other factors than the one described here above that influence the saturation flow. The additional adjustment factors are: road crossing signalization, bus blockage, urban area type, radius of the turning movement, percentage of turning traffic, parking facilities, giving way to pedestrians and bicycles. Each element has its own adjustment factors (Table 4.5).

---

[3]Road capacities; Author John Van Rijn; In development; (2004).

For instance the following table gives an adjustment factor linked to bus stops according to 1, 2 and 3 lanes. According to studies, a single lane road with a typical 40 stop bus line would reduce the overall intensity by 17 %.

So if we subtract from the maximum theoretical road intensity of 1.800 vehicles per lane, the negative impact of all adjustment factors, we can see that the normal capacity is seriously reduced during peak hours, especially in urban environments. Among all factors identified above, light crossing is likely to have the highest impact on intensity (Fig. 4.4).

4.3.4 Road Crossing and Intersection Lights Impact on Intensity

Signaling lights are installed to manage road crossing. Under normal traffic conditions (other formula exists for optimal and extremely low traffic density), the average delay formula is:

- Dav = $I * (1 - Z)^2/(2 * (1 - Y))$; where:
- I = Intensity

Fig. 4.4 Traffic flow on highway will be influenced by lane width and lateral clearance, the mix of heavy and light vehicles, as well as the human factor. *Source* Author

- Z = green light time (in seconds)/complete cycle (i.e., 30/60 s, if green = 30 s; yellow = 10 s and red = 20 s)
- Y = saturation level which can be expressed as Intensity (i.e., number of vehicles per time period)/capacity of leg

Though waiting time per type of color lights is adapted in function of the traffic flow and road characteristics, we could make a simple assumption that on average a main artery will have half its time green and half yellow and red. On a busy road, this would cut capacity by more than half, as in addition to waiting time per se, extra time for car decelerating and accelerating, as well as for driver perception and reaction would be required. With stationary lights, this time is fixed and thus road capacity is diminished independently from the road density.

4.3.5 Intelligent Lighting Systems

To adapt road crossing time to road density, city planners have come up with basically two types of signaling lights:

- **Stationary lights**: lights for which different status have fixed time; and
- **Dynamic lights**: lights for which the status is calculated by computers based on models taking into consideration queuing status.

Dynamic lights combine basically three e-mobility technologies:

- Magnetic sensors in the road that measure the traffic flow;
- Hundreds of IP cameras; and
- Centralized computer system that makes constant adjustments to keep cars moving as smoothly as possible.

For instance, the city of Los Angeles has built over 30 years at a cost of around $400 million and automated traffic surveillance and control system. According to the city's Transportation Department, the average speed of traffic across the city is 16 % faster under the system, with delays at major intersections down 12 %. With synchronization, a 5 miles (8.3 km) drive has fallen from 20 to 17.2 min and average speed has increased from 15 mph (25 km/h) to 17.3 mph (29 km/h).

As seen in Chap. 3, intelligent traffic management could improve further traffic flow. Tests are being conducted to inform in real time the traffic light system and optimize this flow. University research centers are even looking into ways of getting rid of traffic lights altogether.

4.3.6 Maximum and Real Road Capacity

In order to calculate road capacity, we need to multiply the road vehicle intensity by the occupancy rate, taking into consideration all identified limiting factors. However, if we simplify and say that 100 % of the highway is used by cars occupied by five passengers the total maximum capacity for a single lane is around 9000 PPHPD, using 2 s headways. We all know this is far from reality. The average number of passenger is 1.5 in richer countries from Western Europe and North America. During peak hours the occupancy rate is reduced to 1.1. Thus and based on these values, the maximum road capacity during peak hours is around 1900 PPHPD and the average capacity around 2700 PPHPD.

4.4 Road Capacity with Unmanned Cars

We've seen that the maximum car intensity for any given normal lane is limited to 1800 driven cars/h. In this section, we will show that unmanned systems alone can increase such flow rate, especially if traveling in a convoy of cars under the leadership of a first car (or even a truck or bus, which would reduce air lags and improve energy efficiency). Platooning, as it is called, can substantially increase capacity on Automated Highway System. To show how, we need to consider two weaker safety criteria than the "brick wall" principle.

- If a vehicle applies maximum braking until it comes to a stop, the following vehicle should stop without colliding with it, if the interspacing between the two cars is sufficient. Such a hard braking disturbance may arise on an automated highway system in response to an obstacle or a vehicle malfunction.
- If a vehicle applies maximum braking and the following vehicle collides with it, the relative velocity at initial impact should be small. People maintain the same safety distance of around 2 s and thus tend to use the same speed. The low relative velocity safety criterion can be met if the vehicles are either far apart or close enough to one another. In the former case, vehicles have time to stop before they collide while in the latter they collide very quickly and hence the relative velocity at impact is small.

The white paper[4] "Capacity Analysis of Traffic Flow over a Single-Lane Automated Highway System," analyzes the impact of self-driving vehicles on highway capacity. It indicates that if we systematically apply to the leading car, a maximum braking capacity factor lower than the following car, there should be no

[4]White paper: Capacity Analysis of Traffic Flow over a Single-Lane Automated Highway System; Authors: James B. Michael, Datta N. Godbole, John Lygeros, and Raja Sengupta from the California PATH, Institute of Transportation Studies.

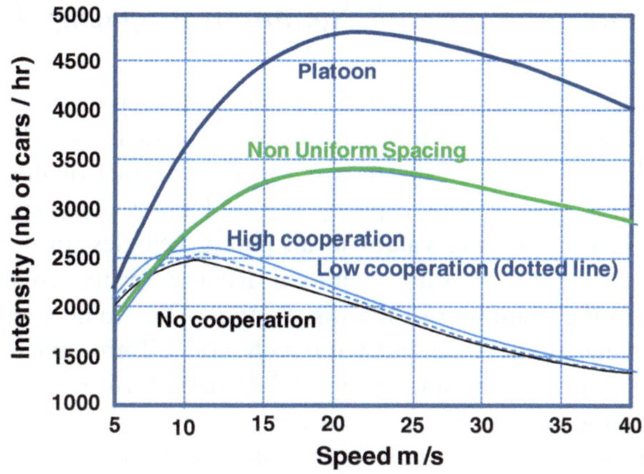

Fig. 4.5 Road intensity as a function of speed for typical vehicle mix (93 % passenger vehicles, 6 % trucks, and 1 % buses) and considering aggressive spacing; data source: white paper "Capacity Analysis of Traffic Flow over a Single-Lane Automated Highway System" (see footnote 4)

intra-collision. The study calculates road intensity taking into consideration platooning, cooperation between autonomous cars with and without uniform spacing.

The graph hereafter sums up the results for several scenarios (Fig. 4.5).

4.4.1 Highway Intensity with Uniform Spacing

The study analyzes the impact of cooperation level between vehicles, with three scenarios, for uniform spacing:

- **Autonomous**: Vehicles do not communicate with each other.
- **Low cooperation**: Vehicles communicate only during maneuvers and emergencies (e.g., hard braking).
- **High cooperation**: Vehicles continuously exchange state information such as speed and acceleration, in addition to maneuver coordination messages and emergency warnings.

The graph here above shows that the benefits of low cooperation between vehicles are minimal. On the other hand, when uniform spacing is maintained for vehicles with a high level of cooperation, car flow can be increased by around 10 %. The study shows also that road intensity for highly cooperative vehicles is extremely sensitive to the vehicle class mix. A high proportion of trucks will impact negatively the results.

4.4.2 Highway Intensity with Nonuniform Spacing Design

Instead of maintaining constant safety spacing, autonomous cars may maintain specific adapted interspacing. Through technologies such as V2V, vehicles can know the braking capacity and distance bumper to bumper of the preceding cars and compare it to its own braking capacity required at any given speed. It can thus, adapt the inter-vehicle spacing values to all paired vehicles on the highway, following the vehicle ahead at a safe separation distance for its actual braking capability, rather than the minimum of the distribution. The Authors of the study (see footnote 4) calculated the road intensity with nonuniform spacing design. The results are also shown in the above Fig. 4.3 (Nonuniform spacing). Using V2V technology on automated highway would increase intensity by around 30 %. (i.e., 29 % at 110 km/h). This is by the way consistent, with headway reduction due to elimination of perception and reaction time.

4.4.3 Platooning Policy

On a highway with platooning capabilities, vehicles could travel in closely spaced groups of up to 20 vehicles. The cars within the platoon through V2V would be able to know the braking action of the other cars and maintain an intra-platoon separation of a few meters. Platoons would need to be isolated from each other by larger distances, in order to avoid inter-platoon collisions.

The same study (see footnote 4) has shown that platooning could double road intensity. The principle is very simple: all cars are running at the same speed, thus avoiding any additional safety distance that drivers instinctively maintain for the 2 s headway. In other words all 20 cars within the platoon can maintain the same safety distance (i.e., 10 m). The inter-vehicle spacing for every paired vehicle is picked up using the lowest and highest values of the distribution to obtain safety in the hard braking sense. It should be noted that maximum intensity is achieved at different speeds (Platooning: 70–100 km/h vs. 25–35 km/h for cooperative vehicles) (Fig. 4.6).

The following considerations should be contemplated when calculating platooning capacity:

- Trucks within the platoon (not as the leader) reduce significantly intensity, as it is highly sensitive to the wide variation in braking capability.
- Intra-platoon collisions may result from hard non coordinated braking in case of mismatch between braking capabilities of the followers.
- Collisions can arise in case of unfavorable pavement conditions (e.g. oil spills or icy patches).

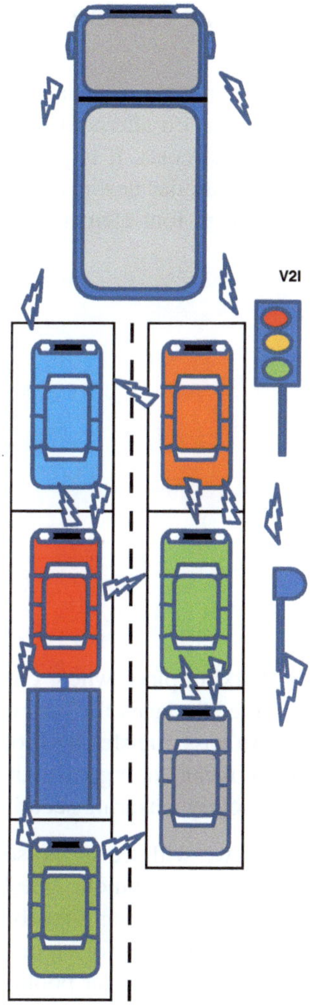

Fig. 4.6 Platooning, using a truck as head of convoy. *Source* Author

- As the platoon size increases, the severity and number of collisions per platoon might increase, if all vehicles in a platoon slam their brakes in response to hard braking by the platoon leader.

Platooning is already being tested successfully. For instance, the EU funded project called Sartre whose aims are to develop strategies and technologies to allow vehicle platooning, conducted in January 2012, its second demonstration in Spain. A lead truck was followed by three autonomous cars at speeds of up to 90 km/h.

4.5 Car Pooling

Carpooling has been around for some time and is probably the simplest of all transportation alternatives to bring additional capacity without creating almost any additional costs. It is just a matter of bringing people together in the same car to share a similar destination and expenses involved in that commute. In order to do so, though four elements are necessary:

- Parking place where commuters may park for free or not much;
- Specific lanes for cars with more than one passenger;
- An easy transfer by foot or by minivan to the car pooling place; and
- A mean to organize the car pooling, such as setting road, paying for such pooling, ensuring security checks on passengers, security records of drivers, etc.

As we will see in the section about business models, carpooling programs are designed to promote ridesharing by identifying riders with similar origins and destinations. Using a database of interested riders, employers, or regional agencies can promote this for an entire region.

The beauty of car pooling is that any increase in car occupancy rate will have a linear effect on capacity. Going from an average 1.1 passenger per car to let's say 3 will almost triplicate the road capacity (Fig. 4.7).

4.6 Bus Capacity

London 2-storey buses, greyhound buses, or even school buses are part of the collective unconsciousness. This image of people boarding or waiting for buses on the corner of the street is shared by many people in the rich world. However, in many developing countries, bus rapid transit (BRT) lines have changed this image of low capacity transport mean, used as a feeder for transporting people from the suburbs to the main metro stations. First introduced in 1974 in the Brazilian city of

Fig. 4 7 Billboard on Italian highway, promoting car pooling usage. *Source* Author

Curitica, BRTs Modern BRT fleets operate on dedicated one- or two-lane lines with special stations allowing up to five buses to load and unload passengers at the same time. For the last 40 years, the concepts have been improved and have transformed BRTs into a direct competitor of metros in medium capacity systems. Thus, understanding how BRTs work and what is their real capacity is fundamental in comparing express roads, metro lines, and buses lanes.

Bus routes, lanes, and terminals capacity is limited by the ability of stops or loading areas to pick up and discharge passengers, by the number of vehicles operated, and the distribution of boarding and alighting along the route. Bus capacity is calculated for three key locations:

- **Loading Area Capacity**: Bus loading areas (berths) are curbside spaces where a single bus can stop to unload and load passengers;
- **Bus Stop Capacity**: Bus stops are formed from one or more loading areas, depending on how many buses can use the stop simultaneously. Bus stop capacity depends on the loading areas' individual capacities; and
- **Bus Facility Capacity**: Roadways used by buses may contain multiple bus stops along their length. Bus network capacity will be constrained by capacity of the critical stop along the facility, which is typically the stop with the highest passenger volumes and the longest dwell time.

4.6.1 Loading Areas

Bus capacity of a loading area is dependent on the following factors:

- **Dwell time**: average amount of time a bus is stopped at the curb to serve passenger movements, including time required to open and close doors;

Table 4.6 Passenger service times with multiple-channel passenger movements

Available doors	Default passenger service time (s/p)		
	Boarding	Front alighting	Rear alighting
1	2.5	3.3	2.1
2	1.5	1.8	1.2
3	1.1	1.5	0.9
4	0.9	1.1	0.7
6	0.6	0.7	0.5

Source Transit capacity and quality of service manual—2nd Edition

Table 4.7 Failure rate

Failure rate (%)	Z
1.0	2.33
2.5	1.96
5.0	1.65
7.5	1.44
10.0	1.28
15.0	1.04
20.0	0.84
25.0	0.68
30.0	0.53
50.0	0.00

Source Transit capacity and quality of service manual—2nd Edition

- **Clearance time**: minimum time required for one bus to accelerate out of and clear the loading area and the next bus to pull into the loading area, including any time spent waiting for a gap in traffic;
- **Dwell time variability**: inconsistency of dwell times among buses using the loading area; and
- **Failure rate**: probability that one bus will arrive at a loading area, only to find another bus already occupying it (Tables 4.6 and 4.7).

Dwell time: Dwell time is the time required to service all the passengers through the busiest doors. This includes people boarding and getting out, as well as the time required for doors to open and close (from 2 to 5 s). The previous table from the transit Capacity and Quality of Service Manual[5] gives default values according to the bus' number of doors.

[5]Transit Capacity and Quality of Service Manual 2nd edition.

It basically considers that the persons are using pre-payment cards and that no tickets are paid onboard the bus. It also doesn't consider any delay due to transporting handicapped people or passengers with bicycles. In this case, typically wheelchair lift cycle times would take 30–60 s on low floor buses. The process to bring onboard bicycles could take approximately 20–30 s.
Clearance Time: Part of the clearing time is fixed, consisting of the time for a bus to start up and travel its own length, clearing the stop. When buses stop in traffic lane (online), this is the only component of clearance time. However, when bus stop offline, time required for a suitable gap in traffic to allow the bus to re-enter the street needs to be added. This re-entry delay depends on traffic volume in the lane and increases as traffic volumes do as well.
Estimating Clearance Time: Various studies have examined the components of clearance time, with total clearance times ranging from 9 to 20 s. The time required for a bus to start up and travel its own length to clear a stop is about 10 s. At offline stops, re-entry delay can be measured in the field or estimated using a 1 s delay for every 100 cars/ hour.
Dwell Time Variability: Based on field observations of bus dwell times in several U.S. cities, the recommended variability is around 0.6 and may vary according to the same factors that affect dwell time.

Failure rate: Bus capacity analysis incorporates the concept of a failure rate that sets how often a bus should arrive at a stop only to find all loading areas occupied. In downtown areas, design failure rates of 7.5–15 % are recommended for estimating capacity. This represents a trade-off between maintaining bus travel speeds and achieving the higher capacities required in downtown areas. The upper limit, 15 %, represents bus stop failure (queues forming behind the bus stop) for about 10 min out of the hour. It also represents the point where bus travel speeds begin to drop rapidly.

4.6.2 Bus Stops

Capacity of a bus stop is based on the following:

- **Number of loading areas**: two loading areas will be able to accommodate more buses than a single loading area, but not necessarily twice as many;
- **Loading area design**: how the loading areas are designed determines how much extra capacity each additional loading area will provide; and
- **Traffic control**: traffic signals may constrain the number of buses leaving or entering a stop during a given period of time.

4.6.3 Bus Facilities

On-street bus facility, capacity is based on the following:

- **Critical bus stop capacity**: bus stop with the lowest capacity along the facility will constrain how many buses can pass through the entire line. This bus stop is usually the stop with the longest dwell time. Heavy right-turning traffic volumes at near-side bus stops or a traffic signal that provides only a short period of green time for the bus facility can also represent the main constraints.
- **Operational procedures**: bus route design that spreads bus stopping activity over a group of stops, rather than having all buses stop at the same set of stops, can greatly increase the capacity of an on-street facility.

4.6.4 Traffic Signal Timing

Traffic signal influences BRT capacity. Traffic signal located in the neighborhood of a bus stop and its loading areas will serve to measure the number of buses that can enter or exit the stop.

$$\mathrm{Bl} = 3600 * (\mathrm{g}/\mathrm{C})/(T_\mathrm{c} + T_\mathrm{d} * (\mathrm{g}/\mathrm{C}) + Z * \mathrm{cv} * T_\mathrm{d}),$$

where

- Bl = loading area bus capacity (bus/h);
- 3600 = number of seconds in 1 h;
- g/C = green time ratio (the ratio of effective green time to total traffic signal cycle length, equals 1.0 for nonsignalized streets and bus facilities);
- T_c = clearance time;
- T_d = average (mean) dwell time;
- Z = desired failure rate; and
- cv = dwell time variability coefficient.

4.6.5 Bus Capacity for One Loading Area

Estimating bus capacity requires incorporating the various limiting factors. To simplify, we will use the following table, which gives the maximum bus capacity per hour (Table 4.8).

For instance, based on it, we can see that with a dwell time of 15 s and a clearance time of 10, 116 buses could use the loading area per hour.

Table 4.8 Bus capacity per hour, assuming a 25 % failure rate and 60 % coefficient of dwell time and no signaling (= 1)

Dwell time (s)	10 s	15 s
15	116	100
30	69	63
45	49	46
60	38	36
75	31	30
90	26	25
105	23	22
120	20	20

Source Transit capacity and quality of service manual—2nd Edition

Table 4.9 Efficiency of multiple linear loading areas at bus stops

Loading area	Online loading areas				Offline areas	
	Random arrivals		Platoon arrivals		All arrivals	Cumulative # of effective loading areas
	Efficiency (%)	Cumulative # of effective loading areas	Efficiency (%)	Cumulative # of effective loading areas	Efficiency (%)	
1	100	1.00	100	1.00	100	1.00
2	75	1.75	85	1.85	85	1.85
3	70	2.45	80	2.65	80	2.65
4	20	2.65	25	2.90	65	3.25
5	10	2.75	10	3.00	50	3.75

Source Transit capacity and quality of service manual—2nd Edition

4.6.6 Bus Capacity for Several Loading Areas

When estimating a bus network capacity with several loading areas, we need to take into consideration the inefficiency resulting from bus loading and unloading. Table 4.9 defines the impact of load area numbers in online and offline areas, with random and platoon arrivals, based on American studies.

It shows that with the increasing number of loading area, the global efficiency falls drastically. Offline areas have a better efficiency level.

For those interested in formulas, the bus stop capacity becomes:

Bl = 3600 * Nd * (g/C)/(T_c + T_d * (g/C) + Z * cv * T_d), with Nd the Number of effective loading taken from the previous table.

Table 4.10 Estimated maximum capacity of online Linear bus stops (bus/h) assuming 10-s clearance time, 25 % failure rate, 60 % dwell time variability coefficient and random bus arrivals, for line with signaling (0.50) and without (1.00)

Dwell time	1 loading area		2 loading areas		3 loading areas		4 loading areas		5 loading areas	
Signaling	0.50	1.00	0.50	1.00	0.50	1.00	0.50	1.00	0.50	1.00
30 s	48	69	84	120	118	169	128	182	133	189
60 s	27	38	48	66	68	93	74	101	76	104
90 s	19	26	34	46	48	64	52	69	54	72
120 s	15	20	26	35	37	49	40	53	41	55

Source Transit capacity and quality of service manual—2nd Edition

The following table gives an estimation of the maximum capacity for online linear bus stops (Table 4.10).

For instance, with 5 loading areas, a dwell time of 30 s and without any signaling, the maximum capacity in terms of number of buses, which can load and unload passengers, would be 189.

4.6.7 Real Bus Capacity at Average Speed

The Transmilenio™ BRT running in the Colombian Capital Bogota offers the largest BRT capacity in the world. Many documents have mentioned its capacity, with a maximum throughput varying from 35,000 PPHPD to as much as 41,000 PPHPD. This BRT system includes 38 km of dual lane, with 59 stations spaced at 500–750 m. These stations have different loading areas, with station length varying from 40 to 193 m (Fig. 4.8).

BRTs can use multiple stopping bays, which increases capacity, especially when two segregated lines (i.e., Transmilenio) are built. The following capacity formula shows parameters used to calculate the multistopping bay BRTs: (Formula extracted from the BRT—Planning guide 2007, source: Steer, Davies, Leave).

$$\mathrm{Co} = \mathrm{Nsp} * X * 3600/(T_{\mathrm{d}} * (1-\mathrm{Dir})/C_{\mathrm{b}} + (\mathrm{Ren} * \mathrm{T1}),$$

where:

- Co = Corridor capacity in PPHPD
- Nsp = number of stopping bays
- X = Saturation level (affects bus queue length)
- 3600 = Number of seconds in one hour
- T_{d} = Dwell time
- Dir = % of bus that are perform express service
- C_{b} = Capacity of the bus
- Ren = Renovation rate
- T = Average boarding and alighting time per passenger

Fig. 4.8 Transmilenio BRT. *Source* (Picture of the Transmilenio BRT Source; http://www.flickr.com/photos/jlascar/4585122508/)

If we apply the formula to the Transmilenio line, we would have:

- 3 stopping bays (Nsp)
- A low level of saturation level: 40 %
- 50 % bus stop limit
- Dwell time: 13 s
- Vehicle capacity: 150 passengers
- Renovation rate: 25 %
- Average boarding and alighting time per passenger: 0.3 s

This translates in a capacity estimated at a little bit more than 37,000 PPHPD, which the operator confirms can be met. If the Transmilenio operation were not to operate express services (i.e.: stopping at each bay), total capacity would be reduced to only 26,721 PPHPD. Mixture of stopping and express lanes is only possible because of the very wide road width.

The following table describes capacity of various BRTs (Table 4.11).

4.7 Unmanned Bus Operation

Can new wireless technology originating from the V2V concepts be applied to buses? Of course, what is true for cars will be applicable to any road vehicle! Obviously specific bus braking characteristics would need to be integrated, but V2V could be applied easily to buses. The question is can it improve conventional bus line and BRT capacity? The theoretical answer is yes. V2V combined with V2I can eliminate perception and reaction time to almost nothing. Managing through V2I road crossing priority for buses could eliminate some of the wasted waiting time for conventional lines. Closer headways would also increase road capacity and reduce journey time.

Table 4.11 Capacity in PPHPD for various BRT systems table adapted from Wikipedia by author

Location	System	PPHPD	Passengers per day	Length (km)
Bogotá, Colombia	Transmilenio	35,000–41,000	2,154,961	106
Guangzhou, China	Guangzhou bus rapid transit	26,900	1,000,000	22
Curitiba, Brazil	Rede Integrada de Transporte	13,900–24,100	2,225,000	81
Belo Horizonte, São Paulo		15,800–20,300		24
Istanbul, Iran	Metrobus (Istanbul)	7300–19,500	800,000	52
Lahore, Nigeria	Metrobus (Lahore)	10	180,000	28
Tehran, Iran	Tehran bus rapid transit		1,800,000	150
Jakarta, Indonesia	TransJakarta		350,000	208
New Jersey, USA	Lincoln Tunnel XBL	15,500	62,000 (4 h peak only)	–
Brisbane, Australia	South-East Busway	15,000		23

However, these potential headway gains for conventional lines could have problems translating into increased capacity for BRTs, as one the main limiting capacity factor is the time taken to load and unload passengers. For instance, could the Transmilenio 2 min departure per bus, which translates into one of the three buses using around 40 s for dwell and clearance times, be improved? It seems difficult but not impossible. As we've seen, the efficiency of the third berth in offline loading area is already 80 %. Perception and reaction time could be knocked out, but wouldn't have much impact. On the other hand, efficiency for the fourth and fifth offline loading berth is, respectively, 65 and 50 % and could be slightly improved by such technology. Probably, the biggest gains though would originate from improving the efficiency of the fourth and fifth berth in platooning (i.e., 25 and 10 %) and random arrivals (i.e., 20 and 10 %). Though no existing study can verify the extra capacity, by using a simple rule of thumb, we could estimate potential BRT capacity gains from V2V and V2I to around 20 %. Indeed platooning could be systematically used instead of random arrival, increasing the overall clearance process. Furthermore, unmanned operations would reduce significantly the number of accidents and errors improving thus the failure rate figures.

However, the biggest winners of the e-mobility revolution would be certainly the more conventional lines. On top of some of the advantages we just described, synchronization of lights at crossing could reduce journey time significantly. We could for instance imagine, waiting time at lights being adjusted, based on higher priority given to buses.

4.8 Mass Transit Capacity

Trains come in all sizes and shapes. Their dimension and resulting capacity depend largely on the application and expected journey distance. For metro and commuter trains with short journeys, capacity will be at its maximum, as most passengers will be expected to stand. For regional or intercity journeys, capacity will be reduced by the need for more comfortable seating and services (i.e., baggage rack, restaurants, toilets, etc.). Another key differentiating feature is the required dwell time. Metros must load and unload people in less than 20 s, while high-speed intercity train may take up to 15 min for that task. Besides the operational requirements, the number of doors per vehicle, their location, size, and design will also impact the dwell time.

Our railway capacity analysis will be restricted to mass transit systems, which can be found in megacities.

4.8.1 Increasing Capacity of Existing Infrastructure

To avoid investing in new networks, Transport Authorities can improve the vehicle and increase the line's throughput. For instance, larger and longer vehicles, running faster and closer to each other can make better use of existing infrastructure. Thus, designers can play with four variables: vehicle capacity, fleet size running on the system, average speed, and headway.

4.8.2 Vehicle Capacity

For any given existing railway infrastructure, vehicle capacity can be optimized by playing with one of the four following elements: physical characteristics, comfort level, safety standard, and occupancy rate.

Physical characteristics Metro, commuter trains, monorails, and tramways vary in size. The external dimensions, which are called envelop, are usually given by the physical characteristics of the network. Tunnel and platform clearance and heights are some of the main variables that impact these external dimensions. As train are moving objects, metro engineers consider the dynamic envelop, which takes into consideration clearance with lateral and vertical movements (Fig. 4.9).

Unfortunately most networks are unique, which prevents metro standardization. The art of the railway manufacturer is thus to take in its product portfolio an existing proven technology, which envelop is the most similar to the network requirements. Once this is done, engineers work on optimizing the seating, door, and equipment arrangements to allow for the highest available useful internal surface.

Fig. 4.9 Network's characteristic impact on trains' physical characteristics. *Source* Author

Comfort level The comfort level is usually imposed by Transport Authorities. For instance, in Great Britain a target for restricting overcrowding was set as "Passengers in excess of capacity" (PIXC). This was defined by the government as greater than 10 % of the vehicle's seating capacity or of an allowance of 0.55 m^2 per passenger depending on the type of vehicle. As a consequence, number of seats and their specific occupying space are normally fixed. Space can typically vary from 0.35 m^2 for metro operation in megacities to 0.55 m^2 for comfortable mass transit seating arrangements (even more for first class intercity train). The surface dedicated to standees is a consequence of the remaining inner available surface of the cars, including door and gangway space (rigid flexible section between two separate vehicles).

Industrial designers are hired by metro manufacturers to optimize seating configuration, loading and unloading flow (e.g., optimizing door size and types) as well as repartition of the standees within the metro. Their objective is not only to make the transit cars appealing and functional, but also to allow for the maximum number of standees. A direct link between seated passengers and standees exist called load factor.

Safety standard As we've seen in the section on signaling, railway operation is very standardized, but standards are not always safety related. For instance, areas reserved for handicapped people, or seats designed for obese are now required for social reasons in most mass transit operations. This has obviously an impact on capacity.

One of the most important safety-related requirements is linked to the train's weight, as it affects structural and performance requirements. Once the standing space is calculated, one must take into consideration the weight of the number of standees per m^2. Usually, 70 kg per passenger is used, though this weight value can change in some cities.

The railway industry takes into consideration mainly 5 weight requirements: AW0 (empty trains), AW1 (weight with seated passengers), AW2 (weight with average peak hour load), AW3 (weight with crush load weight), and some also use AW4 weight measures. In fact, these measures are usually simplified taking into consideration, respectively, 0, 4, 6, 8 and 10 passengers/m^2.

Let's use a simple example to illustrate the impact of passenger weight: a metro car with inner dimensions of 10 m by 2.8 m, with 30 seats required with an area of 0.35 m^2 per seat.

The total available space would be 28 m of which 9.8 m^2 would be dedicated to seating areas. To simplify, we would have 18 m left for standees.

In terms of the seating weight, we would have 30 × 70 kg = 2.100 kg to which we would need to add the standees weight. The number of passengers and additional weight would be:

- 4 passengers/m^2 = 18 × 4 = 72 passengers × 70 kg = 5050 kg
- 6 passengers/m^2 = 18 × 6 = 108 passengers × 70 kg = 7560 kg
- 8 passengers/m^2 = 18 × 8 = 144 passengers × 70 kg = 10,080 kg
- 10 passengers/m^2 = 18 × 10 = 180 passengers × 70 kg = 12,600 kg

For most cities this is just a manufacturing standard, as the vehicle capacity requirement is always calculated at 4 passengers/m^2. However, for megacities of this world, such as Sao Paulo or New Delhi, accepting a comfort level of 6 or 8 passengers/m^2 or even more is a quick and dirty (not to say smelly) way to add extra capacity. Some cities like London, for instance, calculate different load factors such as 8 passengers/m^2 between doorways and 6 passengers/m^2 elsewhere.

The fact of using 4 or 8 passengers/m^2 will obviously impact the structural weight, as more metal will be required with more density. This will have an impact on motorization, which will require more powerful motors, but also on air conditioning, braking system, etc. In other words, though the trains could have the same inner space devoted to standees, the characteristics of a metro car calculated at 4 or 8 m passenger/m^2 would make these two very different beasts.

Occupancy rate This rate gives the real capacity number. In other words, a metro car of the megacity can be designed for 6 passengers/m^2 but the reality is that in some of the megacities of Brazil or India, there are 8 and sometimes even 10 passengers/m^2!

4.8.3 Mass Transit Network Capacity

Capacity is expressed in PPHPD and can be represented as the results of multiplying the number of trains passing by a particular station in one hour by the number of vehicles per train and the number of passengers that could be carried by each vehicle. Similarly to bus systems, we've just seen that there are four main elements defining the mass transit network capacity: train capacity, fleet size, average speed, and headway.

Train capacity Unlike most buses though, metros are composed of several cars linked together by gangways or couplings. Capacity is thus function of the number of cars tied up together. In most modern metro applications, tube shape metros can be composed of up to 12 cars tied up by gangways. The maximum number of cars will be function of the metro station length. The train capacity formula is:

$$\begin{aligned} T_{\mathrm{c}} = {} & \mathrm{train/h} * \mathrm{cars/train} \\ & * \left(\mathrm{seats/car} * \mathrm{passenger/seat} + \mathrm{standing\ area\ m^2/car} * \mathrm{standees/m^2}\right); \end{aligned}$$

Fleet size It is the number of trains available at peak hour to cover the entire line. The maximum number is function of the network characteristics and is defined by speed and minimum headway achievable safely.

Average speed It is mainly influenced by acceleration, deceleration and speed of the vehicles, as well as the distance between stations; and

Headway It is mostly influenced by the authorized time between two vehicles following each other, according to the LMA (see section on safety for more information) and the time required for vehicles to reenter operations after reaching the end of the line.

4.8.4 Railway System Capacity

There is no standard capacity number for railway systems. It really depends on the application from heavy-to-light metro system, commuter, tramway regional, or high speed trains. But even within each category, technology and safety standards will highly influence capacity.

These are a few examples of how other factors beyond the four just identified can influence highly capacity. For instance, commuter train pulled by a locomotive won't carry any passenger while another commuter train based on an Electrical Multiple Unit configuration, will carry passengers in all its railcars. Tramways can be coupled to increase capacity and metro and monorail can see their train composition increased according to maximum station length. Line can be segregated, reducing significantly all non programmable events such as interaction with car, bicycle, bus, pedestrian and road crossing lights and thus reducing significantly travel time.

4.9 Transport Mode Capacity Comparison

In the last sections, we've seen different transportation systems capacity. How can we compare them? The answer is: not easily!

4.9.1 Highway and Road Capacity

The issue is that we cannot really use the maximum capacity for cars expressed in PPHPD, as we do for buses and trains. The reason is that we know that most cars run without any passengers other than the driver. This is why we prefer to use the average occupancy rate rather than the maximum capacity of 5 passengers per car, which in the USA and Europe, is around 1.5 passengers. This means that the theoretical capacity of 5800 PPHPD for one highway lane would be reduced to around 2800 PPHPD at the speed of around 25 km/h (Fig. 4.10).

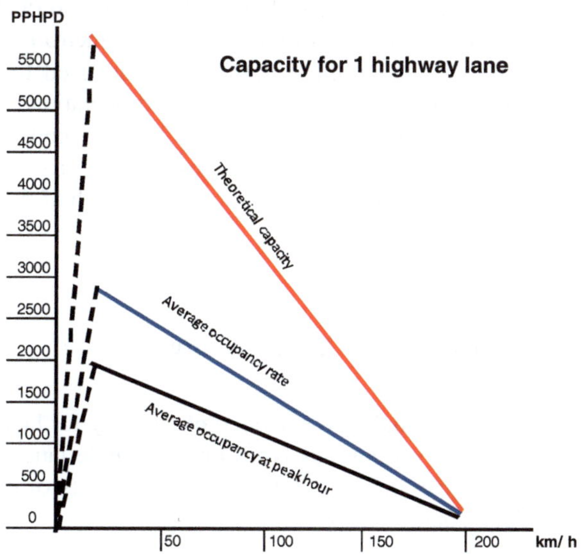

Fig. 4.10 Capacity in PPHPD according to theoretical, average, and peak occupancy. *Source* Author

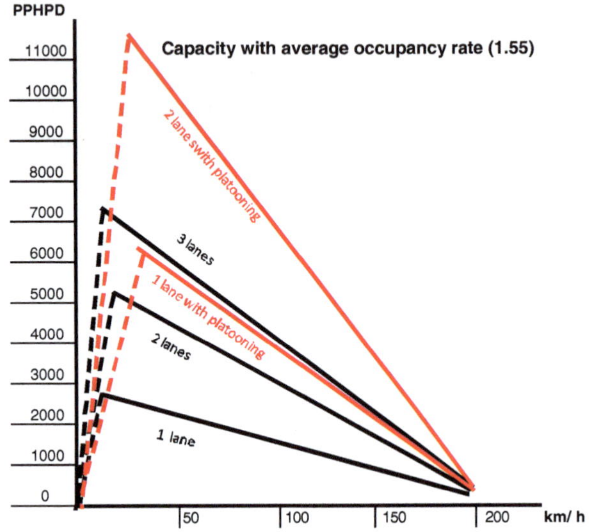

Fig. 4.11 Capacity in PPHPD for 1 and 2 lane with platooning. *Source* Author

This, however, shows the potential of car pooling in addressing capacity issues within the city. Doubling the number of passengers per car, automatically translates into increasing twofold the capacity, without any impact on infrastructure and almost no impact on pollution.

Proposing platooning services of unmanned cars is a formidable opportunity to solve transit issues in gridlock urban areas. Combining platooning with car pooling services (or unmanned taxis with several onboard passengers) offer even more appealing opportunities to solve transport capacity problems. With a two-lane platooning service of unmanned taxis operating at full capacity (i.e., 4–5 travelers per car), the road capacity could reach 20,000–30,000 PPHPD depending on operational speed, or the equivalent of a medium-sized metro line and double the capacity of a tramway! Obviously these numbers would need to be tested, but still show the tremendous opportunities that unmanned operations can unleash (Fig. 4.11).

Now, calculating an average capacity for a conventional road is almost impossible, as there are too many variables, such as crossings, lights, pedestrian, etc. For comparison purpose we could shoot and take half the highway lane capacity: 1600 PPHPD.

4.9.2 Bus and BRT Capacity

The following chart gives a value of capacity for bus and BRT systems taking into consideration various bus types on one single lane. It also shows that operational efficiencies can improve drastically the capacity of the bus network, even for only one BRT lane (Table 4.12).

Table 4.12 Impact of vehicle platform interface and vehicle size on capacity

Vehicle and operation type	Max bus capacity (No.)	Paying on or offboard	Average dwell time (s)	Av. Boarding + alighting time (s)	Corridor capacity (PPHPD)	Bus/hour (No.)
Minibus	15	Onboard	10	3	1140	76
Midi-bus	35	Onboard	11	3	1580	45
Standard bus	70	Onboard	12	3	1870	27
Articulated vehicle	160	Onboard	13	1.5	3780	24
Bi-articulated vehicle	240	Onboard	14	1.5	4020	17
Articulated with level platform	160	Onboard	13	1	5120	32
Bi-articulated with level platform	240	Onboard	14	1	5570	23
Articulated with level platform	160	Offboard	13	0.3	9780	61
Bi-articulated with level platform	240	Offboard	14	0.3	12,170	51

Source Steer, Davies, Gleave: BRT- Planning guide 2007 (Table 8.6)

Table 4.13 Capacity of various transportation means at 6 passengers/m^2

Technology	Description	Car no	Capacity	120 s	90 s
	6 passengers/m^2	No	Unit	PPHPD	PPHPD
Heavy metro	Small body (Rc + M + M) × 2	6.00	1016	30,500	40,600.00
	Medium body (Rc + M + M) × 2	6	1508	42,500	60,300
	Large body (Rc + M + M) × 2	6	1736	52,100	69,400
Monorail	7 car train	7	1002	30,080	40,000
Tramway	30 m	1	270	8100	10,800
	30 m × 2 (coupled)	2	540	16,200	21,700

Source EAMESP and Author

4.9.3 *Metro and Train Capacity*

The following chart based on an EAMESP study (Engineering association of the Sao Paulo metro) and our own calculation gives an idea of the various technologies, using comparable premises. As indicated in the chart, the capacity is calculated with 6 passengers/m^2, which is an operating standard in this city. It also takes into consideration the interval between two trains of 90 and 120 s (Table 4.13).

Table 4.14 Bus capacity at 6 passengers/m^2

Technology	Description	Car no	Capacity	120 s	90 s
	6 passengers/m^2	No	Unit	PPHPD	PPHPD
Standard bus type 1, 1 lane, 1 stopping bay	2 axles	1	85	2550	3400
	Articulated	1	121	3650	4840
	Bi-articulated	2	173	5190	6920

Source EAMESP and Author

The same source also allows us to compare train and bus running on conventional routes (Table 4.14).

It is interesting to note that capacity varies according to frequency and density. For instance, had the frequency been 180 s and the density 4 passengers/m^2, the tramway's capacity would have been 5700 (42 m) and 4200 PPHPD (32 m), and the bus' 2200 (biarticulated) and 1600 PPHPD (normal bus).

4.9.4 Comparing Apples with Apples

Comparing capacity between different transportation means is pretty complex. The following chart coming from a specialist in BRT gives a comparison of light trains, metros and BRTs, which portrays BRTS in a favorable light (Table 4.15).

The following table compiled by Transport for London is interesting, as TfL operates different transportation means and cannot be seen as biased. To give the full picture, we've added to this chart information on conventional road and highway capacity (in italics) (Table 4.16).

Table 4.15 Bus transit systems: the case of Transmilenio

Characteristic	Tram—LRT	Metro	Bus transit
Capacity (Pax/veh)	110–250	140–280	80–160
Vehicles/unit	1–4	1–10	1
Maximum speed (km/h)	60–80	70–100	60–70
Commercial speed (km/h)	15–35	25–55	15–28
Maximum frequency at stops (units/h)	40	20–40	70–210
Capacity at stops (pax/h/direction)	6000–20,000	10,000–72,000	11,000–40,000
Capital costs (€ M/km)	15–50	30–200	1–10

Source Steer, Davies, Gleave

Table 4.16 Capacity and cost per kilometer of different transportation means

Modal characteristics	Unit type	Car road lane	Car highway lane	Bus	Bus max priority	BRT	Tramway	Light rail	Heavy rail
Maximum capacity	PPHPD	*1600*	*2800*	2500	4000	6000	12,000	18,000	30000 and >
Capital cost per route km	M£/km	*<1*	*<1*	<1	1–2	1–20	15–20	10–45	45–250
	M$/km	*<1.5*	*<1.5*	<1.5	1.5–3	1.5–30	20–30	15–70	70–400
Operating cost per passenger	Pence/km	*9–50*	*9–50*	3.8–8.8	2.5–5.8	2.5–5	1–2.1	1–1.4	1.5–1.8
	Cents/km	*15–70*	*15–75*	5.9–13.8	3.9–9.1	3.9–7.8	1.6–3.3	1.6 to 2.2	2.3–2.8
Average speed	km/h	*10–20*	*20–40*	10–14	14–18	15–22	15–22	18–40	18–40

Source Transport for London and Author

As we can see, TfL identifies the main public transport modes' characteristics, focusing on capacity, capital cost, operating cost, and average speed. Not surprisingly, the main difference with the previous chart is concerning the BRT, where capacity is much reduced. The fact is that high-capacity BRTs found in South America aren't really popular in the rich world and probably misunderstood. We will now see why.

4.9.5 Considering Lost Capacity

Not all construction of new transportation lane is neutral from a lost in capacity perspective. For instance, appropriation of an existing road to build a one- or two-lane bus system or even a tramway will reduce significantly car traffic flow in the neighborhood. BRT and tramway capacity can be huge but their impact on the city traffic jams can also be. The reason is that there are two consequential damages of building such systems on segregated lanes.

Bottlenecks By cutting the city into two, such segregated system creates bottlenecks wherever traffic passes over or under the lanes.

Land appropriation These segregated systems either replace existing roads or could be replaced by new roads that would help the traffic flow. The size of the land intake is function of the expected maximum speed, level of traffic, and nature of the separator (can be run through or not). For instance in an urban environment, for a maximum speed of 50 km, the bus must have:

- 6.80 m in double lane for traffic smaller than 60 buses/hour (2 directions)
- 7.10 m in double lane for traffic higher than 60 buses/hour (2 directions)
- 3.50 m in single lane.

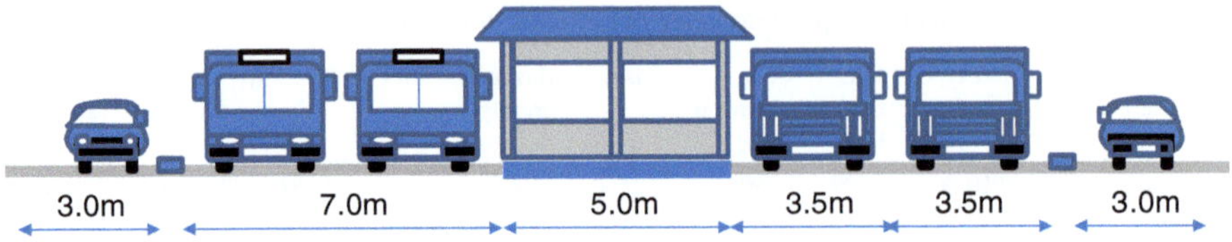

Fig. 4.12 Transmilenio land appropriation. *Source* Author

The following design gives a view of the total space occupancy of the Transmilenio. The total system occupies 19 m. If we consider a 3 m width for a normal lane, the Transmilenio could add 6 extra lanes (Fig. 4.12).

Assuming the minimum occupancy rate of 1.1, the lost opportunity in terms of capacity generated by such system would be at minimum 6000 PPHPD at peak hours and probably more in the range of 10,000 PPHPD.

Furthermore, in the future such segregated lanes could be dedicated to platooning services with car pooling or unmanned taxi services, which we've seen could compete advantageously with BRT or tramway systems (capacity of a 2 lane could go up to 30,000 PPHPD).

On the contrary underground or elevated transportation means such as metros or monorails have very little land intake and almost no negative impact on existing road capacity.

4.10 System Price Comparison

Comparing prices between the different transportation means is even more complicated than analyzing capacity differences. Actually, even comparing the same transportation means across different countries is extremely difficult. To help better understand why, we must obviously look at the capital expenditure (CAPEX) linked to building the system and acquiring all equipment and software, as well as the Operational Expense (OPEX) resulting from operating and maintaining the systems (as TfL has done in their chart), but not only. We also must analyze the impact of financing the projects, which is part of the Total Cost of Ownership. Furthermore, OPEX, CAPEX, and financing costs must be brought back into a value called Net Present Value that takes into account the cost of money over time. However, financing costs for AAA+ rated Governments have nothing in common with nonrated Governments' borrowing cost. Thus, comparing financing costs for a new London or Karachi metro line is really impossible. This already explains in parts why it is so difficult to compare costs between different infrastructure projects, as financial costs which can be substantial over a 30-year period are not technology but mainly investor rating-related.

4.10.1 CAPEX Comparison

Transport CAPEX can be regrouped under two different meaningful categories: system and civil construction work. Under system, we include the vehicles and the signaling system, as well as all the auxiliary equipment necessary to make the network operate (e.g., elevators, telecom, ticketing system, etc.). The construction work will include concrete work for laying tracks or building roads, tunneling as well as all auxiliary building infrastructures, such as stations and control centers. As we've seen in TfL's chart, there is such a big variation in costs (e.g., US$45 M to US$450 M), that these numbers start to be meaningless for comparison purposes. Let's understand why.

Vehicle comparison For cars and buses, prices are today globalized but can be different from one country to the other, mainly due to import or VAT taxes. Furthermore and as we all know, prices between a luxury and a small compact car have nothing in common.

Trains are mainly tailored made and even though the market is globalized there can still be huge differences in prices linked to factors such as quantity, delivery schedule, risks, etc. For instance, a metro car can cost as little as $1 M if produced in large quantity but the same trainset could reach the price of $1.6 M to $1.8 M if produced in small volume. Taxes and import duties can also affect greatly this price. Finally, as train manufacturing is mainly tailor made, cost and productivity of man power will also impact greatly the train's costs. Moreover, frequency and headway will influence the fleet size. Costs of a 20 km line with a 4-car EMU (Electrical Multiple Unit) coming in every 3 min have nothing in common with high capacity systems where 10-car metros must meet 60 s to 70 s headway.

Signaling and other software driven equipment comparison These equipment that are mainly present in the railway industry are also globalized. Price will be mainly affected by the level of safety and the complexity of the integration work required. For instance, a SIL 1 can cost one-fourth of a SIL 4 signaling system.

Construction work General prices for construction work are mainly local. For instance, concrete cost can be affected by the presence of concrete plants nearby or even by the need for special concrete use due to environmental factors. Local man power cost will also greatly influence the overall construction costs, as thousands of blue collars can work on these huge construction sites. Construction work is also influenced by environmental conditions in which the work will be performed. Soil characteristics especially will have a huge impact on tunneling costs even more so wherever lines must run under rivers. Over or under pass, as well as bridges are infrastructure works, which can increase immensely the CAPEX.

All of these factors that influence CAPEX can in fact vary by at least 100 %. This explains, why making any potential investment comparison of transport infrastructure costs across countries is extremely difficult. It is not to say, that CAPEX comparison of different transport means cannot be done. In fact it must be done, but at the local or city level.

4.10.2 OPEX Comparison

Operational expenses can be divided into three main cost drivers:

- **Execution**: the main cost centers are energy consumption, drivers' salary, ticketing, security team salary, administrative work, etc.;
- **Maintenance**: these costs include cleaning, daily maintenance, maintenance regimes for corrective and preventive maintenance, asset replacement; and
- **Soft costs**: they include items such as marketing and gratuities.

Execution costs They depend really on transportation technologies and the size of the operation. Interestingly, bigger operation doesn't always translate into more expensive ones. For instance, energy costs aren't proportional to fleet size. Indeed, kinetic energy resulting from braking can be significantly saved by other vehicles, as long as there is always one nearby train that can recuperate this energy during acceleration. Driverless operations can also get rid altogether of driver and station attendants salary, independently for the network size. Furthermore, ticketing, security, and administrative cost usually benefit from volume as the fixed costs of managing huge or small transport systems are almost the same. When looking at all these elements, we can conclude that operating cost per passenger kilometer transported is usually much lower for big metro networks than smaller ones.

Maintenance costs Thirty years ago maintenance costs were mostly proportional to the fleet's age. Classically, maintenance costs would grow over time. Infrastructure characteristics (ramp, curb, quality of running surface, etc.) as well as operational conditions (driving style, speed, number of stops, mileage performed, etc.) would obviously affect the electromechanical parts, making the need for replacement more or less urgent. However, various maintenance strategies and methodologies were introduced trying to decouple maintenance costs with fleet age. For instance instead, of considering vehicles in its entirety, maintainers optimized the maintenance at the equipment level. As we will see in the chapter on connected cities, new e-mobility technologies are enabling further decoupling by enabling preventive maintenance based on failure predictability rather than operational cycles.

Soft costs Such costs (i.e., marketing, public relations, gratuities) are usually fixed. The operation's size doesn't have much influence.

The Transport for London chart shows OPEX costs of the various transportation technologies expressed in $ per passenger per kilometer traveled. We used extracts from the UK Automobile Association's Motoring cost for 2014, at peak hour for the cheapest and the most exclusive cars). As we used peak hour values, it obviously shows much higher OPEX values than public transport means. Were we to use occupancy rates of 3 and 5 passengers rather than 1.1, it would show that OPEX for small cars would be at par with buses and metros respectively.

Table 4.17 Social benefit costs with an exchange rate of 2.33R$ US$ per US$, not considering inflation; extracted and adapted from the Sao Paulo metro 2009 annual (see footnote 6) report

Social benefits	2009 values		
	Unit	In million	M US$
Reduction in pollutant emission	t/y	783	66
Reduction in fossil fuel consumption	L/Y	407	393
Reduction in bus OPEX	km/y	246	388
Reduction in car OPEX	km/y	1207	245
Reduction in road lane OPEX			15
Reduction in lost time	H/y	556	1038
Reduction in the number of accidents	Number	12	50
		Total	2196

Source Author

4.10.3 Social and Environmental Costs

In order to assess the value for money of any given transport system, one cannot just consider the previous costs, as very few public transport projects can offer a viable return on investment. Social and environmental costs need to be integrated to get the full view on the real costs of any given transportation system. Many metro operators, which need to take into consideration such factors, show in their annual balance sheet figures related to social benefits for the population. The following figures are extracted from the Sao Paulo metro 2009 annual report[6] (Table 4.17).

An interesting feature of such chart is that it shows what would happen to car and bus operations, if the metro would stop carrying so many people. In other words, it shows the opportunity cost of not having such metro network.

The first social benefits, we would like to discuss about are linked to opportunity cost from an environmental perspective. As a matter of fact, we will first describe a cost that is never really considered in most feasibility study.

4.10.4 Appropriation Costs

Road advocates never consider the land occupancy as a major decision-making criterion, whenever comparing transportation means alternatives. However, we've just seen that a BRT system such as the Transmilenio could occupy a 19 m area over more than 50 kms. How much is that worth? Surely it is worth something, even if it is just to transform this lane into a green area. Property value around such a park would most likely double in the midterm and so would property taxes. However, an easy way of estimating such cost exists. If we take the average cost per m^2 within the

[6]Sao Paulo annual report 2009.

neighborhood, we can easily calculate the lost opportunity cost. In cities like London, as there is no more building area available, we could use the 2014 average price per m^2 slightly higher than 10,000 pounds or $15,000. With a 20 km long and 19 m wide area, we would reach an amount of $5.7B (3.8B pounds)! London is not exceptional. For instance, the 2013 price per square foot reached in West Chelsea (New York City) $850 per square foot (US$9150 per m^2)[7]. At such retail price, the same area could be worth around US$3.5B in the Big Apple. With land worth so much, no wonder high capacity BRTs aren't really popular in the rich world!

4.10.5 *Expropriation Costs*

Expropriation shouldn't be mixed up with appropriation. It is basically linked to the confiscation by the Authorities of land and whatever is standing on it, and to the reimbursement of such property usually at fair market price. Obviously the definition of fair market price is difficult to establish and results in numerous legal battles. Expropriation cost varies according to the area in which the line or the road will go through. Cost of land is an indicator of such costs, but can increase tremendously if there are buildings or residences that need to be demolished.

Most transportation projects are affected by such costs. Even metros running in tunnels need expropriation for placement of the TBMs (Tunnel Boring Machines), the construction site and/or the stations. However, stations in many big cities are usually included in buildings or shopping malls, turning a cost into a revenue stream for the metro operator. Monorails can also be integrated within buildings and generate the same type of revenues. They also need expropriation for the depot. However, as their dynamic envelop (maximum limit by which the vehicle moves during normal operation) is very small, it can actually be built between 2 lanes of an avenue, not requiring any expropriation at all (Fig. 4.13).

Bus systems or tramways that run in mixed traffic also don't generate much expropriation cost except for the depot/garage. On the other hand, BRTs such as the Transmilenio require a huge number of land plots to build access ramps to the stations positioned in the middle of the lanes.

4.10.6 *Congestion Costs*

Americans waste 4.2 billion hours in traffic every year or nearly one full week of work for every traveler. When such wasted time is translated into money, traffic congestion represents a $87.2 billion annual drain on the American economy or

[7]Developers play with thin margins as Manhattan land prices skyrocket; Article in NYC's (new) land rush; December 01, 2013 Author: Melissa Dehncke-McGill.

Fig. 4.13 Picture of monorail tracks in the city of Sao Paulo laid between the two lanes of a large avenue, resulting on minimum expropriation and land appropriation. *Source* Author

more than $750 for each US passenger. We also need to add to congestion costs the equivalent in wasted combustible, which represents an extra 15.9 billion liters of wasted fuel. This is in line with the EU's own estimate of congestion costs of nearly $130B (100B Euro) or 1 % of EU's GDP per year. As we've seen before, the Sao Paulo metro quantifies the direct benefits it generates in regard to congestion costs, such as reduction in fuel consumption, in car, bus and road OPEX, as well as the saving in loss time due to traffic jams. The lost time savings are worth around half the social benefits that this city metro system is bringing to the population. This isn't uncommon.

A 2005 US study[8] showed that taking size into account, cities with large, well established rail transit systems perform better in various ways than cities lacking rail systems. These benefits result from the rail's ability to help create more accessible land use patterns and more diverse transport systems. The conclusion of this study is that high capacity mass transit systems have a significant positive impact on congestion cost savings. Benefits to a city operation are proportional to the capacity of its mass transit system and inversely proportional to congestion costs generated. Bus systems, especially when not segregated, produce significantly lower congestion savings.

To translate these costs at the level of a project's feasibility study, we need to integrate the fact that transportation means have different capacity. Thus any lower capacity system should be penalized by the extra amount of congestion costs resulting from such lower capacity to really compare apples with apples.

[8]Rail Transit In America A Comprehensive Evaluation of Benefits (April 2005) Author Todd Litman; Victoria Transport Policy Institute.

4.10.7 Environmental Costs

Unlike electrical traction vehicles, fossil fuel vehicles emit CO_2 and other GHG, which have an impact on global warming, at the point of consumption. In fact buses generate at least 50 times more CO_2 per year than electric metro wheel to wheel per passenger transported. For instance, the average yearly emission of buses running 60,000 km can vary from 87 tons (18 m bus) to 110 tons (24 m bus) of CO_2, when we use a value of 2.63 kg/L of $C0_2$, compared to 0 for electrical metros. However to make a fair comparison, we need to use CO_2 emission values well to wheel (see Chap. 2 for more details).

The price estimation of a ton of CO_2 can vary tremendously depending on the price basis. The current CO_2 price has nothing to do with the challenges of global warming that we are now facing. According to the World Resources Institute, an environmental think tank in Washington DC, bringing carbon dioxide levels down to those suggested by United Nation reports would cost $US20 to $US100 per ton. McKinsey & Co., a consulting firm, estimates that it would cost $US40 per ton of CO_2 to reach the 2030 greenhouse gas levels recommended by the UN studies. Using a price tag of $US40 per ton, we could estimate that solely for carbon dioxide, each bus should bare the annual cost of $4000. This might seem small value, but when we add car pollutant emission, figures start to be pretty impressive. In fact many metro operators calculate the benefits in pollution reduction resulting from their operation. For instance, the Sao Paulo metro estimated in 2009 (see footnote 6), their savings in pollutant emission at 783 M tons, of which 723 M only for CO_2 and estimated the yearly savings for mother earth at US$66 M.

4.10.8 Social Benefits

Public transportation plays a social role. It allows all citizens, whatever their wealth, to get access to services. In megalopolises, poor people are often condemned to live in the city's outskirts, as they cannot afford the high housing price of inner city. Unfortunately, work doesn't exist in these crime ridden regions and the poorer citizens are often condemned to longer commutes and frequently, higher transportation fees to find work. In the megacities of this world, the poorest citizens can lose three hours or more per day commuting from home to the work place and back. Anyone who has experienced traveling in a crowded environment for so long will clearly recognize the burden that it represents on such people's life. In fact, if we look at the 2009 Sao Paulo annual report (see footnote 6), almost half of all the generated social benefits are linked to reduction in lost time due to traffic jams. There are two ways of improving such figures. The first one is to build higher capacity systems which will reduce the traveling time of ordinary people. This is time consuming, expensive and difficult to implement. The second way to improve such figures is to make the travelers' journey more productive by enabling them to work or get access to the internet. The benefits and the way to enable this alternative solution will be described in detail in Chap. 5.

4.10.9 Health Cost Linked to Air Pollution

From an health perspective, where the pollution is emitted is probably more important than how much pollutants' is send in the air, as many inhabitants live in cities where pollutants are trapped.

Air pollution impact on OECD countries health costs (including deaths and illness) was estimated at around $USD 1.7 trillion (2010 values). In a 2014 publication[9] the OECD suggests that road transport would account for roughly 50 % of these costs. Obviously yearly health cost in the range of US$850 billion imputable to air pollution is no small number.

Estimating how much health cost bills would be reduced by introducing a new metro or bus line is complicated but doable. As we've seen in Chap. 2, we need to take into account the pollution impact well to wheel to really understand the complete cost—benefit of removing fossil fuel vehicles.

To finalize this chapter on megacities, we would like to talk about a last important feature, which is the quality of ride.

4.11 Quality of Ride

In the previous section, we clearly identified that poor people of megacities can spend a significant portion of their spare time traveling between their work place and their home. Obviously middle class citizens or even the Richs can also spend hours in traffic jams but the burden on someone life's isn't the same. This reality is that journey time is clearly influenced by the perceived comfort and quality of the service provided, which are usually better in the city center, where these social classes live. To quote the French writer Françoise Sagan "If you need to cry, you might as well cry in a Rolls Royce rather than in a deux Chevaux (an old popular French car)."

There are in fact various ways of improving the quality of ride.

4.11.1 Average Speed

The most obvious one is to increase the network's average operational speed. To give an example, bus speed can reach 25 km/h, however during peak time average

[9]Cost of air pollution; health impact of road transport—OECD (2014).

speed can fall to values as low as 12 km/h on busy routes. Putting in place measures and equipment to come back to the average speed outside of peak hours, would immediately cut journey time by half! Changing the transportation mean would also do wonder. The following figures give a view of the worldwide average commercial speed for other transportation means:

- 27 km/h for heavy metro;
- 20 km/h for tramways on segregated lines;
- 17 km/h for BRTs; and
- 34 km/h for automatic Light Rail Trains.

4.11.2 Comfort

Improving the conditions in which people travel is a good way to enhance the quality of ride. Air conditioning is a feature of modern transport that comes immediately to mind when one thinks about comfort in megacities, mostly located in hot climates. Wider seating space, good lighting conditions, low level of noise, nice design, are a few of these additional features that make traveling more pleasant. Reducing overcrowding is also a measure that greatly enhances the traveling experience but unfortunately decreases capacity.

4.11.3 Quality of Service

Facilitating the life of passengers can also improve the quality of ride. For instance reducing, the queuing to enter or exit metro stations can be greatly appreciated. Improving payment convenience by using for instance rechargeable electronic payment cards or even in the future smart phones, adds flexibility to passengers who don't have to queue to buy paper ticket or monthly pass. Higher bus or metro frequency can also greatly improve quality of ride, as people can avoid wasted connection time.

4.11.4 Access to Information

Passengers want to get quick access to information concerning their trip. As we will see in the next Chapter, people want to know immediately if there is any problem on their itinerary that will affect their arrival time. Passenger information systems, which provide in real time such information, are plebiscited by passengers. Decent internet access to surf on the web is also extremely important, especially for the younger generation.

4.11.5 Transportation Modes' Integration

The last element that improves journey time, which is maybe the most important of all, is the integration of transportation means in a coherent manner. People don't travel from point A to point B. They travel from one activity to the other. For instance, they go from resting at home to working in an office or plant. If there is no direct link between activities A and B, they will either cancel such activity or if they don't have the choice, such as driving their own car, will go from location A to B through point C and even D. This adds not only to the journey time but also to the exhaustion and stress of these trips. Thus, the issue for most city planners is to ensure that as many people as possible can go from point A to B with the less connections possible. The consequence of this is that they need to improve transport coverage in order that passengers avoid losing time walking to or waiting for the next vehicle. This unfortunately translates into higher public transport costs.

In today's world of manned vehicles, this means that you need to provide a wide spectrum of transportation means that connect easily. In transportation, the image of a Fishbone is often used to describe the best possible integrated network. Cars or small buses feed into secondary bus or light train lines. These secondary lines integrate with primary lines such as heavy metro that constitutes the backbone of the city transportation network, at connection stations (Fig. 4.14).

In a driverless car world, this fishbone image could change drastically. Indeed this current ideal integration is based on a point to point transportation model, where metros and buses need to stop at all stations. In a driverless world as we will see in the next chapters, multi-point to multi-point transport will be possible. In other words, we could envision that through transport apps, unmanned taxis pick up and drop off several passengers according to an optimized journey time, at a very

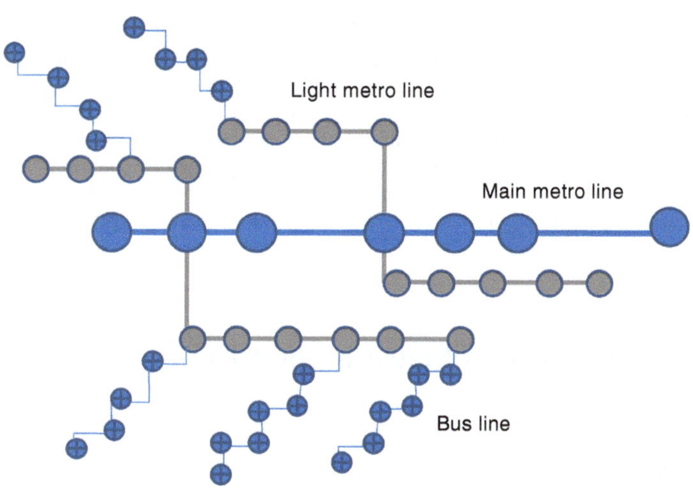

Fig. 4.14 Transport means integration taking the shape of a fishbone. *Source* Author

Fig. 4.15 Picture of a multimodal integration platform in the German city of Unna, where regional and city trains, station, as well as several bus platforms and an adjacent parking are located together in the city center. *Source* Author

low fare cost. This will obviously disrupt the public transportation industry as we know it, but will improve drastically the quality of ride of millions of poor citizens (Fig. 4.15).

Acknowledgments and disclaimer The use of pictures or references made to studies or companies and their brand does not in any way suggest that the authors of such studies or the mentioned companies endorse in any way this book or its content. The author endeavors in respecting the copyright subsisting any of the graphics, and texts that he uses, to use graphics and texts he has himself created or to use graphics and texts not covered by copyright. All trademarks and brand names quoted in the book including those protected by copyright of third parties are subject unreservedly to the provisions of current copyright law and the rights of ownership of the registered copyright holders.

Companies and Brands Stated in the Chapter

- Transmilenio S.A.
- Transport for London
- McKinsey & Co.

Chapter 5
Connected Cities

Acronyms

ABS	Anti-lock braking system
API	Application program interface
CAN	Controller area network
CCTV	Closed circuit television
EDA	Event-driven architecture
HVAC	Heat ventilation and air conditioning
(IPv4 and IPv6)	Internet protocol version 4 and 6
LIN	Local interconnect network
M2M	Machine to machine
Mbps/Gbps	Mega bytes per second/Giga bytes per second
MOST	Media-oriented system transport
NGN	Next generation network
PDA	Personal digital equipment
PnP	Plug and play technology
PTC	Positive temperature coefficient
RFID	Radio-frequency identification
SDP	Service delivery platform
SOA	Service oriented architecture
TCP/IP	Transmission control protocol/Internet protocol
URI	Unique resource identifier
V2I	Vehicle to infrastructure
V2V	Vehicle to vehicle

5.1 Introduction

It is trendy for big companies to talk about intelligent cities. It brings a notion of high tech and green image that big corporations love to be associated with. However, the notion of intelligence can barely be applied to any city at all. As far as

S. Van Themsche, *The Advent of Unmanned Electric Vehicles*,
DOI 10.1007/978-3-319-20666-0_5

we know, there isn't yet any mega computer supervising the various infrastructures and harmonizing the residents' life accordingly. What we believe exists and is being developed further in many cities is the infrastructure to improve the connectivity between all things. Already in 1999, the concept of the "Internet of Things" was introduced to society. The main idea behind such concept was that all things could be interconnected. Initially, it was thought that all devices would be connected through a unique electronic product code (i.e.,: RFID) but under the advances of internet, this same concept has evolved into objects having an IP address or Unique Resource Identifier (URI). This means that any device—computer, phone, or machine—connected to the internet is associated through a URI. Originally, this URI was a 32-bit number (IPv4: Internet Protocol version 4) but as the number of connected devices is expected to grow exponentially, a 128-bit identifier was launched in 2013 under the IPv6, allowing for trillion of devices to be interconnected. With IPv6, the Internet of Things growth won't be restricted anymore by software limitations.

Changes in hardware have also happened to allow for this increase in communication capabilities. On the fixed network side, most city residents have now access to Ethernet at home or at work. But until now, these Ethernet accesses were limited in download speed and data transport capacity. This is now changing in most big cities, as cable companies are adding the last few meters of connection, enabling fast internet services. These fast Ethernet medium will not only enable video and music on demand, they will allow for more sophisticated service providing, fuelling even more the shift from retail buying to online purchasing. These fast Ethernet connections will also enable data transfer at a reasonable cost and speed, fuelling the machine-to-machine (M2M) communication revolution. But the main game changer in the equipment area is the advent of smart phones or tablets connected to 4G telecom technology. What is actually happening in front of our eyes is the laying of the infrastructure foundation enabling everyone or everything, everywhere in the city, to get access to whatever information or services he, she, or it needs 24/7.

This chapter looks at the various opportunities and challenges linked to these new Internet-based technologies and how they will affect public and private transport. In order to get a glimpse of the impact of the internet on transportation, let's first go back to the main communication modes. Basically, there are three modes (a fourth one would be one to nobody but this isn't hopefully the focus of this book…):

- One to one;
- One to many; and
- Many to many.

In the past, communication was mainly "one to one" (i.e. phone) or "one to many" (i.e. newspaper or television) types. However, with internet and new software technologies "many to many" communications has become increasingly popular. A good example of this technology is social media such as Facebook or Twitter, which we are all familiar with. New technologies are also blurring the distinction between communication types. For instance Skype, an online phone service software, allows smart phones or tablets to be used in groups or on a one to one basis. Media network such as Facebook allows people to show a message on their board or to send individualized messages.

These technological changes are now at the heart of the news (or at least the users' news) and can immediately transmit worldwide information or picture of events. Besides the social aspect of being connected 24/7, there is also a cultural impact to get peers' information and seeing the world through the eyes of our connections. Furthermore the "Internet of Things" is changing our daily life and the economy by for instance changing the buying decision process. We will now see how this will influence transportation.

5.1.1 The Mobile Environment

For many years, the internet experience has been achieved through a PC in the comfort of one's home. Home wireless routers have helped diffuse wireless technology through the mass market but it is only with the introduction of smart phones and tablets that the ordinary users have been able to connect on the move. For the IT world, the possibility of getting in touch with potential clients outside of a building opened new business opportunities. The problem for many years was still the cost of data transfer. However, in this last decade mobile operators have started offering fixed rate packages with unlimited download capacity, opening the way for 24/7 internet access. With the introduction of 4G networks, mobile operators are now able to improve download quality and speed, fuelling even more this e-mobility revolution. Be it wirelessly or through fixed lines, there is now almost no limit to quality and quantity of bits of information transmission. This opens the constant connectivity revolution within the cities.

5.1.2 Acquisition of Intelligent Thermostat Manufacturer

In January 2014, Google acquired Nest, a manufacturer of intelligent devices, such as thermostats, which can through clever algorithms understand energy consumption patterns and optimize the turning off and on of boilers and air conditioning to reduce energy consumption.

Imagining a future where home automation would be integrated within Google's own traveling application platform, we could envision smart phones adapting their user's daily environments through M2M communication, according to anticipated time arrival and departure. To understand the benefits of such acquisition, we need to make the link with the expected exponential growth in M2M communications likely to occur in building infrastructure. Indeed, it is estimated that by 2022, 37 % of all connections will be in the area of intelligent building, mainly coming from communication between HVAC and security devices (2013 white paper)[1].

This acquisition illustrates how big IT players are positioning themselves around the "Internet of Things" opportunities. They propose and will continue to add services and technologies to benefit from the growth of M2M communications [2 billion in 2013 to the anticipated 18 billion in 2022, of which 8 % will be in automotive sector alone, according to the same study (see Footnote 1)]. As mobile devices will have grown during this same period from 146 million to 2.6 billion devices, with M2M communication accounting for 22 % of all cellular connections, we understand better the need for a company such as Microsoft to buy the mobile activities of once number one mobile manufacturer Nokia.

An outsider to this book, could ask herself, in what way is all this information relevant to transportation. The fact is that the M2M revolution is already reshaping the public transportation world and will impact the automotive industry as well. To understand how a world without human interference can work or even as a matter of fact write about e-mobility, connected cities, connected electrical grid, transport Apps, etc., it is in our view impossible not to describe the main technological factors behind this revolution: the Internet and its big players. We will very briefly explain what the main IP technologies are, and how all this works together, before showing how this is affecting transportation means.

5.2 The "Internet of Things" Technologies

We realize that for non-IT specialist, this section will seem a little dry, but we will try to keep it as simple as possible. For those that are really not software driven and wish to skip the section, we will resume what is necessary to understand for a M2M world where decisions must be made without human beings:

- **Rugged, proven software technology**: in order for everyone to be able to communicate, people must speak the same language. The Internet Protocol (IP) technology has been this universal "lingua franca" for the last 40 years and is likely to continue to be so for many more, regardless of its current limitations.

[1]M2M market in 2013 white paper published by Machina research (2013).

- **Fast wireless and fixed networks**: There is no communication without a medium through which such information can be exchange. Like the strings of the throat that create sounds that travel through air, there needs to be a way to transmit and transport messages from one side of the world to the other. New broadband IP networks developed by communications carriers can provide such capacity.
- **Open software architecture**: If people don't know whom to talk to, we basically have a situation where everyone receives every conversation, ending up in cacophony. Equipment must also know what information is pertinent and which other devices should receive what kind of data. Service-Oriented Architecture, with its open service approach, enables a harmonious exchange of information.
- **Service delivery platform**: It typically provides services (control, creation, and orchestration) and execution environments. In other words, it sets the rules of how things work, especially for multimedia services.
- **Event-driven architecture**: In a world without human beings, there needs to be triggers for machines to send information to other machines, in order for the system to take precise actions. These triggers are function of defined events, which can for instance, be as simple as time based or as complex as a camera identifying someone in an area where he or she shouldn't be.
- **Plug and Play**: A world where hard programming doesn't need to be done for each machine to communicate simplifies the introduction of thousands of M2M products. In this M2M world, auto-discovery is the way to go.

5.2.1 Internet Protocol Definition

IP is a data-oriented protocol used for communicating data across packet-switched networks. It is a standard that controls or enables connection, communication, and data transfer between two computing endpoints. More precisely, it is a network layer protocol usually associated with another communications protocol called Transmission Control Protocol (TCP) and is encapsulated in a data link layer protocol (e.g., Ethernet). The TCP/IP defines a set of rules to enable computers to communicate over a network. It provides end-to-end connectivity specifying how data should be formatted, addressed, shipped, routed, and delivered to the right destination. It specifies protocols for different types of communication between connected devices and provides a framework for more detailed standards (Fig. 5.1).

5.2.2 TCP/IP Layers

Layer 4: Process or Application Layer This is where the "higher level" protocols operate and include all processes involving users or device interactions. The application determines the presentation of the data and controls the session.

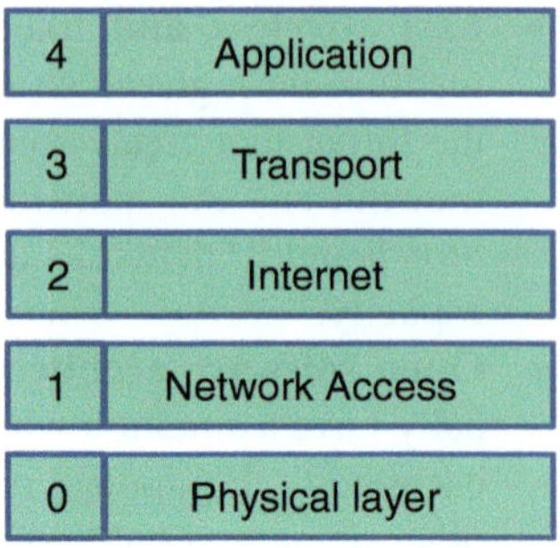

Fig. 5.1 TCP/IP is generally described as having four 'layers' (or five if you include the bottom physical layer)

Examples of these protocols are:

- Simple Mail Transfer Protocol (SMTP, the de facto standard for email transmissions across the Internet);
- File Transfer Protocol (FTP);
- Hypertext Transfer Protocol (HTTP, a communications protocol for the transfer of information on intranets and the World Wide Web); and
- SOAP (a protocol for exchanging XML-based messages over computer networks, normally using HTTP/HTTPS).

Layer 3: Transport Layer This is where flow-control and connection protocols exist. This layer deals with opening and maintaining connections, ensuring that packets are in fact received.

Examples of these protocols are:

- Transmission Control Protocol: TCP;
- User Datagram Protocol (UDP). enables programs to send short messages (called datagram);
- Real-time Transport Protocol (RTP) defines a standardized packet format for delivering audio and video over the Internet;
- Wireless Access Protocol (WAP), etc.

Layer 2: Internetworking Layer This layer defines IP addresses, with many routing schemes for navigating packets from one IP address to another. The IP is the primary protocol of this layer.

Layer 1: Network Access Layer This layer describes both the protocols used to mediate access to shared media, and the physical protocols and technologies necessary for communications from individual hosts to a medium (which, as mentioned, can also be considered a fifth layer). As new hardware technologies appear, new network access protocols must be developed so that TCP/IP networks can use this new hardware. Consequently, there are many access protocols—one for each physical network standard.

Examples of these protocols are:

- **WLAN** is a wireless computer network that links two or more devices using a wireless distribution method;
- **WIFI** is the name of a popular wireless networking technology that uses radio waves to provide wireless high-speed Internet and network connections, based on the IEEE 802.11 standard;
- **WiMAX** (Worldwide Interoperability for Microwave Access) is a telecommunications technology aimed at providing wireless data over long distances in a variety of ways, from point-to-point links to full mobile cellular type access. It is based on the IEEE 802.16 standard; and
- **Ethernet** is a family of frame-based computer networking technologies for Local Area Networks. It defines a number of wiring and signaling standards for the physical layer, through means of network access at the media access control (MAC)/data link layer, and a common addressing format. Ethernet is standardized as IEEE 802.3.

Layer 0/Physical Layer This layer describes the physical protocols and technologies necessary for communications from individual hosts to a medium.

Examples of these technologies and physical protocols are:

- Ethernet physical layer;
- RS-232;
- Optical fiber;
- Coaxial cable; and
- Twisted pair.

IP is at the basis of the "Internet of Things" revolution, but it isn't sufficient to make it work. A proper environment is also required, which can foster easy connectivity between devices. One of the examples is the V2V network that we have seen in a previous chapter where a mesh network topology is required. The choice of the network topology isn't neutral. For instance, it allows for the different network elements to communicate, even when one of the elements is failing.

5.2.3 Network Topology Description

Unmanned cars will require constant communication. Thus, selecting the right network's topology enabling failsafe communication will be crucial for the advent of self driving cars.

Network topology definition Network topology is the study of the network arrangement of the elements (links, nodes, etc.), especially the physical and logical interconnections between nodes. Any particular network topology is determined only by graphical mapping of the real configuration and the virtual connections between nodes. The arrangement of the elements of a network gives rise to certain basic topologies, which may then be combined to form more complex topologies.

Basic types of network topology The following design shows the basic network topologies: point-to-point, bus, star, ring, mesh, and tree.

Point-to-point It is the simplest physical network. It can be an established permanent link between two end points or can be established by circuit-switching or packet-switching technologies;

Bus network This topology can be described as a backbone (or trunk), where all of the nodes of the network are connected to a common transmission medium and where data can be received by all nodes in the network virtually simultaneously.

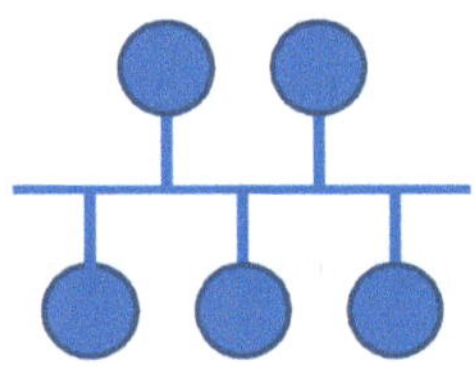

Star network It categorizes networks in which each of the network nodes is connected to a central node with a point-to-point link. All data transmitted between nodes in the network is transmitted to this central node, which is usually some type of device that then retransmits data to some or all other network nodes;

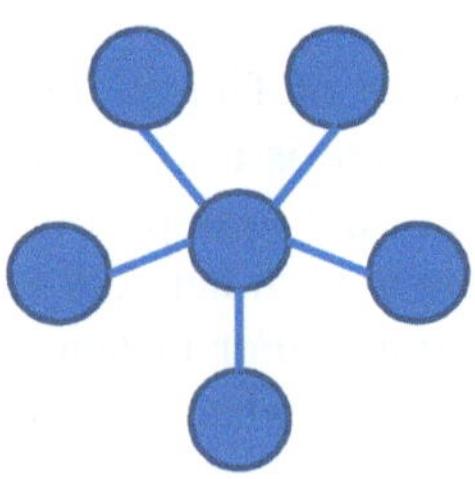

Ring network It can be described as a topology in which each of the network nodes is connected to two other nodes in the network and with the first and last nodes being connected to each other, forming a ring. All transmitted data between network nodes travels in a circular manner and the data generally flows in a single direction only. Redundant systems which are common in public transport often use a dual-ring network where each of the nodes of the network is connected to an extra other network node;

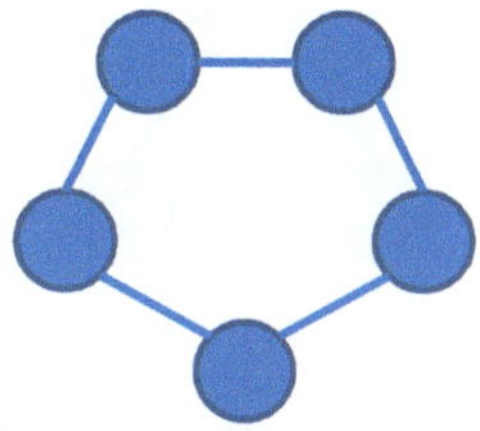

Mesh network It is a network topology in which some of the nodes are connected to more than one other node with a point-to-point link. This makes it possible to take advantage of some of the redundancy that is provided by a physical fully connected mesh topology, without the expense and complexity required for a connection between every node in the network;

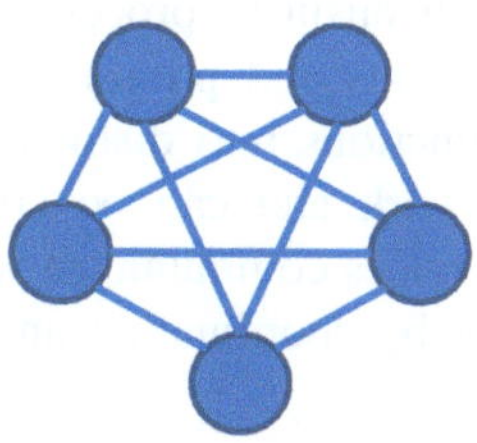

Tree network It is a hierarchical network topology where the top level of the hierarchy is connected with a point-to-point link to other nodes that are one level lower, while these intermediary level nodes also are connected via a point-to-point link possibly to one or more other nodes that are one level lower in the hierarchy.

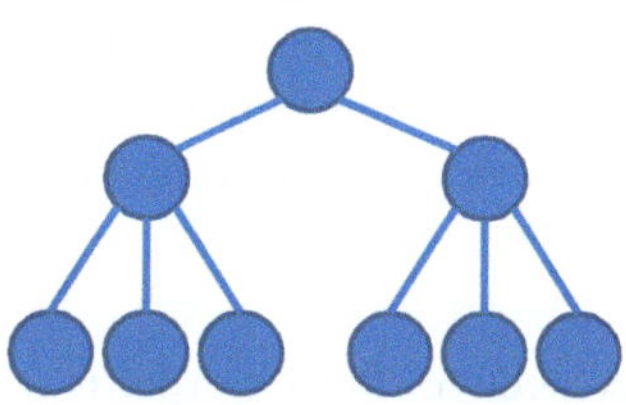

Hybrid network topology It is a type of network topology that is composed of one or more interconnections of two or more networks that are based upon different physical topologies. Hybrid networks use a combination of any two or more topologies in such a way that the resulting network does not exhibit one of the standard topologies (e.g., bus, star, ring, etc.).

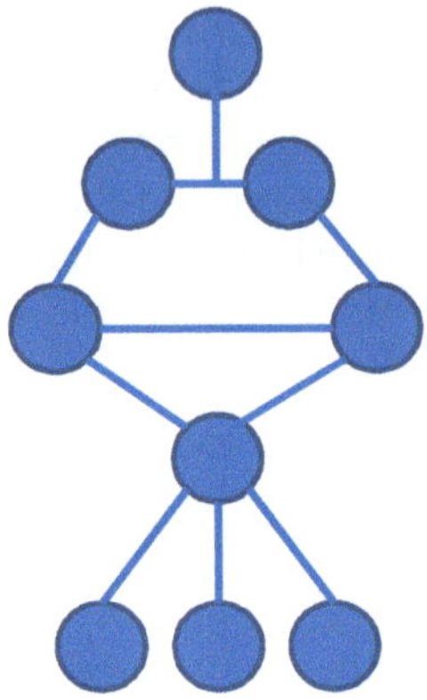

5.2.4 Service-Oriented Architecture (SOA)

In a world without man interfacing, machines or cars need to know who to talk to and what to listen to. SOA, which is a computer system's architectural style for creating and using business processes packaged as services, allows doing so. It defines and provisions the IT infrastructure enabling different applications to exchange data and participate in business processes. These functions are loosely coupled with the operating systems and programming languages underlying the applications. SOA separates functions into distinct units (called services), which can be distributed over a network and can be combined and reused to create business applications. These services communicate with each other by passing data from one service to another, or by coordinating an activity between two or more services (Fig. 5.2).

Open architecture SOA architecture is not tied to any specific technology. It may be implemented using a wide range of technologies, (such as the newer technologies: SOAP and Web Services). SOA can be implemented using one or more of these protocols and, for example, might use a file system mechanism to communicate data conforming to a defined interface specification. The key is independent services with defined interfaces that can be called upon to perform their tasks in a standard way, without the service having foreknowledge of the calling application, and without the application having or needing knowledge of how the service actually performs its tasks.

From a programming perspective, SOA can be regarded as a style of information systems architecture that enables the creation of applications that are built by combining loosely coupled and interoperable services. These services inter-operate based on a formal definition that is independent of the underlying platform and programming language. The interface definition hides the implementation of the language-specific service. SOA-based systems can

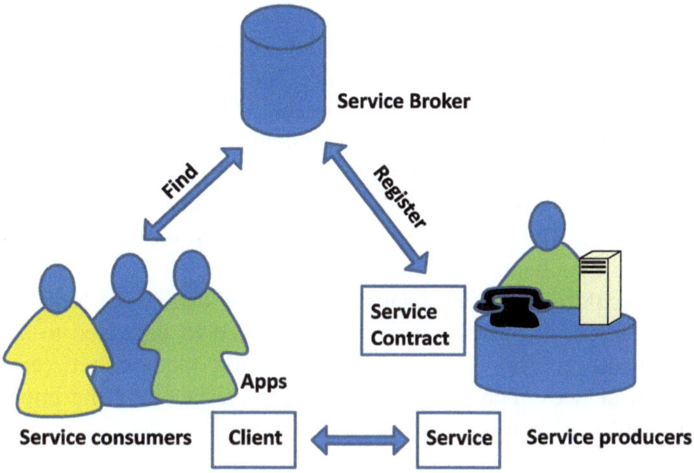

Fig. 5.2 Illustration of a SOA architecture with a service provider (primary engine underlying the Web services), service broker (ensuring service information is available) and service consumers (issuing one or more queries to the service broker to locate a service and determine how to communicate with that service). *Source* Author

therefore be independent of development technologies and platforms (such as Java, .NET, etc.). For instances, services written with Microsoft's C# and running on .NET platforms can be consumed by a client application. Applications running .Net or Java platform can also consume services running on the other as Web services, which facilitates reuse.

- **SOA benefits**: A flexible, standardized architecture, which can better support connections of various applications and sharing of data coming from various sensors or systems. It achieves the goals of increased interoperability.
- **Programming reusability**: SOA achieves this by allowing fairly large chunks of functionalities to be tied together to form new ad hoc applications, which are built almost entirely from existing software services. The larger the chunks, the fewer the interface points required to implement any given set of functionalities.
- **Composability**: SOA is a highly composable architecture (that brakes complex problems into simple ones), which provides recombinant services that can be selected and assembled in various combinations to satisfy specific user requirements.
- **Information exchange**: SOA unifies business processes by structuring large applications as an ad hoc collection of smaller modules. These applications can be used by different groups of people.
- **Plug and Play**: The key to a successful standard lies in its usefulness and easy adoption in a particular field/industry. Whenever adding new

devices, SOA approach is enabling plug-and-play enhancements without the need for binary code programming and subsequent recompilation of software code by computer programmers.

- **System upgradability**: Newer technologies such as SOAP and Web Services enable the separation between the interface and the implementation. In other words, the invoker of a service needs only to understand the interface. Thus, the implementation can evolve over time, without disturbing the clients of the service.
- **System scalability**: Traditionally, when a new device (i.e. sensor such as a new camera or photoelectric switch) needed to be added to a system, APIs would be the common communication method between vendor supplied devices. The separation between interface and the implementation enables APIs to be eliminated all together.

5.2.5 Service Delivery Platform (SDP) and Access Networking

Speaking the same universal language (IP) and knowing who to talk to and about what subject (SOA) isn't sufficient for a world without human to work. Something needs to set the rules of how things work and this is where SDP plays its role.

Thanks to new SDP standards and network architecture, telecommunication companies are driving SOA implementation to:

- Decouple applications via middleware from IT server/storage/network resources;
- Compose new services using standard-based technology and protocols;
- Reuse architectural components to lower costs and time to develop;
- Improve configuration and change management.

Telecom and network providers are putting their R&D efforts on evolving the current generation of SDP to converge Web and Telecom domains. This increases demand for communication-entertainment devices and services from internet users. As the telecom and Internet networks evolve to 4G, the benefits of the converged Web and Telecom will become even more pronounced.

SDP refers to an architectural style applied to telecommunication infrastructure. SDPs typically provide service creation, execution, and control environments, a service orchestration and abstractions for media control, presence/location, integration, and other low-level communications capabilities. SDPs applied to business applications often focused on the integration of telecom and IT capabilities.

Newer SDP technology: New standards for voice and image transfer have been developed recently to support new technologies, such as Voice-over-IP (VoIP) and Session Initiation Protocol (SIP).
IP Multimedia Subsystem (IMS): The IMS is an architectural framework for delivering IP multimedia to mobile users. To ease the integration with the Internet, IMS as far as possible uses IPs such as SIP.

5.2.6 Next Generation Networks (NGN)

Speaking the same language (IP), knowing who to talk to and on which subject (SOA) and understanding what are the rules to communicate (SDP), isn't sufficient for a world without human interference to thrive. Indeed without a medium through which such information can be exchanged, we literally have discussions between mute equipment. Broadband IP networks, which provide the carrier capacity to transmit and transport messages from one side of the planet to the other, are also necessary.

New broadband IP networks developed by communications carriers (called Next Generation Networks. NGNs) incorporate real-time multimedia communications and service quality management functionality, providing high-quality video, videoconferencing, and high-reliability communications services in addition to existing fixed and mobile telephone service. NGNs also support SDP functionalities mentioned earlier, allowing communications carriers to offer application-building functions to third parties.

NGN definition: The general idea behind NGN is that one network transports all information and services (voice, data, and all sorts of media such as video) by encapsulating these into packets, like it is on the Internet. NGN is able to make use of multiple broadband and Quality Of Service (QoS) enabled transport technologies, in which service-related functions are independent from underlying transport-related technologies.

NGN involves three main architectural changes in the:

- **Core network**: NGN implies a consolidation of several transport networks each historically built for a different service into one core transport network (mostly based on IP and Ethernet);
- **Wired access network**: NGN implies the migration from the 'dual' legacy voice to a converged setup in which the voice port is integrated on IP (VoIP), allowing to remove the voice switching infrastructure; and

– **Cable access network**: NGN convergence implies migration of constant bit rate voice to CableLabs Packet Cable standards that provide VoIP and SIP services.

Benefits of NGN: In a NGN there is a more defined separation between the transport (connectivity) portion of the network and the services that run on top of that transport. This means that whenever a provider wants to enable a new service, they can do so by defining it directly at the service layer without considering the transport layer—i.e. services are independent of transport details. Increasingly applications, including voice and images, will tend to be independent of the access network (de-layering of network and applications) and will reside more on end-user devices.

5.2.7 Event-Driven Architecture (EDA)

In a world of human being, using a "lingua franca" (IP), knowing who to talk to and on which subject (SOA) under established rules (SDP) and through a medium (NGN) would be sufficient. However, in a world of machines or cars, there is one element that is missing: when should they communicate. Without the notion of event, all the equipment would have the capability to communicate but would be paralyzed by the fact that they wouldn't know when to talk or act. In a world of machine to machine communication, the "when" is triggered by events.

EDA is a software architecture, which promotes production, detection, use of, and reaction to events. This architectural pattern may be applied by the implementation of applications and systems transmitting events among loosely coupled software components and services. It is clearly divided into two sections: event selection and handling.

EDA typically consists of event consumers and event producers. Event consumers subscribe to an intermediary event manager, and event producers publish to this manager. When the event manager receives an event from a producer, the manager forwards the event to the consumer. If the consumer is unavailable, the manager can store the event and try to forward it later.

Building applications and systems around EDA allows these applications and systems to be constructed in a manner that facilitates more responsiveness, because event-driven systems are, by design, better adapted to unpredictable and asynchronous environments.

In EDA the flow of the program is determined by sensor outputs (such as video cameras), user actions (mouse clicks, key presses) or messages from other programs.

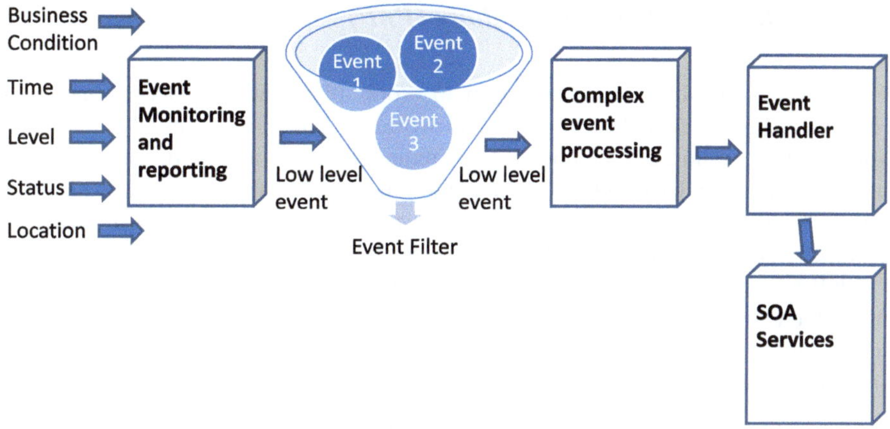

Fig. 5.3 EDA and SOA are complementary architectures. SOA 2.0 in fact, is a design approach that deals with complex event processing. *Source* Author

> Event-driven programs can be written in any language, although the task is easier in languages that provide high-level abstractions to support it.

5.2.8 *EDA and SOA Together*

EDA complements SOA architecture because services can be started by triggers such as events. Business systems benefit from the features of both a SOA and an EDA, since an EDA can trigger event consumers as events happen and loosely coupled services can be quickly accessed and queried from those same consumers (Fig. 5.3).

For systems to be highly responsive, they must be able to quickly determine the necessary actions when events are triggered. To this end, events should be published and consumed across all SOA boundaries, including the layers of the architectural stack and across physical tiers.

> Events can be defined as any published change in a system, platform, component, business, or application process. They can be high-level and business-oriented, or low-level and technical in character. Because events can be transmitted and received, event-aware applications and services can respond to the underlying changes as needed. The secret to understanding any given event is to know its cause at the time the event occurred.

5.2.9 *Plug and Play Technology (PnP)*

For a world of M2M communication to work, plug and play technology isn't required. Having said that, for a M2M world to grow exponentially, one must simplify to its minimum the connectivity requirements of all equipment. Indeed, if an object costing let's say 5$ must communicate with the network, it needs to automatically self connect or the cost of installing such device will make the connectivity requirement unviable. The secret is auto-discovery on the network.

PnP is a concept that addresses integration's interfacing aspects. Solutions are assembled from components designed and delivered using common platform architecture. The idea is that the platform architecture specifies an interface pattern that allows new components to:

- Be defined, providing specialized services based on the common interface pattern—often in the form of an API;
- Plugged in easily, using existing available component services;
- Immediately substitute existing similar components from other vendors, whenever supporting the same interface and services.

Universal Plug and Play (UPnP) is a set of computer network protocols promulgated by the UPnP Forum. UPnP goals are to allow devices to connect seamlessly and to simplify the implementation of networks (data sharing, communications, and audiovisual content) and corporate environments. UPnP achieves this by defining and publishing UPnP device control protocols built upon open, Internet-based communication standards. UPnP architecture allows peer-to-peer networking of PCs, networked appliances and wireless devices. It is a distributed, open architecture based on established standards such as TCP/IP, UDP, HTTP and XML.

Any UPnP compatible device can dynamically join a network, obtain an IP address, announce its name, convey its capabilities upon request, and learn about the presence and capabilities of other devices. Devices can leave the network automatically without leaving any unwanted information behind.

5.3 M2M Communication

All the IT concepts running together around the TCP/IP technologies we briefly described before can form the basis for the M2M infrastructure. This M2M buzz word refers to technologies that allow both wireless and wired systems to communicate. It is a broad term that doesn't refer to any specific technology but only to exchange of information and performs actions accordingly without the use of any human interference. Typically device to device communication network connections have evolved from "one-to-one" to "one-to-many" communications. In the past, equipment would talk to each other via dedicated APIs, supervised by a monitoring system. In this new IT environment, all connected device will inform about and will only use the services that are relevant to it. The worldwide Web expansion and the emergence of these new IT technologies have made it easier for M2M communication to take place. We will in this section try to explain what constitutes a M2M network, besides the software and telecom equipment we've just presented.

5.3.1 Sensing Devices

Sensors come in all forms and have different measurement purposes. They will include simple equipment such as photo-electric or electromagnetic sensors that can detect the presence or passage of objects, but also smarter equipment that can detect shapes or colors (i.e.,: cameras used in people profiling). Sensors can be embedded within equipment (i.e.,: PTC sensor which measures motor heat) or not (i.e.,: HVAC's thermostats hanging on the wall). Sensors are fundamental because they are the link with the real word and provide the basis for machine decision making. In a world without human interference they will identify the events that will triggered a set of services and actions linked to these services.

5.3.2 RFID

It is typically a chip that equips non-powered equipment or non smart equipment that allows for small data recording and updates. There are typically two types of RFID:

- **Passive**: data is recorded or read when an object powers on the RFID; and
- **Active**: it can send data according to a programmable logic.

The RFID cost is dropping quickly allowing for the simplest objects to be RFID tagged, providing the basis for data storage and transfer of information of all objects.

5.3.3 Algorithms

Computers are as intelligent as their algorithms are sophisticated. These algorithms need to interpret data and make decisions based upon events or information supplied by sensing devices. They are at the heart of the M2M revolution, because a world where only events would be gathered without any action decided would be pointless. The secret of M2M lies in these algorithms, for which companies like Google are willing to pay big money.

5.3.4 Communication Links and Networks

Wi-Fi, cellular, or fixed links allow for data exchange between connected sensors and computers. There is currently no standardized communication platform throughout the M2M equipment providers but as seen, many IT advances are fueling this evolution. Furthermore, with the growth in the M2M market, it is likely that vendors will agree upon a standard or adopt directly these new TCP/IP technologies, which will continue simplifying the life of businesses wishing to develop new equipment operating around these concepts. The market being potentially huge, many companies are trying to push their proprietary solution, teaming up with complementary technological partners.

Some of these most prominent consortiums are:

- AllSeen Alliance integrating companies such as Qualcomm, Microsoft, LG Panasonic, Haier, etc.;
- Industrial Internet Consortium, composed of companies such as General Electric, IBM, Cisco;
- Ipso Alliance promoting the use of IP technology; and
- Open Interconnect Consortium, with Broadcom, Dell, Samsung, etc.

As for many new technologies, in the end the industry is likely to adopt a common standard. As a matter of fact, the Institute of Electrical and Electronic Engineers (IEEE) is already building such a consensus among several consortiums, working on defining this common architecture.

5.3.5 Transportation Within the M2M Market

We need to recognize that the adoption of these standards and technologies will not be dependent on, nor be specific to the transportation industry. As we've seen, M2M applications touch various areas of society. Currently, non transport related activities are mainly related to warehouse and supply chain management, logistics services, telemedicine and robotics, where tracking and tracing applications play a big role. In the transportation area we can identify a least three areas were M2M applications is bringing added value:

- Security (see section on security for more information);
- Traffic control and signaling (see section on traffic control); and
- Predictive Maintenance.

Furthermore SOA and EDA technologies have also started being implemented in the public transport. In particular, projects financed by the EU such as InteGrail, Protectrail, and SECUR-ED, have taken up the challenge of testing across multiple product platforms from various vendors, these technologies in the real environment of various European city railway operations.

5.4 M2M Applied to Public Transport

We've discussed already a lot about driverless trains, where mobile and way-side environments are connected through a control center. This is obviously an area where the M2M buzz word could be used, though at all time someone in the signaling control center can intervene. The technology being extremely specific, related to safety and not IP based, it is usually not defined within the M2M technology portfolio. This can however, be seen as semantic, as in the end the objective of new moving block signaling technology is to use device to device interface to avoid human errors. Human interference is usually reduced to intervening whenever there are problems on the line.

5.4.1 Railway Onboard Networks

The TCP/IP revolution that transformed the "wired world" has not left the public transportation world untouched. Obviously, on the way-side, the same trends as in other utilities could easily be seen. However, what is interesting is that in the past few years, railway manufacturers have strongly focused on implementing IP-technologies onboard vehicles.

Deployment of local area networks onboard rolling stock goes way back to the 1980s. Technologies such as CAN bus, Profibus or FIP, which were originally developed for the automation and automotive industries, were widely adopted by the public transport sector. Strangely, an old technology such as CAN bus is still widely used throughout the automotive industry. However and though it is true that IP technology is being implemented in the railway industry, the prevailing railway technology still remains the Train Communication Network (TCN), defined in IEC 61375 and in IEEE 1473 standards. Though different in their characteristics and performance, all these old technologies have a common bus topology, which allows for limited data transmission in the range of 1 KBit/s up to 10 MBit/s. Indeed, all these bus topologies have physical limitation with respect to achievable bandwidth, with technological limitation around 40 MBit/s transmission rate.

In order to overcome the old technologies' limitation, railway manufacturers are changing to point-to-point connections by using modern switching technologies, such as Ethernet. The introduction of these switching technologies also offers the possibility to adopt standardized and open layer protocols like TCP/IP. However, contrary to IP networks in the office or on the factory floor, train IP networks are mobile and thus are dynamic with respect to their topology, because train compositions can change manually or automatically during operation, by adding or removing train cars.

5.4.2 New IT Technologies that Affect Transportation

Dynamic addressing requires defining a railway-specific approach for the deployment of IP. This is especially true for concepts, such as addressing, network topology, redundancy, and network operation. In particular, it requires sophisticated networking technologies such as ring redundancy topology or wireless bridges (Fig. 5.4).

These more complex configurations need correspondingly advanced addressing technologies. Two technologies are particularly useful in railway operations:

- **Automated ring coupling**: After train consist re-configuration, the status of ports that are active or blocked can change. In order to have the system automatically auto-configure, switches must determine by themselves which ports to set as active, redundant, and inactive.
- **Automated wireless bridges**: going wireless brings many advantages, especially in refurbishment projects where coupling pins (equipment that will connect electrically two train-consist together) are usually mostly used and where electrical shielding issues are constant between car-sets. Wireless bridges using standards such as 802.11n exist with auto-carriage configuration.

New IEC standard (IEC61375-2-5) for Ethernet Train Backbone network, has also been introduced to simplify communication across multiple train consists. The ETB can use the Train Topology Discovery Protocol (TTDP), to identify the elements of the train starting from the lead vehicle. Based on the definition of the lead train (only remaining input given by the driver), TTDP will extrapolate the other car consists. Then the ETB maps all onboard

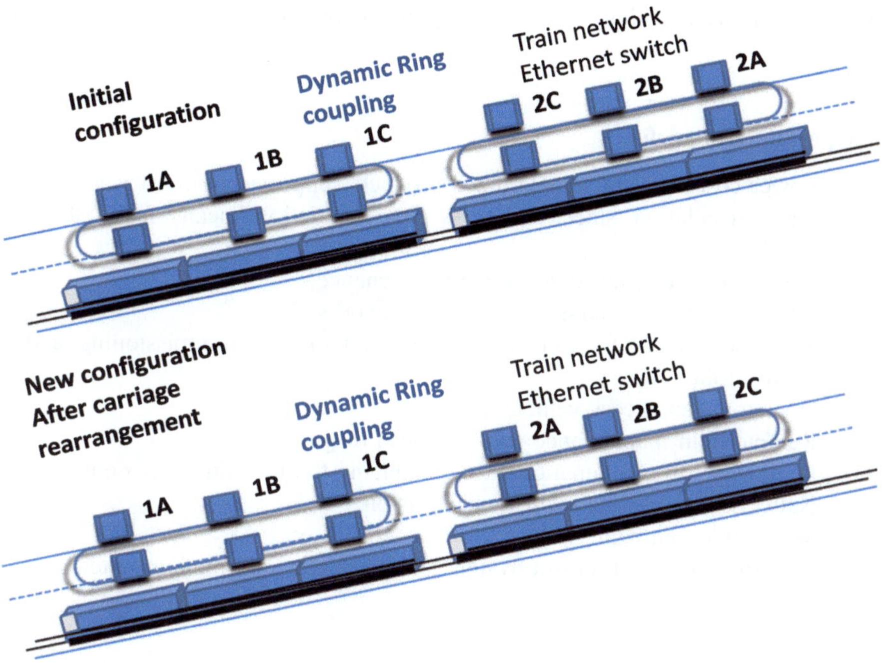

Fig. 5.4 With dynamic coupling each port of the ETN will automatically identify if it should be active or inactive. *Source* Author

IP equipment throughout the various consists and uses a table to give then a dynamic unique IP address. To work, the ETB also needs a third element, which is a technology that allows fiber optic connections to define the primary and backup links in order for ETB to have redundancy capability.

Other technical working bodies are trying to define the sets of rules and concepts necessary for the train to ground communication to work. The EU founded project InterGRail worked on standard definition such as IEC 61375-2-6 within environments with heterogeneous technologies, (i.e.,: IPv4/IPv6; WIFI/GPRS; WEB Services/FTP; etc.) enabling the railway community to proposed a unified approach to train to way-side communication around TCP/IP technology.

5.4.3 *Benefits of IP Networks Onboard Trains*

In the past years, railway manufacturers have strongly focused on implementing IP technologies on vehicles, which benefits from the fast technology improvements in office and network technologies. This approach provides many benefits for the operators.

Operational benefits:

- Lower cost by using standardized products supported by most vendors;
- Easy installation since higher system integration means less cables, smaller couplers, etc.;
- Easy to maintain through remote maintenance;
- Integration with ground system data networks;
- Subsystem suppliers can access their systems for commissioning and quality follow-up;
- Operators can access vehicle data;
- Remote train maintenance and advanced diagnostics.
- Integration of new functionalities enhancing the traveling experience:
- Network cameras for security supervision;
- Video entertainment; and
- Real-time information to travelers and passenger information system.

5.4.4 SOA Applied to the Railway Environment

One of the biggest current technological trends in the railway environment is the evolution from older concepts of distributed computing and modular programming toward a "built-upon service" concept, such as that found in SOA concepts. On the way-side these technologies can already be found.

For instance NS, the Dutch railway operator has implemented in 2011, a single network based on SOA for most operational services. This wireless network covers 49 rail stations throughout the Netherlands. At each of these stations public hotspots can be found for passengers but also for the 10,000 NS employees. Through PDAs, train conductors, platform staff, and operations teams have access to different application services depending on their access rights, regardless of where they are located. These services include business-critical data such as train traffic conditions and schedule, employees working shifts, maintenance information, etc. Security staff members are also alerted via their PDA when any abnormal behavior is observed and can get real time streaming of CCTV images from within the trains to identify potential hazardous situations. Train departure and arrival information is easily accessible online in the station and on the trains for passengers. Inspectors check through their PDA, personal information of customers trying to avoid paying tickets and, if necessary, ask security personnel to intervene. NS is also contemplating using mobile phone technology for paperless proof of payment.

5.4.5 *InteGRail*

SOA networks, which can be found on the way-side, are rare within the onboard environment. In 2011, the EU founded project called InteGRail allowed for integration tests, based on SOA onboard network.

It demonstrated that SOA could be implemented at the subset of rolling stock, infrastructure, train operation, and traffic management levels. InteGRail was successful in demonstrating that cooperation between different applications was possible and integrated technologies, in order to achieve working prototypes. The wide scope and size of the project allowed it to tackle all aspects of integration of railway information systems, defining an overall framework from which many specific, more focused solutions could be derived. One of them is related to predictive maintenance applied to railway subsystems.

5.5 Predictive Maintenance

Predictive maintenance has been for years, part of airplane manufacturers strategy to improve equipment reliability and increase aircraft overall performance. Recently, predictive maintenance has become a hot topic of railway manufacturers. Indeed, more and more private operators and Government authorities are looking at the total cost of ownership in their rolling stock tender process. Cost of maintenance, but also down-time of material, are important cost drivers of such concept (Fig. 5.5).

Traditional asset management strategy focused primarily on reactive maintenance management, where train operators only responded to failures. Predictive maintenance, in comparison, takes asset management many steps further by helping users monitor equipment condition and identify vulnerabilities before the equipment failure, based on historical data and predictive models.

By predicting equipment failure, predictive maintenance helps reduce equipment downtime, unplanned maintenance, and schedule disruption. Other advantages of such concept include increased asset utilization and enhanced productivity, both of which result in tangible benefits and accelerated return on investment. Furthermore, predictive maintenance increases reliability, reducing by the same token hefty

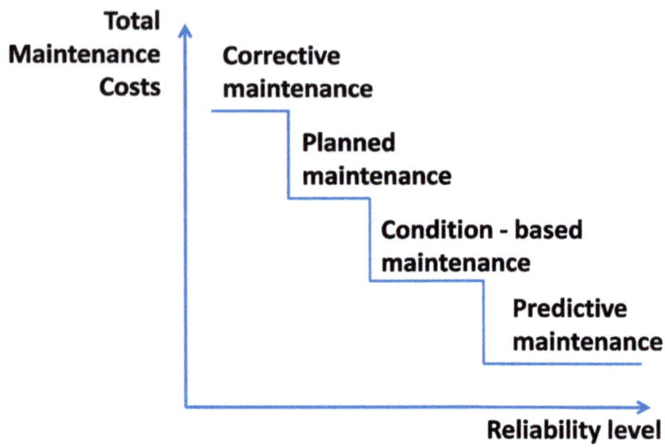

Fig. 5.5 Choice of maintenance strategy and impact on cost and reliability levels. *Source* Author

penalties that have been pushed to the railway manufacturers to ensure that the trains' performance doesn't affect the operator's QoS.

Several KPIs have been introduced to asses train performance, such as:

- **Mean Time Between Service Affecting Failures**, which basically identifies the time between two failures that have an impact on operation, even if not impacting traffic flow; and
- **Mean Kilometers Between Failures**, which identifies the number of kilometers performed between two failures.

With penalties and cost of ownership being so critical, no wonder why manufacturers have started investing heavily in predictive maintenance. This maintenance methodology relies on the possibility to get precise status of different critical train components and to make useful correlation between equipment status and the probability that they will fail according to past experience, taking also in consideration the impact of such failure on operation.

5.5.1 *Constant Monitoring*

The first key element of predictive maintenance is the possibility to monitor on a constant basis the various connected equipments. This seems obvious in a fixed environment where IP-based networks connected allow easy "many-to-many" connections. This is much more complex in a mobile environment. Thus, the first brick that train manufacturers had to lay to create a predictive maintenance engine was the possibility to create an onboard network using modern rugged network technology, such as Ethernet (Fig. 5.6).

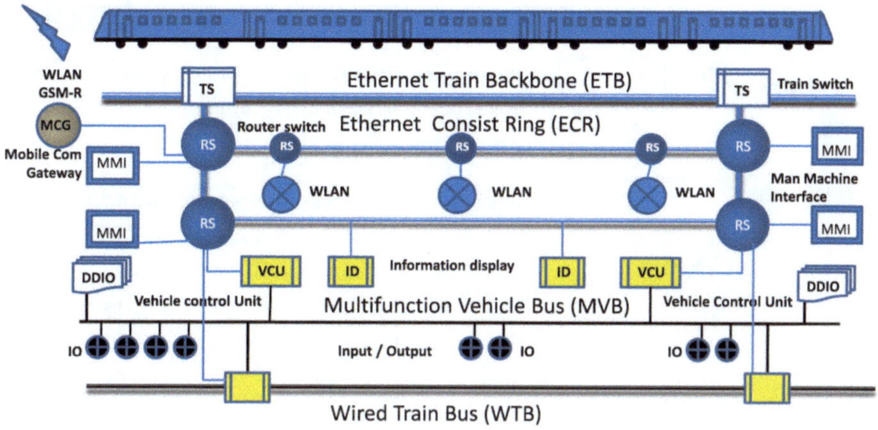

Fig. 5.6 Illustration of train onboard networks. *Source* Author

> Many issues had to be solved such as automatic configuring of the various car-trains composition, electro-magnetic compatibility and shielding of data packets (especially whenever connecting separated train sets through automatic coupling), train distances of more than 100 mbit for gigabit Ethernet, integration of CCTV with its high bandwidth requirements, etc.

This network technology is now mature and many trains are equipped with fast Ethernet networks for non-vital information. However and as mentioned, safety-related information is mostly still being sent through other network technologies (Fig. 5.7).

The second brick was to convince equipment and component suppliers to use the IP technology. Such equipment involves engine, gearbox, traction system, brake, door, battery, HVAC, etc. This was reasonably easy as these OEMs could see a direct benefit in terms of return of experience. With most vendors now using Ethernet connections, all onboard equipment can now send status report and communicate together.

The problem for railway manufacturer didn't stop there. The information supplied by the thousands of onboard equipment needed to be transferred to the way-side for integration, in order to enable predictive maintenance. This means that information needed to be uploaded and downloaded between train and based stations positioned along the tracks to relay the information to the predictive maintenance engine. This mobile environment meant additional issues, such as latency time, bandwidth capacity, QoS, etc.

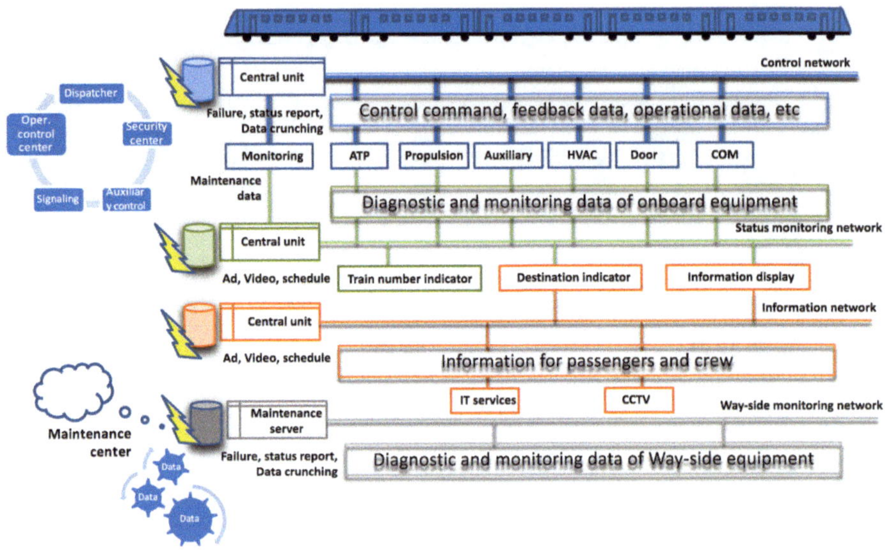

Fig. 5.7 Illustration of data acquisition and treatment. *Source* Author

Latency problem: it is due to the fact that internet data is carried in small packets. In M2M communication the issue is less about receiving smoothly the information to ensure a good user's experience than the fact that information packets can be lost while traveling from one base station to the other. The higher the speed of the train and the further apart the based stations are, the more critical becomes this problem. In high speed train for instance, communication needed to use specific satellite technology to upload and download internet content.

Bandwidth capacity: Providing metadata information is usually not that demanding in terms of bandwidth capacity. The issue is more linked to the fact that other information such as audio or video stream must often be downloaded simultaneously from the train.

Quality of Service: whenever partial information is sent, the system must figure out which data wasn't sent, so that it can be done in priority.

Though real-time communication is ideal for predictive maintenance, some operators have decided to do the download at night, whenever the trains are in the storage yards.

5.5.2 *Data Crunching*

The second critical element of this predictive maintenance is the capacity to gather billion bits of information coming from thousands of equipment installed on hundreds of trains, operating on many lines and make a sense out of it all. This means that in the cloud, servers hosting databases, having constant access to various fleets, must receive and organize data in a meaningful way. This implies not only comparing values using historical data but also reorganizing it in a way that alarms and new thresholds can be established.

5.5.3 *Event-Driven Information*

Some old railway folks might say: "well the industry always had alarms that informed the operator whenever an issue was detected. Thus, nothing's new under the sun." The reality is that going from alarms to event driven environments, is a complete change of paradigm. Alarms or logs give a picture of the reality at a given time. In order to make useful correlations between equipment states, data must be organized according to the same master clock and stored together.

To help understand this point, let's take an example. If a door fails within the usual maintenance environment, a message will be sent that at a given time, a specific door had a malfunction. There is nothing we can learn from such information and someone from maintenance will have to search the causes of that malfunction to solve the problem. Now, in an event driven environment, for instance opening and closure time, as well as electric information (door voltage and intensity), number of opening and closure already performed, etc. can be identified. This not only helps understanding the root cause of the problem, it allows also to build historical data that can be reuse for future malfunction prediction.

For instance if the correlation between events shows that closure and opening time was 5 % above the normal time and the number of opening and closures was inferior to 1 % of the limited recommended number before going into maintenance inspection, new procedures could be established with new thresholds to take this malfunction into consideration.

5.5.4 Useful Algorithms

Making useful correlations between events is good. Making intelligent decisions upon such correlations is much better. By integrating past experience, these algorithms can be improved over time, either through the work of analyst or through self-learning algorithms. This area is still new but operators and railway manufacturers will want to increase the intelligence of their system to benefit fully from predictive maintenance.

5.5.5 Failure Criticality

Supposing that through an algorithm, the system can detect that one component is likely to fail or as a matter of fact already failed (i.e.,: the air conditioning compressor), the system may decide that it is not worthwhile stopping the train. Instead the train will continue running and will wait for the train to be back to the maintenance station outside of operating hours, to replace the defective equipment. On the more advanced systems integrated to maintenance software such as Maximo™ from IBM, the system could automatically send a maintenance order to perform such substitution after operating hours.

5.5.6 M2M Public Transport Market Estimation

M2M business characteristics in public transport are primarily driven by cost reduction. For those who come from other industries this is a concept that can be difficult to understand. After all, any new business generates revenue growth. The reality is that added value and its associated markup is usually given up by railway manufacturers in the lowest bidder tenders, as it is perceived as a risk reduction enabler or as a way to reduce cost of ownership.

In other words, solutions such as predictive maintenance aren't sold for what they can bring but rather as a tool to lower the total cost of ownership so that rolling stock manufacturers can sell more trains and maintenance contracts. Furthermore, the cost of such solutions (software, a few telecom and network equipment) is usually very low in relation to the value of the global contract, representing less than 1 % of the overall contract value. It is true, that some manufacturers have been able to sell such predictive maintenance services as a stand-alone contract, but once again the railway industry, isn't keen on buying such services independently from the complete maintenance management of their fleet. The other M2M services such as Security and traffic controlling, are also cost driven and the M2M solutions are being used to lower the costs of the tender proposals.

5.6 M2M Applied to Cars

M2M applied within the automotive industry is still in its infancy but picking huge momentum as it is being pushed by EU and American initiatives to reduce fatalities after car accidents. Indeed new EU regulations supporting the implementation of eCall requiring car owners to have M2M technology enabling automatic notification to the nearest emergency center whenever drivers have been involved in a crash, is likely to be mandatory by 2015. The cars will be required to send a minimum set of information to the nearby emergency service. As a result of this, technology experts estimate that response times could improve by up to 40 % in urban areas and 50 % in rural areas, saving many lives. A consequential benefit will be that M2M telecom infrastructure will be available on all new cars and will be able to transmit wirelessly whatever type of data, via a secure mobile network. Two options are being considered for connecting cars to the network:

- From the SIM that is inside of the driver's mobile phone, through the car's Bluetooth or USB connection; and
- From an embedded SIM card inside the vehicle.

Many people within the industry are predicting that the EU Commission will soon make the embedded SIM mandatory, as it removes uncertainty over mobile availability and would lead to more consistent responses.

5.6.1 Re-programmable SIM Cards

In 2013, the association of mobile operators agreed on a standard for reprogrammable SIM cards. Though these SIM cards could also be used for voice transmission (Apple is pushing for such a solution), the main benefits of such technology is to connect devices.

By doing this, SIM cards could be fixed to the car and allow switching from one telecom provider to the other according to data transmission quality or even price. For instance, if automakers were to issue their own SIM cards and rent spare capacity of operator's network, they could save $billion a year on operating cost and propose an attractive package to car owners.

By forcing car manufacturers to work together for once on driverless functionalities around a common onboard network platform around the Ethernet/IP technology, car manufacturers could also save billion on harness and cable design.

5.6.2 *Existing Onboard Car Networks*

Probably drivers don't know that high end cars integrate around 1000 wires, totaling a few kilometers of cables. Modern cars have dozens of different electronic units, such as engine management system, power steering, electric windows, HVAC, lighting, etc. all controlled by chips that use wiring harnesses to communicate. In fact, cars have been using for years many onboard network technologies. The most widely used technology—CAN Bus—was created as early as 1984.

At least three network topologies have been used by the automotive industry.

- **Linear topology**: It is used in drive train and body network systems;
- **Ring topology**: It is used for multimedia; and
- **Star topology**: This network is low cost design single wire interconnect network for use in on-off equipment such as seat control, door locks, mirrors, etc.

Today most vehicle innovations involve electronics and software, and the number of control units (e.g., Dynamic Stability Control) has increased significantly over recent years as a result. For instance, up to 70 control units are fitted in premium-class cars. The latest models boast several hundred electric and electronic functions, and up to 115 MBytes of program code and data on board. Up to four different bus systems, such as CAN, LIN, MOST, and FlexRay, work side-by-side to transport electronic data around today's vehicles. All are equipped to meet demands of their individual areas of application, but all speak their own automotive languages (Fig. 5.8).

5.6.3 *Onboard Car IP Networks*

The automotive industry hasn't yet embraced the TCP/IP revolution. It is expected that the first all Ethernet-based IP network will only hit the market around 2015– 2016, enabling 100 Mbps throughput. The reality is that it wasn't really needed till today. It is only with the advent of cameras, infotainment services and more generally of all the technologies required to support driverless cars that such capacity became a necessity. As the new safety-related technologies will require

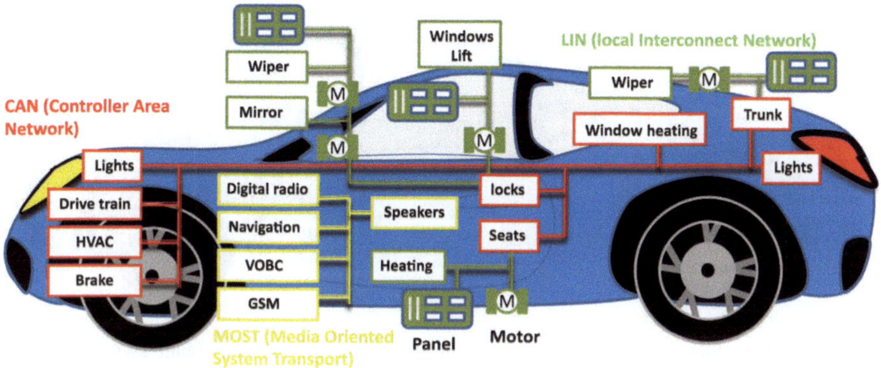

Fig. 5.8 Illustration of the various car networks. *Source* Author

transparent communication across the automotive industry, the big market players have decided to team up their efforts to create a common network platform.

Open Alliance: This group is formed by an alliance of around 200 car manufacturers and OEM companies (i.e.: Ford, GM, Daimler, Honda, Hyundai, BMW, Toyota, Volkswagen, Renault, Bosch). Its aim is to create a single vehicle Ethernet standard around the IEEE 802.3 working group, to establish 100 Mbps and 1 Gbps onboard Ethernet networks.

Working groups are still establishing the requirements for the different sub-systems. Currently the groups are assessing several scenarios with as much as 35 Ethernet nodes. These scenarios include sophisticated driver assistant systems such as 3D and infrared cameras, radar sensors, telematic transceiver systems for GPS and vehicle-to-vehicle communications.

IP benefits IP networks with their faster network connections will improve performance (for instance, increasing response time between ABS braking and drivetrain systems). A collateral benefit will be a reduction in weight due to Ethernet ports connection, which requires fewer cables. It is estimated by some industry experts, that by around 2020, 100 million of such Ethernet ports will be found worldwide (Fig. 5.9).

5.6.4 New M2M Added Value Services

By making M2M (and most likely V2V communication infrastructure) technology mandatory, Governments are helping the automotive industry get access to other beneficial information. Automotive manufacturers and their OEM suppliers can get

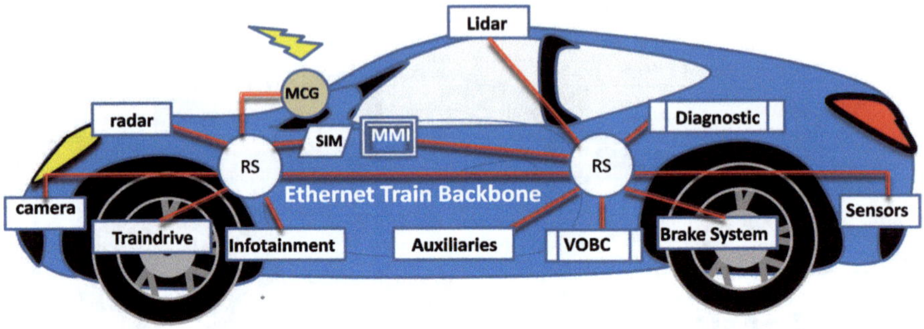

Fig. 5.9 Connecting all equipment through an ethernet backbone and router switches. *Source* Author

access to onboard equipment status. This allows better understanding of the vehicles' performance and driving behaviors. Furthermore, this enables car manufacturers to develop and evolve relationships by having continuous exchange of information with their key-stake holders (dealers, suppliers, and consumers) during the lifetime of their product. This should bring to market new innovative value-added services, but also create new service offering from other automotive players such as:

- **Fleet managers**: they will be able to know in real time the status of their car fleet, helping them manage better their assets;
- **Insurers**: they will be able to track the portfolio of their insured cars (Fig. 5.10).
- **Emergency call centers**: A major M2M benefit is to enable immediate automatic contacts with emergency call centers. EU's initiative called eCall was voted on March 2014 and will impose to all car manufacturers to equip after October 2015 (this date could be slightly postpone), an emergency call system.

The minimum set of information that will be sent through eCall includes:

- Accident information: the exact crash site location, time of incident, travel direction;
- Car information: car's category, fuel type; and
- Safety information: number of buckled safety belts.

It is anticipated that this extra device will cost a few hundred euro.

- **Safety and security service providers**: automatic dialing to roadside services or vehicle assistance will be possible, informing about potential hazardous situation such as cars stopped on the side of the lane or in tunnels. Car thefts immediate reporting to police station will also be possible.

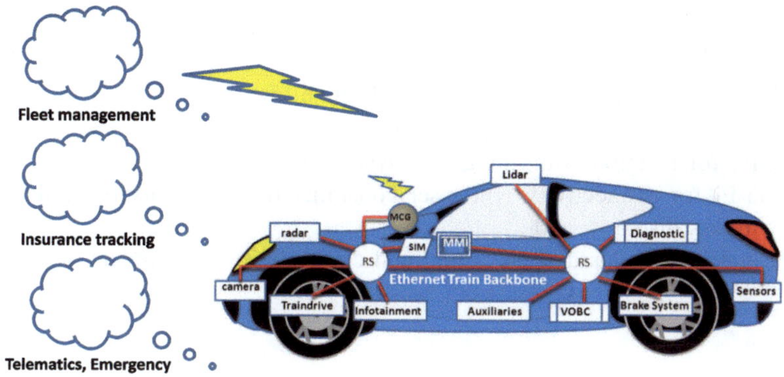

Fig. 5.10 Illustration of new service offering

- **Electric energy providers**: Some service providing will also include the capacity to manage the connection between home or energy providers for electric vehicle recharging.

As in the railway industry, new innovative services will be in the area of:

- **Remote monitoring and diagnostics**: Basic remote monitoring of vehicle equipment is already possible. Some simple diagnostic such as a malfunction of equipment can be made. The advantage of such services is that drivers will be able to be assisted from the comfort of their home or on the side of the road, by experts that could detect quickly the cause of a car failure. It could also enable the same revolution that has already started in the railway environment of predictive maintenance. Some high end services already provide data about the car's state to identify problems before they happen. For instance, analytics can diagnose engine or braking problems that would in the past been caught only at the garage maintenance check-up. However and as seen for railways, predictive maintenance will require synchronized event driven information, supplied in real or quasi real time. Besides the issue about privacy (i.e.,: geo-localization, speed, etc.), communication cost might also end-up reducing the attractiveness of such services.
- **In-car infotainment and entertainment**: As people spend more and more time in their car, stuck in traffic jam, they wish to create an environment that feels like home. Car manufacturers are providing passengers with internet access to radio and over-the-air content delivery. Drivers can also use their system to play music from their smart phone, while other passengers can simultaneously connect laptops, tablets or smart phones via the vehicle's WIFI system. DVD or Blu-ray disk players can also be found to play movies on rear end or even dash-board screens.
- **Safe telecom network**: The embedded SIM will enable 3G/4G-based devices from common design platforms to connect and communicate through multiple mobile broadband networks offered by different wireless operators.

The SIM card will provide:

- Enhanced coverage, redundancy and reliability;
- Application processing capabilities;
- GPS for location with navigation system update;
- Wi-Fi for connectivity with user consumer devices (hotspot mode) and wireless displays; and
- Bluetooth for hands-free calling and connectivity with tablets or PCs.

5.6.5 M2M Versus V2V or V2I

As indicated for trains, specialized communication between vehicles (V2V) or between vehicles and the road infrastructure (V2I) are in our view separate segments of the M2M market. This can be seen once again as semantics as the final objective is to have no interface between cars and human beings. As mentioned, the main difference is that V2V and V2I are safety-related and use-specific communication technology (i.e.,: Dedicated Short Range Communications, based on WiFi-like 5.9 GHz technology without preidentification required).

5.6.6 M2M Private Transport Market Estimation

M2M in private transport is price driven and thus can be more easily assessed. In order to do so, we need to analyze three different market segments: equipment, communication, and added-value services.

M2M Equipment We've seen that the cost of an embedded SIM card with connection to other navigational systems will cost around 250 euro per car. In the last years, the number of cars sold topped 80 million. As it should become mandatory within the next years, we could estimate this market for the equipment to reach euro 20 billion before 2020, depending on adoption rates outside of Europe and North America.

M2M Communication In 2010, around 450 million European had 660 million SIM cards. Included in this, 4.3 million cars had already onboard telematic system, such as theft prevention, emergency call, and motor insurance tracking. Despite reduction in price, higher volume of data transported allowed the telecom industry to grow. In 2010, the mobile industry generated 175 billion euro. Thus, and all things being equal, yearly revenue per SIM card was 265 euro that year. Using this amount, which is probably acceptable to most car owners, we would again identify 20 billion euro in market opportunities.

Big telecom providers have tried to close deals with car manufacturers to position themselves in regards to this market. Vodafone announced a new agreement with Volkswagen and Audi in Europe to provide connected car services through embedded global SIM cards, starting with new Audi models from 2015. With a strong European presence, Vodafone wishes to provide high-speed internet services throughout that continent. Telefónica a Spanish mobile provider has also teamed up with U.S. carmaker Tesla to propose an in-car TV demonstration. Ferrari, Mercedes-Benz and Volvo also demonstrated a new in-car user interface from Apple, named CarPlay, at the 2014 Geneva International Motor Show.

5.7 E-Mobility Technology Limitations

M2M, V2V, and moving block technologies, as well as the electrification of transport have the possibility to change completely the landscape of the transportation business. Thousands of new companies will create new services that will make our trips safer, cheaper, more environmentally friendly, productive, and comfortable. New business models and software applications are already being developed to benefits from such disruptive technologies. However, in order to rip the full benefits from these technologies, society will need to solve a crucial problem: the lack of investment in new infrastructure. The attractiveness of these technologies is indeed somewhat limited if city residents can only drive at 10 km/h or need to wait for three passing-by metros to be able to commute. As we will see in the next section, there is a growing gap between the infrastructure funding and financing needs and Governments or private sector capacity. We will propose solutions that, with courage, could solve problems not only in the rich world but also in poor countries of Asia and Africa.

Acknowledgments and disclaimer The use of pictures, or references made to studies or companies and their brand does not in any way suggest that the Authors of such studies or the mentioned Companies endorse in any way this book or its content. The author endeavors in respecting the copyright subsisting any of the graphics, and texts that he uses, to use graphics and texts he has himself created or to use graphics and texts not covered by copyright. All trademarks and brand names quoted in the book including those protected by copyright of third parties are subject unreservedly to the provisions of current copyright law and the rights of ownership of the registered copyright holders.

Companies and Brands Stated in the Chapter

- Facebook Inc.
- Twitter Inc.
- Google Inc.
- Nest Labs
- Skype is division of Microsoft Corporation
- Microsoft Corporation
- Nokia
- .NET and C# are Trademarks of Microsoft
- Java is a trademark of Oracle Corporation
- Qualcomm
- LG Corporation
- Panasonic Corporation
- Haier Group
- General Electric Company
- IBM/International Business Machines Corporation
- Cisco Systems
- Broadcom Corporation
- Dell Inc.
- Samsung
- Nederlandse Spoorwegen/NS
- Maximo is a trademark of IBM
- Apple Inc.
- Ford Motor Company
- General Motors Company
- Daimler AG
- Honda Motor Co., Ltd.
- Hyundai Group
- Bayerische Motoren Werke AG/BMW
- Toyota Motor Corporation
- Volkswagen group
- Renault S.A.
- Robert Bosch GmbH
- Vodafone Group plc
- Audi AG
- Telefónica S.A.
- Tesla Motors Inc.
- Ferrari S.p.A.
- Mercedes-Benz is a division of Daimler AG
- Volvo Group/AB Volvo
- CarPlay is a trademark of Apple Inc.

Chapter 6
New Transportation Business Models

Acronyms

BRICS	Brazil, Russia, India, China, South Africa
CAPEX	Capital expenditure
IEA	International Energy Agency
OPEX	Operational expenses
PPC	Pay-per-click model
PPHPD	Passengers per hour and per direction
PPP	Private public partnership
ROI	Return on investment
SPE	Special purpose enterprise
SWF	Sovereign wealth fund
TCO	Total cost of ownership
V2V	Vehicle to vehicle

According to the World Economic Forum, the yearly global spending on infrastructure is around \$2.7 trillion while the needs are around \$3.7 trillion. This gap of almost 30 % is likely to increase as population of Africa and India continues growing quickly, putting even stronger pressure on transport infrastructure. In the past, most of this money came from Governments. Transportation was seen as a public service, which wasn't supposed to be subject to market forces.

Today, few countries can afford funding such expensive projects and fewer are doing it. China and Middle East Countries such as the GCC and Saudi Arabia are probably the current exceptions. Governments from the developed countries that could still afford improving their transport means by using the current very low borrowing rates prefer to use their budget to fund recurring costs rather than invest in new electric grids, water treatment plants or transportation projects. If borrowing rates are low and Countries like the US didn't even invest in times of high unemployment in labor intensive projects, what will happen to their crumbling infrastructure? Will they wait for their bridges to fall apart before they take the necessary measures? We've stated

S. Van Themsche, *The Advent of Unmanned Electric Vehicles*,
DOI 10.1007/978-3-319-20666-0_6

the USA, but the problem isn't limited to this Country. Developing countries like Brazil, Argentina or India have systematically under-invested in infrastructure. In poor countries the problem is even worst. There, Governments just can't afford to build and maintain basic infrastructure, such as power station, roads or railway lines.

The issue is that the global financial resources, which would be needed to be allocated to infrastructure projects, are so huge that Governments won't be able to solve the problem all by themselves. For instance, $16.6 trillion will need to be invested worldwide in roads only, between 2013 and 2030 to meet the growing population needs, according to the McKinsey Global institute. Furthermore, $4.5 trillion will have to be allocated to railway projects during this same period. If Governments won't or can't fund this level of investment, how and who will finance a large portion of these $21 trillion? The obvious answer is the private sector will. However, this isn't a good answer. If the investment opportunities were attractive enough, the current situation wouldn't show such a gap.

Furthermore, the problem humanity is facing is even more drastic when we take into consideration the investments needed in green solutions to avoid global warming. According to the IEA, $500 billion in yearly green energy investments are required. Public transport, which could reduce significantly carbon emission and pollution levels, could be part of the solution. By investing in railway networks, society would kill two birds with one stone, reducing CO_2 emission while improving the life of billion of citizens. New business models that the e-mobility revolution can bring such as car sharing, car pooling, or convoying on dedicated lanes have the possibility to solve this problem by reducing the need for more infrastructures. This chapter will describe the likely financial players and new business models that can help bridge that gap.

6.1 Project Finance

Understanding how to solve this yearly $1 trillion gap means knowing:

- Which financial players could be attracted to such opportunity;
- What are the legal conditions that make such project attractive;
- What are the risks that the private sector won't accept;
- What are the additional subsidies or tax exemptions that would increase the projects' attractiveness; and
- Finding new financial solutions that could help boost the investors' confidence in such projects.

We will firstly explain the different project finance concepts to which these privately financed investments usually answer to.

6.1.1 Corporate Versus Project Finance

The simplest way for a project to happen is to find a corporation that will accept to finance a project from its own balance sheet. For small investments, this is usually possible. However, Finance 101 will tell us that money could be better used for other activities related to the core business of the corporation.

For instance, a high tech company could make better use of the share-holders' money than to buy its head-office building. Besides the notion of return on investment (ROI) and immobilization, many companies' stock value is also assessed on the ratio of revenues versus equity. In other words, many financial analysts will sanction companies that have too many assets. This is why few companies like tying their money to big investments and will only do so if they don't have the choice. Railway manufacturers that invest in concession usually will do it because this is the only way for them to sell their own rolling stock technology. If they can't avoid it, these companies will want to be sure that their participation in the project and the money that they will earmark for it, as well as the maximum liability they will be exposed to, will be limited to the amount of their participation in the project. This is typically when project finance kicks in.

Project finance is a method of financing in which the lender looks primarily to the revenues generated by a single project, both as the source of repayment and as security for the exposure. The definition given by the Basel Committee on Banking Supervision, International Convergence of Capital Measurement and Capital Standards ("Basel II") is:

Project finance may take the form of financing the construction of a new capital installation, or refinancing an existing installation, with or without improvements. In such transactions, the lender is usually paid solely or almost exclusively out of the money generated by the contracts for the facility's output, such as the electricity sold by a power plant. The borrower is usually a special purpose entity (SPE) that is not permitted to perform any function other than developing, owning, and operating the installation. The consequence is that repayment depends primarily on the project's cash flow and on the collateral value of the project's assets.

By limiting their total financial exposure to a defined level, companies can accept risks that could otherwise bankrupt the corporation and would make the investment unattractive.

To illustrate a structured Project Finance, let's say a railway manufacturer invests $10 million in a $200 million project to secure a rolling stock sale of $100 million. The other investment partners would have to foot in $40 million, if we consider the remaining $150 million to be financed by bank loans.

Thus, the manufacturer would own 20 % of the equity. We could suppose a net profit of 10 % on the equipment and a linear $2 million yearly profit on the results of the railway operation over 20 years after debt repayment. If the SPE would pass the entire $40 million profit as dividend, the overall profits for the manufacturer would be $18 million ($10 million for equipment and 20 % of $40 million) over that 20-year period. Through financial calculation such as net present value (current value of money after discounting the cost of money), the manufacturer would need to estimate if the ROI would be sufficiently attractive. The issue is when profit isn't there or when manufacturers need to add more equity. In our example, the manufacturer could have to invest in theory up to $200 million if all the other partners went bankrupt just to finalize the project and could have to absorb operational losses for 20 years. In other words, with Project Finance, the manufacturer could hope for $18 million in profit with its maximum $10 million exposure on equity. With Corporate Finance, the profit level would be the same but with a maximum exposure of $200 million, with operating risks over 20 years. Furthermore, the manufacturer would most likely have to book on his balance sheet the $200 million assets and the loan which would deteriorate its financial status (under special conditions, the manufacturer could avoid to integrate the SPE assets in its balance sheet).

Thus one of the first conditions for the private sector to become an active player in the infrastructure financing market is to have laws enabling Project Finance on a final recourse basis, where the lender or the Government cannot go after the corporation. Having such laws isn't sufficient though. The market needs also financial institutions that accept to play by these rules. For instance in Brazil, BNDES one of the few players which mission is to finance project at reasonable interest rates, won't accept nonfinal recourse project financing. Projects still happen because other players will take the risk that the banks or the Government won't. Thus understanding who the likely players are and what makes them accept projects is key for creating a vibrant private market.

6.1.2 Private Financial Players

There are several players involved in transportation projects with different roles:

- Lenders: usually private or public banks
- Institutional investors
- Corporate investors
- Manufacturers
- Civil companies.

We will first analyze the potential roles of financial institutions and try understanding if there is an appetite for infrastructure projects.

Potential bank role In the past, infrastructure financing was dominated by European banks (i.e., Paribas, Deutschbank, UBS). Due to the subprime crisis, many of these banks have become more conservative and are focusing on improving their balance sheet. Furthermore, under the "Basel 3" banking rules, banks have less incentive to get involved in long-term financing. As typical big infrastructure project financing duration is 15–25 years, the financial resources for Project Finance have all but dried up.

Banks as we've just said are penalized to offer long-term loans but have also a much lower appetite for risk. In the past and on projects with reasonable predictivity and visibility, banks would finance up to 90 % of the project needs. This would be the case for instance on highway project finance in rich Countries. Today for these same projects, banks will only offer up to 70 % of the financing needs. The rest must come from private owners or from other institutions, which cost of borrowing is usually higher.

If there is less appetite from banks to finance projects, who could replace banks and bridge this 20 % difference in level of financing?

Long-term investors With so much money required, infrastructure is now attracting long-term investors such as insurers, pension funds, and sovereign wealth funds. It is estimated that these institutions have around $50 trillion to invest and that not even 1 % of their investments are in the infrastructure sector. Were they to increase it to 10 %, a huge portion of the gap would be filled. This level should be reasonably achievable as there is a clear correlation between their long-term financing obligations and the long-term profitability of such project financed investments. Furthermore, once the risks from the construction phase are over, the revenue inflow usually offers high predictivity and is usually protected from inflation and economical crisis.

A bigger role from these institutions would bring three major advantages:

- More long-term capital allocated to transport would not by itself lead to much more projects, but the increased access to capital would lower capital cost to the sector and hence make more projects bankable.
- These long-term investors are interested in buying back the assets of projects already developed, freeing the capital of private players that don't wish to immobilize for the length of the concession, such as manufacturing and civil companies.

– Having onboard such players also offers a protection against Government interference, as it is more difficult for a Government to penalize let's say a pension fund than a private equity firm.

Private equity firms Corporate buyout remains the core business of private equity companies such as Blackstone, Apollo, KKR & Co., and Carlyle Group LP. However, as competition is becoming fiercer, many private equity firms have started analyzing possibilities to invest in infrastructure and some have already started doing so.

For instance Blackstone, one of the most well known private investor has invested $125 million in a $900 million hydroelectric project in Uganda, ensuring a 2/3 stake in the project. Besides power projects like the dam in Uganda, Blackstone has used billions of dollars of its investors' money to buy ships and gotten into ocean transport. Blackstone is said to pursue similar dam projects elsewhere in Uganda, in Tanzania and along Rwanda's border with the Democratic Republic of the Congo.

Unfortunately, for the transportation area most of these infrastructure projects are in the energy business. The ROI of transportation projects would need to be higher (typically around 10 %) to attract such players.

Sovereign Wealth Fund (SWF) Total wealth of the world's SWFs amounts to more than $5 trillion, of which 60 % belong to the world's oil and gas exporters. A few of these SWFs, like Singapore's Temasek and Abu Dhabi's, have already started investing in infrastructure project. However, the great majority are only allowed to invest in listed equities, bonds, and real estate. For instance, this is the case of Norway's SWF, the world's largest SWF.

However, there is a move for Norway to allow their SWF to invest directly in projects, though only renewable energy ones. If this SWF was given the green light to expand its portfolio to include direct investments in renewable energy assets and ideally other green infrastructure project (i.e.: railway though this isn't even considered yet), this could potentially have a significant change on infrastructure financing. For instance, if the Fund was allowed to invest up to 5 % of its total assets into infrastructure, the Fund could on average allocate in the range of 10 $Billion USD per year to the green energy investment market from 2015 onwards. As Norway's SWF is seen as a peer leader among SWFs when it comes to governance and social responsibility, a decision to let Norway's Fund seek low-carbon direct investment opportunities, could be followed by other SWFs.

Private companies In order to find potential investors, that will take over the construction and operating risks, Governments have very often drafted private public partnership (PPP) contracts that made it interesting or compulsory for companies that would be taking over the construction and the system to be involved in.

For most of these companies blocking for a 30-year period their precious assets is something they do reluctantly. Indeed, technology and civil companies are usually financially assessed against the level of equity they use and often punished by the financial markets for using too much equity in regards to debt or asset. Furthermore, the amounts invested in the SPC's equity are usually better used in developing new products or new markets. In other words, these companies could achieve a much better return on investment allocating this money to other needs, than the 5–7 % that PPP opportunities usually advocate. Even though by doing this, they could win a portion of the construction or system scope, they still resent blocking equity for such a long period. In some case, they add extra provisions in their cost to compensate for lower ROI or rush after entry into service operation to sell their equity to long-term financial investors, which are happy to lock long-term asset for a 5–7 % steady income flow.

For this and other reasons, concessionaire companies, usually composed of long-term investor shareholders or the financial institutions' specialized divisions in concession, often prefer not to have the construction or system companies within the operating entity. By doing this, they usually can benefit from their lower rate of borrowing and avoid any additional costs that construction and system companies might decide to add within their equity.

6.1.3 Private Participation in Transport Infrastructure

Governments are always highly involved in transportation projects. This involvement ranges from direct ownership and management to elaborating a regulatory framework that defines operational standards. The trend toward greater private involvement in transportation started with the privatization in the 1980s of existing transportation firms. New relationships started to be established with financial institutions since public funding and subsidies were substantially reduced and new competitors entered the market. Then, many transportation firms were able to expand through mergers and acquisitions into new markets.

In the many countries where Governments' responsibilities and roles are well defined, the private sector will accept to get involved in all phases of transportation project, from the early stage of the project (project financing and feasibility studies) to the building phase (design, construction and project management) as well as the operation phase (operation and maintenance). Contemporary transportation infrastructure financing is facing the following challenges:

- Lack of funding
- Divergence of purpose between political and commercial interest;
- Uncertainty in the outcome on overall transport system performance; and
- Time frame misalignment between financing and operation.

6.2 Transport Infrastructure Privatizations

In many countries, government and public authorities are under a lot of pressure to improve mobility, while at the same time must cut public deficits to bring debt level in line with their reimbursement capacity. In such environments, privatization can be seen as an attractive solution. Higher efficiencies that can be brought by the private sector, resulting from lower operating and maintenance costs, are usually an important reason invoked to justify privatization. In railway and road transport, there are very few privatizations where Government assets are sold to private investors (contrary to airlines or maritime sectors). In general, Governments decide to maintain land ownership and grant a concession to the private sector.

6.2.1 Concession Agreement of Existing Transport Facilities

The first dominant form of privatization is to grant the right to operate road or railway lines for a given period of time. As road concessions are usually profitable, the concession is mainly awarded without any subsidy. In general, the winner of the road concession is the company that will require from road users the lowest pay-toll amount possible. Usually, investments such as lane duplication or road re-asphalting will also be required. In public transport, existing transportation operations are rarely profitable, except for regional and high speed concessions. In most cases, the winner of the concession is the company that will make to the Government the most attractive offer, resulting from the lowest subsidy and the lowest fare level possible.

6.2.2 New Project Concession

The second popular form of privatization is through project financing. This strategy brings new sources of financing but with a higher cost of borrowing (typically most companies will pay around 2–3 % base over the Government's borrowing cost) that will need to be compensated by better technical and managerial expertise during construction and operational phases. Several types of tendering process exist.

The one with the most demanding private sector involvement is design, finance, built, operate, transfer (DFBOT), where the private sector basically runs the show from the start of the project till a period where it basically transfers the infrastructure back to the Government, usually after a period of 25–30 years. Other forms exist with less involvement such as design, built, operate, and transfer (DBOT), built operate and transfer (BOT), or design and built (DB). Governments will decide on the model depending on the level of risk that they wish to take and the financial attractiveness of the concession to the private sector.

6.2.3 Management Contract

The last dominant model is the management contract. While infrastructure asset ownership remains with the Government, the operation and maintenance is transferred to the private sector.

In many countries, privatization is commonly favorably perceived by Governments and population. This is especially the case wherever new projects must be built and less so when existing infrastructure are leased. As in DFBOT, DBOT or BOT models, the risks are mainly bared by the private contractor, there are usually few voices raised against such models. However, this isn't the case for management contracts, were assets are taken over by a company that hasn't really made the initial investment.

Furthermore, not all privatization have been successful and fear is higher that the provided services will be downgraded. Three main reasons are usually linked to such failures:

- Government doesn't honor its part of the contract, relying on the private investors to solve all issues, such as land expropriation or obtaining the necessary licenses, delaying the construction phase;
- Private contractor are badly performing and unable to meet commitments made; and
- New cost structure is perceived to be unfair by users since privatized infrastructure must reflect more openly the real cost structure to repay the investment made. This happens often in cities where customers are used to low subsidized costs.

6.2.4 *Private Public Partnerships (PPP)*

PPPs are contractual agreements between a public agency (Federal, State, or Municipal) and a private sector entity. PPPs take place in environments where stakeholders find the project too risky or need subsidies in order to construct or operate the project's assets. Governments thus help leveraging the position of the private sector, which commonly results in a better allocation of resources than if they would have done so independently.

6.3 New Potential Financial Instruments and Incentives

Creating the right financial instruments to add liquidity to the market, while protecting investors is something that should be analyzed urgently by financial authorities. Financing needs of infrastructure are so huge that there is a real incentive for all parties to find solutions to bridge the mentioned gap. Governments could find alternative sources to their own funding. Financial markets could create new products. Long-term investors, such as pension funds, would find projects with a duration meeting their long-term maturity obligations. Manufacturers and civil companies could find more easily other private investors willing to take over their investment in an SPE after the riskier first years of construction and operation.

Creating these financial instruments might not be sufficient though. Giving also tax incentives would help. In the next section, we will see a few emerging financial instruments and will suggest a few tax models that should reduce the gap between infrastructure financial needs and means. We will start with three green financial instruments that could effectively be used extensively for public transport, as long as railways start rightfully being considered as a solution for carbon problems by Public and International Institutions.

6.3.1 *Green Bonds*

Green bond is a financial instrument that ties proceeds of bond issue to environmental activities.

It is always hard to say what is green and what isn't. A metro line will undoubtedly take out of the road thousands of pollutant cars. However, how many is always the question, as it is difficult to assess on the long term. Because it is already difficult for transportation specialists to answer such question, it is even harder for investors to know what is green and what isn't. This is why green bonds were usually associated with the World Bank's environmental department. When they were first launched in 2008, they were associated to small emission values and weren't very popular. In 2013 this changed with the emission of a $1 billion green bond by the International Finance Corporation. Later on that year EDF the French energy concessionaire emitted a $1.9 billion green bond and Toyota issued a $1.75 billion to help finance the sales of hybrid cars.

The President of the World Bank estimates that the market in 2014 could rise to $50 billion from $11 billion. At that growth pace, green bonds could represent up to 15 % of the corporate bond market by 2020.

6.3.2 Carbon Credits

In order to reduce global warming, several Governments have come up with a financial instrument called carbon credits, which relies on regional financial markets and tax incentives. But carbon credit prices have crashed to $4 levels owing to the financial crisis in Europe and lack of consensus on the future of the Kyoto Protocol beyond 2012.

Carbon credit It is basically a title bought by companies in exchange for the right to pollute. This title is traded openly on a specific market. Though several countries have tried pushing such type of markets (i.e., Australia), the main market remains European.

Carbon credit price evolution When it was first introduced in Europe in 2000, the price was set around 20 euro per ton of CO_2. In order to avoid penalizing too much their own industry, European Governments provided too many exemptions. As a consequence, the price has been falling down to a value lower than 5$ per ton, a price which makes carbon credit too cheap to be worthwhile the hassle.

Impact on transportation financing Even before the carbon credit value crashed, carbon credit had a limited impact on the financing of public transport projects.

These are some of the reasons explaining why carbon credits haven't been used extensively:

- **Cost of new metro line**: The cost of a metro line can reach $100–$200 million per kilometer, which means $3–$5 billion for a normal 20 km line. One must add on top of the CAPEX, OPEX over a 30-year period, which usually doubles this amount. CO_2 savings resulting from the reduction in cars due to a new metro line will at most generate one to two million dollar in yearly revenue at a medium carbon price ($10 per ton).

This is still worth around $10 million for a seven year period but the process needs to be renewed each year, requiring lots of efforts.

- **Difficulty to assess carbon reduction**: calculating how many CO_2 tons can be reduced by people switching from private to public transport can be tricky, time consuming and expensive. Potential ridership studies along the future line must also be done, adding extra difficulties.
- **Ownership of carbon credit**: revenues from carbon credit would need to go to the SPE to be meaningful. However, this cannot be done as credit demand needs to be done prior to the constitution of the SPE. Indeed, carbon credit requirements are that they should be included in the project's financing calculation. However, being in an open bid, the potential SPE cannot use it, as theoretically the Government needs to authorize such use of carbon credit, which would then be available for each bidding SPE. This is a chicken and egg issue.
- **Rules of operational cost**: In order to receive carbon credit approval, potential users must usually follow already approved methodologies. The other option is to create a new one, which can be difficult, very expensive and uncertain in terms of approval. Once the project benefits have been calculated, it must then be registered by the United Nations and validated. However, the main methodology which could be used for financing new transport concessions is very difficult to use as it validates only projects for which the benefits coming from the carbon credit can contribute significantly to the concession's OPEX. Though the railway industry has asked successfully to relax such rules, it still is almost impossible to meet, as carbon credit prices have dropped to an all time low. Another carbon credit offsetting methodology based on mounting regenerative system onboard trains has been used successfully to overhaul train and improve energy efficiencies. But with carbon credit prices being so low and this system being generalized it is unlikely that it will be used commonly in the future.
- **Poor Country rules**: These credits have in the past been used for poor and developing countries projects. The EU has changed the rules in 2013, to rule out all developing countries, such as BRICS Countries. The remaining Countries are basically too poor to afford a metro line, thus defeating the purpose of such tool for mass transit project financing.

6.3.3 Voluntary Carbon Market

As indicated in its name, this market is composed of companies that decide for social responsibility purposes to buy carbon credits. This is quite a dynamic market, with a yearly growth in the range of 5 % in volume of carbon offsets (100 million tons in 2012), though in value the amount fell significantly due to the price of carbon credit (slightly higher than $500 million in 2012).

The "State of the Voluntary Carbon Markets" 2013 report, estimates that the total market value could reach a value of $1.6 billion (projection based on historical rates) to $2.3 billion (projection based on predicted rates) by 2020. If proven true this would mean that the market would double the current market size, offsetting around 200 million tons. However, with a price of 4 $ and not knowing if Governments would allow project developers to issue and monetize their own offsets, it is still not clear if such offsetting amounts could be earmarked towards transportation projects or be mainly used in reforestation project in poor African country, as it is today.

6.3.4 Corporate Donation and Tax Exemption

Even if all these measures would be applied, it would still not solve the problem of citizens of the megacities of poor African Countries. As seen no mass transit system can survive without getting subsidies. In poor Countries, Governments have problems financing the project and offering such subsidies. By taking away the Governments' obligation to fund the project but keeping a certain level of subsidies, such projects may happen. However, most companies will not invest as they see big infrastructure projects as too risky. In order for the private sector to invest in these countries, there needs to be a sense of doing good as well as the possibility to obtain immediate gains from such action. Why not allow corporations throughout the world to consider investment in a SPE of such countries, as a nonprofit action and allow the SPE to apply for a qualified charitable organization classification? In the eyes of tax authorities such as the IRS in the USA, charitable donations are gifts made to qualified charitable organizations. These include churches, nonprofit organizations, a foundation or trust fund or any other organization "operated exclusively for charitable, religious, educational, scientific, or literary purposes." If you define charitable as the possibility to do good to thousands of people, such a Transportation SPE should evidently qualify.

The interest of enabling these SPEs to be characterized as charitable organization is that the contributions could qualify as tax deductions against the company's annual tax liability and not as an expense or worst an investment. Monetary contributions through share acquisition in SPEs could be tax deductible up to a certain level of the company's yearly profits (e.g.: 10 %).

Some level of protection for potential smaller investors would be required to ensure that there are always serious committed active investors. Profits generated by such SPEs would be taxed in the corporation's home country as it would be for any other business investment. Corporations would be incentivized to contribute during the early construction phase as the expected gains would be the highest. Possibility to have shares buyback could make these contributions even more attractive. In other words, in the best case scenario, the corporations would be entitled to get a profit proportionate to the level of contribution in the SPE as a dividend and potentially to sell its share at a profit. In the worst case scenario it would have been able to deduct a portion of its taxes and still done a good action.

6.3.5 Certificate of Potential Increase in Construction (CEPAC)

For once, one of the most interesting financial instruments doesn't come from the Northern but from the Southern Hemisphere. The Certificate of potential increase in construction (In Portuguese CEPAC or Certificado de Potencial Adicional de Construção) is a financial instrument, whose objective is to obtain financial resources dedicated to public work. Created in Brazil in the early 2000, this financial instrument has already been able to finance several re-urbanization projects in big cities such as Rio de Janeiro and São Paulo.

Capturing speculation benefits for the collectivity It works on the principle that real estate speculation that would inherently result from the increase in attractiveness of a region due to work improvements, can be anticipated and promoted to finance public work. To illustrate how it works, let's take the example of Rio de Janeiro's Porto Maravilha.

Rio's Porto Maravilha project: In Rio de Janeiro, a 4 km elevated inner city overpass called Perimetral was inaugurated in 1960. The Perimetral ran through the heart of Rio's historical past where many beautiful historical buildings shared space with the central market. Unfortunately, to build such infrastructure, many old buildings and the central market itself had to be demolished. With the years, this part of the city became dirty, polluted and dangerous. Middle and upper class left and the region started to be deserted fueling even more a no man's land.

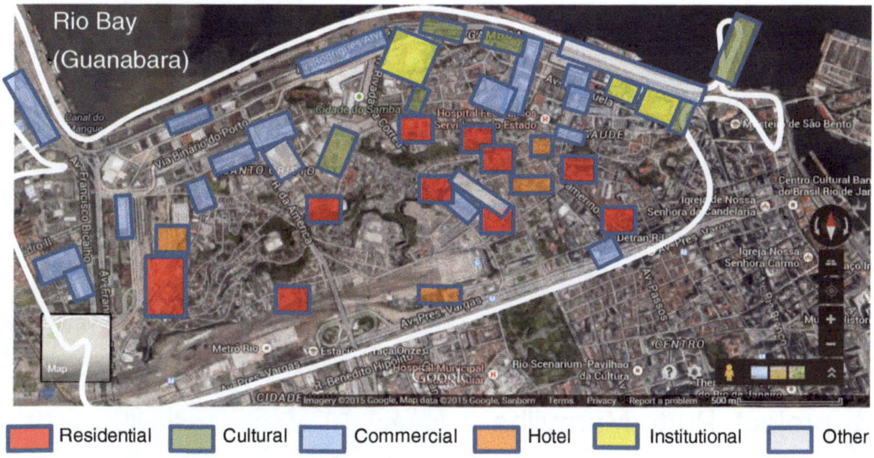

Fig. 6.1 View of the various areas allocated to different construction types in the project Porto Maravilha in Rio de Janeiro Brazil. *Source* Author

In June 2010, the city launched the project Porto Maravilha, which objective is to re-urbanize a region near the inner port of Rio (see map hereafter) (Fig. 6.1).

At the heart of this re-urbanization lies the demolishing of the 4 km overpass and its substitution by the construction of three tunnels running directly under the overpass. Other infrastructure improvements include the re-urbanization of 70 km of roads and 650,000 m^2 of walkway, as well as the reconstruction of 700 km of urban infrastructure network (water, sewage and drainage, telecom) and 3 water treatment plants. The project also includes the implementation of 52 km of a free catenary operation tramway. All these transportation infrastructure work should allow for an increase of 50 % in traffic flow (Fig. 6.2).

To improve the quality of life of this neighborhood, the project plans to implement 17 km of bicycle lanes and to plant 15,000 trees (the green area should raise from 2.5 to 11 %). The city's objective is to revitalize the region not only by creating offices but my motivating Rio residents to move back to the area. The city objective is to make the region attractive enough and supply the necessary apartments to increase the residential density from the 22,000 inhabitants today, to 100,000 within a 10-year period.

All these works are estimated at R$7.6 billion (around US$ 2.5 billion) and will be financed by CEPACs and executed within a PPP between the company responsible for the RIO port development (CDURP) and a private consortium (Concessionária Porto Novo S/A constituted of the civil companies OAS Engenharia, Norberto Odebrecht and Carioca Christiani-Nielsen Eng.).

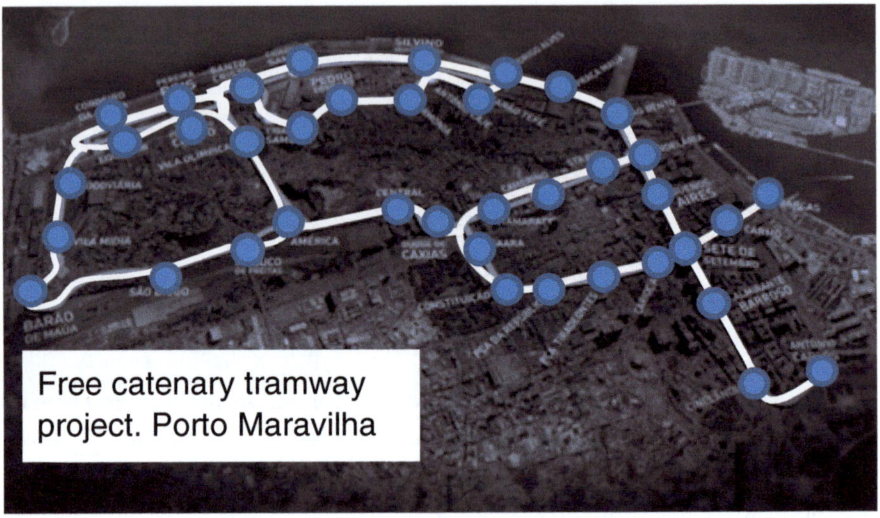

Fig. 6.2 Illustration of the free catenary tramway line of 52 km to be funded entirely by the CEPAC financing scheme. *Source* Author

Legal framework In order to create such successful financial instruments, Brazil needed to vote special laws that could create a legal and financial framework.

Based on articles 182 and 183 of the Brazilian Federal constitution, the Federal Government voted in 2001 a special law (10,257), which defined the municipal status. This law allowed Cities to promote and vote special laws enabling the financing of re-urbanization projects for a specific city area (called a OUC/Operações Urbanas Consorciadas).

For instance, the city of Rio de Janeiro, passed a law in 2009 (i.e.: complementary Law 101/2009 for the OUC Porto Maravilha), authorizing the construction above the existing city restriction on land occupation. This law defined the perimeter of the region where the works would happen.

Limit on surface to be built It also imposes a ceiling for new surface to be built. For instance, in Porto Maravilha the limit was set at 5 million m^2. The project also defines how these m^2 should be distributed according to office, commercial, and lodging necessities. Areas already protected such as natural preservation and cultural patrimony buildings or areas, as well as public service offices cannot be touched by construction.

What is a CEPAC? It's a financial instrument emitted by a municipality (i.e.: Rio, São Paulo) via a SPE (i.e.: CDURP in Rio Porto Maravilha). This CEPAC is used as a counterpart for the right to increase the construction level in a dedicated area of the city (i.e.: 5 million m^2 in Porto Maravilha). Each CEPAC is equivalent to a value of m^2 for construction area or for a modification in the use of land. A table is established for each project giving the CEPAC m^2 cost according to the type of use or region within the defined urban area. This price is updated on a monthly basis, according to a construction index taking into consideration inflation. The CEPAC is specific to a project and cannot be used to finance infrastructure improvements in other city areas. The CEPAC doesn't have any time limit. It lasts as long as the law voted for this specific project is still active or until all CEPACs are sold.
Purchase of CEPAC: Investors interested in getting access to real estate space buy CEPAC, giving the right to increase construction levels from the municipalities of the said region. CEPAC are ruled by the instruction 401 of the Comissão de Valores Mobiliários (CVM) financial market. The CVM market will oversee the project making sure that the CEPAC investors are protected throughout the execution of the project. CEPAC can be bought from two different sources: In auction organized by the main Brazilian financial market called BOVESPA; and on secondary markets through financial brokers. They can be bought and resold to other Parties freely on these secondary markets, until they are specifically associated to a dedicated lot in the region.
Control of CEPAC: A SPE (i.e.: CDURP for Porto Maravilha) must be established by the municipality to manage and supervise the works to be performed. The SPE must participate in the approval of the real estate projects, during technical meetings with the municipality's secretary of Urbanism. It is also the entity's role to make land available for the market. The money received through the sale of CEPAC is separated from the municipality's budget, which cannot use it for any other reasons than the said project.

Two financial institutions are in charge of monitoring the entire process. As mentioned, the Brazilian market authority CVM will control the allocation of the CEPAC to ensure that investors are protected. The Federal financial institution Caixa Economica will oversee that the resources generated by the CEPAC sale are indeed allocated to the specific project purpose and will follow the different construction interventions throughout the project.

Benefits for the Municipalities Most cities experience budgetary restraints, which make it difficult for them to front money for a project or earmark a significant amount of their budget to a re-urbanization project. Another interesting financial aspect of CEPAC is that they don't generate any rights of credits against the municipality. In other words they are treated off balance sheet with no impact on the city's indebtedness level. With money secured, contracts are easier and cheaper to negotiate with potential civil construction companies that can execute the works, without the risk of seeing the money disappear with changes in the Mayor's agenda priorities or changes in the city's political team.

Benefits for the investors CEPACs allow investors who believe in the real estate potential of the region to buy the rights of increasing the urbanization level, within a strong legal framework. It creates a real estate market in a place where usually no home or office owner would dare investing, either by fear of the re-urbanization project not going ahead or because the minimum transport infrastructure wouldn't be there to attract residents and workers.

6.4 Reducing Infrastructure Construction Needs

One way of bridging the gap between infrastructure investment needs and the money available is obviously to reduce the need for more investment in transport infrastructure. We've seen in the previous chapters a few of these solutions.

6.4.1 Encourage Working from Home

The easiest way of reducing the need for new infrastructure is to motivate people to work from home and enabling companies to legally do so. With internet, there is no technological limitation to working from home.

Some practical restrictions may exist related usually to worker's position. Indeed, some jobs require interaction with clients, suppliers or even other colleagues and thus may involved from time to time physical meetings. But this is more the exception than the rule and even more so in global companies where people cannot easily meet. However, in most cases, corporate culture is at stake. Top management in most companies believes that coming to work enhances employees' productivity. Bosses also feel more confident when they can oversee what their subordinates are doing. Thus, many employees believe that their yearly appraisal is positively influenced by their long hours spent at work, very often regardless of if real work was performed or not. Additionally, culturally many employees accustomed to working with colleagues or under monitoring feel uneasiness whenever working from home. This feeling is what many freelancers call "free floating anxiety".

6.4.2 Encourage People to Stay in Their Neighborhood

By tying real estate development to job creation and available commercial facilities, city planners can reduce the need to commute or travel.

6.4.3 Higher Occupancy Rate

Increasing the number of passengers per existing transportation mean, by using for instance car pooling is a good way of achieving such result quickly. With 1.1 passengers per car at peak hours in the main American and European cities, doubling or triplication of the occupancy rate is an instant win for the infrastructure, the environment (any additional passenger will only consume between 2 and 5 % extra energy) and the passengers' wallet, if they were to share the cost. In public transport, increasing the number of passenger per m^2 or adding more spacious trains or buses is also a good way of increasing nominal capacity without changing the infrastructure and the cities environmental footprint.

6.4.4 Higher Transportation Means Density

Increasing the density of existing transportation networks is an interesting way of reducing the needs for more road or railway infrastructures. It can be achieved by adding new signaling system (i.e., moving block) allowing for lower headways. Road platooning technology allied to V2V technology should also reduce spacing between vehicles to a minimum and increase safely the road capacity.

6.4.5 Car and Parking Space Reduction

New alternative car ownership and rental models are being created, such as peer-to-peer car sharing, or rental by the hour. These models will have a durable impact on infrastructure needs as most cars sit idle on the street or in parking lots 95 % of the time. Through new Apps, the e-mobility revolution will optimize their usage, reducing significantly the needs for car parking space.

6.5 Electrifying Transport

Global warming is one of the biggest threats the future generations will face. Many governments have decided in theory to reduce their CO_2 footprint, but results so far have been disappointing. In fact, CO_2 emission continues to rise inexorably and an increase of a minimum of 2 °C is now accepted by most scientists. Limiting this increase to only 2 °C will already be a challenge and few countries have taken real steps to reduce their CO_2 emission.

Probably one of the most active Countries in this area is the UK. The British Government has made some difficult political choices, such as restarting the construction of nuclear power plants. Though some green activists in Germany would see that measure as completely inappropriate, environmentalists in the UK have embraced such policy, as they consider that nuclear is the lesser of the two evils. They also recognize that a responsible energy strategy cannot rely in the short term only on renewables and that the UK needs a source of stable energy, till there is plenty enough of other renewable sources. We don't want to start a debate on the benefits of nuclear versus its negative impact because this isn't the purpose of this book. However, what is important with these measures is that the UK Government puts its money where its mouth is. By supplying stable energy with zero CO_2 emission, it ensures that its important decision to transform the current transport industry based mainly on fossil fuel to an industry driven by electric technology has teeth. In its report "Making the connection: The plug-in Vehicle Infrastructure strategy" the Office for Low Emission Vehicles sets the strategy and the steps forward for the Government and all industrial players to provide the necessary recharging infrastructure alongside improvements in vehicle range. It also provides favorable incentive for the switch to electric automobile, such as reduction in vehicle excise duty, company car tax exemption as well as enhanced capital allowances. Planned investments in new train or metro networks such as Crossrail should also help reduce significantly UK's transportation CO_2 footprint. As we've seen in Chap. 3, the collateral benefit of all this is that it will also bring huge financial benefits for UK's trade balance.

6.6 Changing Public Transport Business Model

Building a new train network or metro line from scratch is hugely expensive, especially at a time when most Governments have problems meeting their debt commitment. Unfortunately, such infrastructure project cannot be done without public support and money. As we've just seen, PPPs have been introduced as a ways to share risks between the private and public transport to finance or fund CAPEX. However, even mast transit operations are likely to run a deficit and must rely on government subsidies. During lean time, there are basically two ways of reducing the reliance on government money: reduce the operational costs and enable higher revenues.

Here also the PPPs have been used to lower the amounts flowing from the governments to the transport authorities. Some governments have outsourced the entire operation and others have decided to maintain it, while subcontracting the maintenance usually to the technological provider. However, what should the basis for such subcontracting be? Based only on the lowest acquisition price? The point is that what seems the cheapest isn't always what will cost the less to the asset owner. This is why some governments have introduced the concept of Total Cost of Ownership, which basically takes into consideration, the cost of an asset over its entire life. This is an important change to the basic tendering model of many countries, tying the technological provider to the long-term performance of its asset. Backed by hefty penalties and sometimes bonuses, it ensures that the cost estimated over a 30-year period will be met. It also passes the risk from the buyer to the manufacturer, enabling the Governments or investors to know before committing to the asset acquisition, their long-term base costs.

6.6.1 Total Cost of Ownership (TCO)

The TCO of an infrastructure or metro installation is the total expenses incurred since its acquisition by the operator, including acquisition, financial, operation, maintenance, and obsolescence costs (Table 6.1).

Typical TCO distribution of a metro vehicle is shown in table above. This is obviously benchmark information, as prices can vary extremely depending on the technology used, the solvability of the asset owner and even the acquisition model (i.e., some contract are based on rental hours where the customer is charged by the km traveled). Roughly, 30 % of total life cycle cost (LCC) can be assigned to the initial purchase price, while customers typically spend about 70 % during the rest of the life cycle, mainly in operation and railway maintenance activities. The financing costs may on average represent around 15 % of the TCO. This number can obviously be higher in developing or poor countries where the cost of financing is high.

An important element that doesn't show that table is that investments that are carried out at the start of the life's asset reduce subsequent upkeep costs, even though it may seem as more costly initially. For instance, low initial

Table 6.1 TCO breakdown of a metro

TCO phase (%)	Initial purchase price (%)	Operation (%)	Failure cost (%)	Financing (%)	Maintenance (%)	Spare parts (%)
100	30	20	5	15	10	20

Source Author

Table 6.2 Elasticity per transportation type in Europe

Transportation type	Price peak	Price off-peak
Vehicle travel—essential trips	−0.16	−0.43
Vehicle travel—optional trips	−0.43	−0.36
Bus, metro, tram passenger per km	−0.19	−0.29
Rail passenger per km	−0.37	−0.43

Mayeres 2000 study

investment (such as heavy standardized carbon steel trains) will end up increasing various operational costs, such as maintenance, energy, warring of parts, etc. On the other hand, investing in new moving block signaling technology, which is still more expensive from an acquisition cost perspective, will offer good payback, as it will reduce staff cost over 30 years (as well as strike risk) and maintenance costs, as many way-side equipment are eliminated.

Knowing your costs and maintaining them to the bare minimum is a good step to reduce reliance on Governments, but doesn't allow public transport to survive without subsidies. Thus, one must work on increasing revenues. Increasing fare prices would seem an easy solution. However, there are three main issues with such strategy. First, public transport is generally used by the poorest citizens. Increasing price too heftily will penalize those that public transport is actually supposed to help. Second, mass transit users represent valuable votes. No mayor in his right mind will want to confront million of potential electors. Lastly but not the least, by increasing prices too much, people will vote with their feet and will start buying cars, defeating the purpose of promoting public transport. The following table compares elasticity for different transportation means. It shows that an increase of 10 % in price, reduces the number of km traveled in buses by 1.9–2.9 % (Table 6.2).

One other way of increasing revenue is to enhance vehicle and network capacity. Putting more people without changing much the network's characteristics will increase revenues with little effect on the cost structure. But here also, if the comfort level declines too much, people will opt to go back to their car.

We suggest two out-of-the-box ideas, enabled by e-mobility technologies that could bring extra revenues: cross-subsidizing of fast lane with localized congestion charges and geo-localized ads.

6.7 Private to Public Transport Cross-Subsidizing

Life is often not fair and the Government's role should be to make political choices that are the best for the greater whole of their population, in particular giving people equal opportunities. Such choices never satisfy everyone but as long as they please the vast majority, politicians are happy to make such decision, as they increase their chances of being reelected. This is why for instance, politicians like to invest in new railway networks and highways, as ribbon cutting ceremonies can usually translate into additional favorable vote intention. Unfortunately, in a democracy such events are becoming extremely rare as many interest groups will have different views of what is best for society. Furthermore, in society prone to litigation, these groups have the means to delay or kill projects, even if they could bring great benefits to the majority.

Advocating that Governments should introduce initiatives to cross-subsidize public transport with private transport money is likely to be seen a priori by the car industry lobby as a no go. These big interest groups have deep enough pockets to kill such initiatives easily. They would probably argue that car owners pay enough taxes when buying their cars, register their automobile or refuel. Without entering into a debate, one could probably answer, that the car industry is one of the most subsidized industry when considering that road work and their maintenance is paid directly by income and property taxes, regardless of the fact that the tax payer might not even be a car user. For instance, a 2005 EU commission study[1] estimates that every year local Governments in Germany spend 15 billion euro on their car traffic. Only between 15 and 45 % of this aide is actually recuperated through parking fees, taxes, etc. In other words, each German citizen pays around 100–150 euro for the car industry to be compared to around 84 euro subsidies for public transport. The report states that these subsidies are in line with other European countries. The other issue is that the appropriation of road is taken for granted but, as we've seen, is a lost opportunity cost, as it could be used for housing or green areas. Furthermore, the automotive industry doesn't foot the bill for the million of car deaths and injuries resulting from car crashes nor do they pay for health costs resulting from pollutant emission.

The fact is that if the automotive industry wishes to continue selling high end cars in the big cities of this world, they will need to promote initiatives that will allow their clients to drive faster than walking during peak hours. This can only happen, if a sufficient number of citizens ride on the high capacity transportation means. Thus initiatives that could provide money to fund Public Transport and consequently reduce the number of cars within the inner city roads should be welcomed. As a matter of fact, our first suggestion should be a no brainer for the automotive lobby but a politically not correct proposal for left wing Parties.

[1] Hidden subsidies for urban car transport; Public funds for Private Transport ICLEI (2005 study).

6.7.1 *Fast Lane Cross-Subsidizing*

Large monorail project can carry around 40,000 passengers per hour and per direction (PPHPD). This is almost comparable to the large mass transit system that could go up to 60,000 PPHPD. High-capacity BRT and tramway projects can generally carry around 20,000 PPHPD. Without rejecting BRTs that can be an effective, quick to implement and cheap solution or tramways that can rehabilitate entire inner city region and cut pollutant emission, both technologies require "expropriating" housing and or "appropriating" land. Metros and monorails on the other hand, have a very low environmental footprint and usually will improve the car flow in the neighborhood of these new networks. Our idea is based on a concept similar to CEPAC, which enables remuneration of housing expropriation based on the future increased real estate value of a neighborhood, due to infrastructure improvement. Why not bundle with a private concession, the future benefits of road appropriation. In other words, circulation improvement to a large avenue resulting from a new mass transit system could be used to cross-subsidize the infrastructure works and mass transit operations.

1800 and 3600 PPHPD for one and two exclusive lanes respectively, could generate big money. Rich people or even taxis would most likely pay twice the price of a bus or metro ticket to avoid spending hours in traffic jams. The same e-mobility technologies that exist today on the highways of this word to pay tolls without stopping, would allow people to get in and out quickly from these exclusive lanes. At 5 $ per access, an exclusive two lane express road could maybe generate between $50 and $100 million per year in revenues with not much cost. Furthermore, people not willing to pay would have fewer options not to hop on the network, increasing the concession's ridership. Both revenue impacts would reduce significantly the need for Government subsidies.

The problem with this approach is obviously that left wing Parties might claim rightfully that this creates two types of transport: for those able to afford private roads and the others. However, it is already the case as the Rich stay in their car while the Poor use public transport. Furthermore, one could argue that it gives an attractive transportation choice to poor people in an area where they either would have to use a car that they couldn't afford or nonsegregated bus lanes.

6.7.2 *Localized Congestion Charges*

For those who systematically don't want to create cross-subsidies by mainly improving "ostentatiously" the life of the Rich, there is the possibility of using congestion charges within the new line's neighborhood to subsidize the mass transit operation.

As we've seen several cities, such as London or Milano have put in place these charges to discourage residents from using cars within certain area. There is no technological limitation in installing the same video technology that reads your driver's plate and checks if you have paid your charges, in smaller area alongside the network. Such congestion charges that can go up to 11.50 pounds ($18.70) in an expensive city such as London city could easily be implemented in a restricted area benefiting from better traffic for a fee of $10.00 per day with weekly discounts to make it attractive enough. This would penalize local residents but technology would probably allow for modular charges, according to residency criteria. It should however, be noted that congestion charging, road pricing and tolls have an impact on road usage. For instance, a 2010 study[2] concluded that elasticity to road charging is between −0.1 and −0.45.

Another way of generating extra revenues for public transport is to turn time wasted in transport into productive hours. In fact, the average one hour lost by commuters in the rich world represents around 20 % of their free time. If only a fraction of such time could be transformed on online searches to buy products or services with smart phones and this money be channeled through the public operator website, millions of dollars could be generated each day.

6.8 Internet Connectivity

Be it at home, on the metro, at work on one's computer, walking on the street with a smart phone, with Google glasses or with an Apple or Samsung watch on the wrist, as well as driving in a car with a Google or Apple onboard platform, anyone can now be connected 24/7 anywhere. Inversely, it means that the big IT companies have unlimited access to customers' information as well. Until recently, these companies wouldn't be in a position or willing to consolidate consumer across multiple platforms. However, this isn't the case anymore and they can follow consumer behaviors across multiple platforms and physical locations. We will analyze which are the potential IT companies that are likely to play an important role in the public and private transport e-connectivity market. In order to do so, we need to understand what the powerful business models and trends behind this push are.

[2] 2010 study done by Spears, Boarnet and Handy.

6.8.1 Pay-Per-Click Model (PPC)

Companies like Google or Facebook, get their revenue mainly from advertising. The PPC model, basically directs traffic to company websites. These companies pay the publisher a fee for all redirected connections. Websites using PPC ads display these advertisements when a keyword search matches an advertiser's keyword list, or when a website shows relevant content.

Google AdWords: It is the most popular PPC advertising system. It enables businesses to create ads that appear on Google's search engine. In AdWords advertisers bid on keywords and pay for each click on their ad. Every time a query is initiated, Google digs into the list of AdWords advertisers and chooses companies which products will appear in the ad space of Google's search result page. Selection is made according to a combination of factors, including quality and relevance of their keywords, the advertiser's rank and the highest amount an advertiser is willing to spend.
Cost-per-click: The average Adwords PPC cost was around 0.90 $ per click in 2013 (Facebook average PPC was $0.80). As Google' advertising revenues were around $32 billion, we can easily understand the importance of increasing the overall number of PPC opportunities. If these PPC were achieved mainly by people at their desk in the past, more and more they are now generated through mobile devices.

6.8.2 Time Spent Online

Around the world, 2 billion people are connected. On average they spent online each month around 16 h (2013 values). The monthly connection time of 153 million Google visitors was around 1 h 45 (7 h 45 spent by the 137 million Facebook members).

Among the favorite internet activities relevant to this chapter, as seen recently in a survey, search engines topped the lists (over 90 % favored that activity). Information search to buy a product and the real buying act were also activities well appreciated with an overall assessment in the 70 %. Furthermore, the highest growing trend seen online was location based services being sought out.

6.8.3 *Mobile Shopping*

According to a McKinsey report, 50 % of consumers spent 75 % or more of their total shopping time conducting online research. In the past it was mainly linked to electronics but consumers are more and more searching or buying directly furniture, health and beauty products, clothes, and even food. Furthermore, the presence of mobiles in this shopping experience is growing. The last years have brought major shifts on how users employ their mobiles to shop. Almost half of their time is now spent on non voice activities.

For instance, European smart phone users are spending more than one fifth of their time doing online research on the go. 40 % of them actually mentioned doing research in stores, using various research tools, such as professional review, coupons, price comparison tools, stores website, taking picture, scanning bar codes, queries, social network.[3]

6.8.4 *Mobile Activities*

Why is all this online activities relevant to transport? The fact is that people spend valuable time in transport and that time is worth a lot of money, if passengers can do something useful out of it. On average American spend roughly 600 h in their car per year. Over their average lifespan that equals out to be roughly 5 years of their life or roughly 7 %. The following tables give a view of the time structure of European men and women aged between 20 and 74 of an average day. This table is based on a seven-week schedule and includes men and women not working. Thus, when all this is factored in, European men and women spend at least 5 % of their time traveling (Table 6.3).

Now, if we consider that people spend one-third of their time sleeping and more than another third working or doing domestic work, sharing meals or taking personal care, we can conclude that the traveling experience if connected correctly, could generate an extra 20 % opportunity for people to search or buy online. This could generate significant extra revenues for the public transportation players if only a portion of these revenues could be captured by them.

[3]Mckinsey IConsumer, Europe 2011.

Table 6.3 Time spent in Europe on main activities according to gender (*Source* 2004 EU report)[a]

Activities	EE		UK		FR	
	Women	Men	Women	Men	Women	Men
Free time	4.36	5.28	5.05	5.30	4.08	4.46
Meals personal care	2.08	2.15	2.16	2.04	3.02	3.01
Sleep	8.35	8.32	8.27	8.18	8.55	8.45
Travel	1.06	1.17	1.25	1.30	0.54	1.03
Domestic works	5.02	2.48	4.15	2.18	4.30	2.22
Work or study	2.33	3.40	2.33	4.18	2.31	4.03
Total	24.00	24.00	24.00	24.00	24.00	24.00

[a]How Europeans spend their time; Everyday life of women and men (2004 edition) source EU Commission

6.8.5 *Generation Gap*

A massive generation gap is fueling the migration of advertising revenue from old to new media and telecom revenues from fixed landlines to mobile. The generation that is under 35 isn't buying fixed technology but going directly to tablets or smart phones. Furthermore, they are up to 50 % more likely to use social media and VOIP/video chat than older generations. For youths, social media has displaced the use of email. For them, social media has become information and entertainment portals. In the long term, this massive shift in how consumers are informed and communicate with the word cannot be ignored by all industries. This includes the private and public transport players that will need to adapt their offer to these changes.

For instance,[4] reports that around 50 % of people aged 16–34 identify internet as the most useful media compared to only around 15 % for population aged 55 and over. On the contrary, only 10 % of this population find print most useful versus 32 % for the over 65.

6.8.6 *Mobile Advertising Market Estimation*

The total online advertising revenues according to price Waterhouse was $42.8 billion in 2013 for the USA only (IAB internet revenues 2013 full year

[4]Mckinsey Media and Entertainment practice.

report) of which $7.1 billion was mobile (an increase of 110 % over 2012). Experts estimate that global online ad spending will most likely overtake newspaper expenditures in 2014 (USD $100 billion vs. $93 billion), and will equalize the combined total revenues of newspapers and magazines in 2015 ($132 billion).

At such a growth rate, the traveling experience be it on the bus, metro, train, or in the cars will be worth $25 billion before 2020. Furthermore, this rate will continue to increase under the phenomenal growth in car ownership, the increasing time spent in private and public transport and the demographic tendencies, where younger generation or poorer citizens opt to go entirely wireless. Where can these numbers stop? The mobile search market is experiencing huge growth (i.e., 200 % in 2012). Some industry experts predict that within a few years mobile will become the primary way people access Google's services. If we consider that cars will double by 2030 worldwide, it isn't too optimistic to think that advertisings revenue generated during transport can reach $50 billion.

6.9 Transport Information Supplier of Choice

To put things into perspective, this $50 billion value is as big as the entire yearly revenues generated by the total railway industry. If we want to understand better this advertising market, we need to divide this between public and private transports. The following 2004 table gives a view of weekly European traveling patterns and shows that though women used as much private and public transport, men spent twice the time in their car (almost ten times more in Germany) (Table 6.4).

We will come back to these numbers and how they impact transport. But before, we would like to discuss about the passenger information revolution in public transport, as the two issues will become more and more interlinked. This is also an important subject because most passengers want to know when they will arrive to their final destination. They will also want to experience the best quality ride during the lowest journey time possible and to achieve this will need to know:

Table 6.4 European traveling patterns (*Source* EU 2004)[a]

Transport mode	EE		UK		DE	
	Women	Men	Women	Men	Women	Men
Car, motorcycle	0.13	0.30	0.49	0.57	0.48	1.00
Public transport	0.17	0.14	0.12	0.12	0.12	0.12
On foot, bicycle	0.28	0.25	0.18	0.16	0.19	0.16
Unspecified mode	0.03	0.04	0.05	0.06	0.01	0.01
Total	1.01	1.13	1.25	1.30	1.17	1.26

[a]How Europeans spend their time; Everyday life of women and men (2004 edition) source EU Commission

- When the next train or bus will arrive. Ideally, they will try to be there at the stop immediately before it arrives at the station, to reduce their lost time or avoid spending time in the cold or under the rain;
- Where to embark and get out, to get the most direct access from one activity to the other (i.e., home to work);
- Any potential road or network disruption due to unplanned events (i.e.: accidents or union strikes);
- How crowded is the road or bus, to postpone departure or select an alternative transport solution (i.e., wait for the next bus, select another train compartment or use a different highway).

Most railway or road operators will invest to provide this information via information displays, in the stations, trains, and buses as well as on the road. However, passengers with smart phones or navigational systems will be able to get on the go such information. Along with this information, these connected passengers will want to be as productive as possible and most likely want to find potential services along their way to simplify their life or decide to search for items to buy. By doing this, they will grow exponentially this mobile internet advertising market.

6.9.1 Public Transport Information System

Public transport in many countries is known for late arrivals and departures. The main sector associated with such late arrivals is obviously the airplane industry. Huge arrival and departure panels are there to remind us of such frequent delays. The railway industry has also been prone to such delays. Except maybe in a few countries like Switzerland and Japan, late arrivals have been more the rule than the exception. However, things have improved lately, with the entry into services of new signaling systems. Time information displays have now moved from the intercity stations to the commuter, metro, tramway and even bus lines. Two technologies have allowed for these information displays:

- **GPS-based information**: typically buses can indicate their position and based on traffic information, systems can calculate the time of arrival and update such information on real-time basis; and
- **Signaling based information**: all modern signaling systems have real-time schedule based information (i.e., train location, time to go, direction, etc.) which can easily be sent to wayside information panels (Figs. 6.3 and 6.4).

In onboard environments, dynamic maps based on signaling or GPS information can also indicate the direction and the next stops.

Fig. 6.3 Information system on bus line in Luebek Germany, GPS based. *Source* Author

Fig. 6.4 Information system in Berlin's U-Bahn (Muckernbrucke metro station) based on signaling system. *Source* Author

6.9.2 *Battle for Passenger Information Access*

In North America, public transport is mostly managed by Public Authorities with a nonprofit orientation. Cities want their local operator to stay just that way: local. A few private operators have been successful in winning bus, metro or commuter line operations, but these contracts are more cost than revenue based. Mostly underfunded, American public operators have been all too happy to give big IT providers access to their critical schedule information. By transferring this responsibility to content providers, they've been able to provide such valuable service without the need to invest in passenger information technology. These same big IT providers have tried to replicate such advantageous business model in Europe, but with much less success. There, big railway and metro operators have denied such access, as they felt this was part of their core business. Holding a strong grip on their clients was perceived as fundamental to avoid being commoditized and maintaining a strong brand.

Why is such train and bus schedule, maps and traffic disruption so important? Well, as mentioned, people want to improve as much as possible their quality of ride. They will connect to get such information and then might decide to use their mobile phone to do other things such as web searches. As most players in this industry will recognize it, one out of three searches is currently localized base and as we've seen this trend is likely to increase significantly. Thus it is fundamental to motivate potential users to connect and be associated with this first connection. Furthermore, as information at the point of purchase is increasingly important, getting information on nearby shops or services is likely to generate extra revenues with higher margins.

Application using mobile maps to which are added data specific to users' interests will help public transport passengers find locations, click to call, get the address and show how to get there: by walking, driving, or public transport.

6.9.3 Social Public Transport Apps

Moovit is proposing a social network Apps, based on crowdsourcing, but centered on public transport users. Present in 11 countries and 65 cities the application gives real-time traffic information on buses and metros. With more than 2 million users it has become a must for all public transport passengers. More than 3.5 million researches are done each day and more than 10 million interaction between users. This application gives real-time information on the next displacement. It tells the passenger if any disruption happened on the bus line and can provide her or him with an alternative transport solution. It will recalculate the time necessary to reach your final destination with this alternative trip. This platform relies on direct sharing of information between passengers. Users may indicate information, such as delays on the line, metro or train dysfunction, etc., or even send alarms to other users in case of labor strikes, accidents, works, etc. All this data is integrated and fused with official data coming from the operator information system. The aim is to give a better traveling experience (Fig. 6.5).

Another interesting public transport application called Tranquilien™ was created by the French start-up Snips. This application indicates in which car set to hop on, based on the number of passengers per car. This App uses

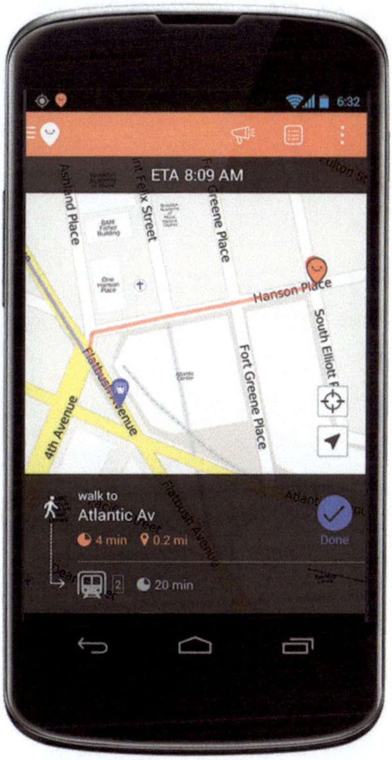

Fig. 6.5 Picture of Moovit App. *Source* [CC BY-SA 3.0 (http://creativecommons.org/licenses/by-sa/3.0)], via Wikimedia Commons

geo-localized data on smart phones to calculate the density of people in a given place. The start-up doesn't record the personnel data but analyzes it and transmit this analysis to other companies for use by application companies.

By being the primary source of information, these Apps can motivate passengers to use the site to look for advertisement but also to do online searches. To avoid being a hostage to the attractiveness of the big online search engines or social apps, the big mass transit operators can hope to develop location based application. There are basically two types of technology for geo-localized ads: GPS and beacon based.

6.9.4 GPS Location-Based Information

This technology is based on smart phones' GPS capabilities. Location based information works on the principle that, knowing where the consumer is physically located, it can deliver an ad for a product/service that can be found nearby. It can actually even send an ad for a shop that is further along, competing with a nearby

Fig. 6.6 Picture of a geo-localized shop in Berlin. *Source* Author

shop. Ads or messages will be triggered whenever a mobile user crosses a predefined area on a map, also called geo-fence. This geo-fence can vary in size according to the advertisers' goal, from a few meters to a few kilometers. It is important to note that not every service will be advertising based (Fig. 6.6).

6.9.5 Beacon Location Based Information

Beacons are equipment located in a certain area, which broadcast messages that are detected by a receiver, usually a smart phone.

The major transmitting signal is based on Bluetooth (Low Energy) technology. Bluetooth based beacon broadcasts very simple information: a Unique identifier and two information (major and minor), enabling mobiles to calculate their position in regards to these two information. Unlike GPS locations that are forever fixed and public worldwide, a beacon is something that can be placed anywhere and has its location changed at anytime. However, without the Unique Identifier, the smart phones will never know the beacon is there. Furthermore, a beacon's reception is not precise and the distance between this broadcasting device and the receiver will be greatly affected by interferences (150 m maximum with best coverage around 30 m).

With the broadcasted information, Apps can detect it and tell how close the phone is from it and perform actions, such as display alerts to the user, offer discounts, or trigger automated events such as turning lights on or opening doors.

6.9.6 Legal Use of Geo-tagging Space

Geo-tagging, which is at the core of geo-localization ads, is based on the physical position—latitude, longitude and even in some cases height for GPS based technology—and relative position in regards to beacon—of the user. To whom belongs the physical location of a geo-tagged position? To the service provider of the ad or to the owner of such position?

> This is much more than a discussion on the sexes of angels. Would it be acceptable for a company to see its customers receive ads in its own building pushed by competitors? Will homeowners of an apartment over a sex shop need to put parental guidance on their child's phone? In fact this is a multimillion dollar question, if people and companies need to protect their geo-tagged position like they do for their web addresses, unless laws and courts decide who geo-tagged space belongs to.

For the public transport industry, this is a key issue. Million of passengers use every day, the metro and commuter systems of the big cities. Well-positioned geo-tagged ads could bring lots of extra revenue every day. Will the lost making transport operators be able to tap into this lucrative market or just see their billboard and multimedia revenues shrink as they are replaced by these ads? Could a sharing of revenues between content providers and landowners be possible on the basis of a rental contract? Advocates of freedom of the internet and of the right of land ownership are likely to clash. Furthermore, can one of the lesser important IT player of the online advertising market team up with the public transport operators for GPS based geo-fencing? The risk for the public operators isn't only to see their advertising revenue shrink. It is also the risk of seeing their brand being commoditized, especially under the threat of specialized social networks such as Movit or other networks providing real-time information on transport conditions.

6.9.7 Cross-Subsidies of Mobile Advertising Revenues

How much online revenue could be generated in a public transport environment? If we take an important metro such as London, there are 970 million passenger-journeys per year or 2.7 million passenger-journeys per day. If only 10 % were to do online searches and click a PPC twice, this would still represent around $200 million per year in extra revenues. Though we haven't found specific research indicating what percentage of the total $132 billion (2015 estimated value) mobile ad revenue could be generated in public transport, it is still likely to be an important value, not even considering geo-tagged advertisement revenue, a technology still in its infancy. If Governments were to decide to pass a law making the right of property more important than intellectual property rights for such technology, it would in the long run generate a complete new source of income. Such law would be probably challenged by big online search groups but at least for beacon based ads there would be a good legal foundation for public operators.

6.9.8 Road Transport Information System

On many autoroutes or important traffic lanes, infrastructure providers are able to transmit in real time any road disruption. Based on traffic flow models, they can also calculate and indicate to drivers, the average time a trip is likely to take between two locations. However, such source of information isn't as important as for public transport, as most cars have now onboard GPS. In the next section, we will look at the disruptions generated by the new e-mobility Apps that will change the private transport picture.

6.10 E-mobility Software Applications

In transportation, social networks have been playing an increasingly important role. Be it through advertising revenues or through impact on the younger generations that are tomorrow's transportation consumers, social networks can both represent a threat or an opportunity. Depending on which side of the barrier the companies are positioned—first access provider of critical traffic information or remote alternative—companies and operators will be able to create loyal followers or face the risk of being commoditized. Social network dedicated specifically to private transport have been recently extremely successful. So successful in fact that they have been acquired by the big IT players or are likely to be so very shortly.

6.10.1 Social Car Apps

The Tel Aviv start up Waze, which had almost 70 million users in 2013 throughout 110 countries, has finally been bought by Google for a price of almost $1.3 billion after being pursued as a potential acquisition by IT groups such as Microsoft, Apple and Facebook. This success story which started in 2006, around an open source application to localize Israeli road radars had launched in 2012 an advertising platform to monetize its services. This start-up had edited since 2008 a free GPS App, which instead of relying on expensive map suppliers, was based on social members' information. Members would be invited to share their trips and signal any disruption on the road (i.e., accident, road work, radars and policemen presence).

Furthermore, members were also invited to improve existing maps. Through this acquisition, Google was able to get access to an advertising platform proposing to businesses to pay to be shown on the maps during the nearby trips.

6.10.2 *Car Pooling Apps*

As seen, one of the best way of fighting city congestion and pollution is to favor car pooling. Transportation agencies especially in the USA have developed infrastructure to support car pooling with high-occupancy vehicle lanes, discounted tolls for registered carpools, and support for areawide ridesharing programs. New Apps now exist to improve the car pooling experience. For instance, Internet service providers such as Carpool World, proposes free ridesharing services worldwide. After registering contact information with privacy controls, the users indicate a trip origin and destination, a weekly schedule or dates for such trips, and then can search for potential car ridesharing partners using interactive map displays.

6.10.3 *Car Sharing Initiatives and Apps*

Ten years ago, no one knew whether car sharing could work. Yet today car sharing initiatives have popped up all over the world. In many cities, car sharing has made a strong impact on how city residents travel. Even in strong car ownership cultures such as North America, car sharing has gained momentum. In the USA and Canada, membership now exceeds one in five adults in many urban areas. Strong influence on congestion and pollution can be felt from Toronto to New York, as each shared vehicle replaces around 10 personal cars. City authorities with downtown areas struggling with congestion and lack of parking slots are driving the growing popularity of car sharing, encouraging people to give up their cars and use alternative transportation.

Car-sharing-rental Apps: A few innovative companies have appeared throughout the world, using such Apps. Zipcar, the biggest car sharing company in terms of membership (760,000 members), was one of them, recently acquired by the Avis budget group for $500 million. This company follows the faith of several other innovative start-ups snapped up by big car rental companies. The Avis acquisition has shown synergies. Zipcar members have gained access to Avis' massive fleet during weekends, when car-sharing companies are usually short of cars and renting companies short on customers.

Peer-to-peer car-sharing Apps: it allows city residents to rent out their personal car to people who need transport, usually through Apps. To participate in a car-sharing service, people must become members. Some companies charge an annual fee for membership while others don't. Users receive a card that unlocks and locks the rented vehicles, which are conveniently located to allow easy access. Reservations for the vehicles are made online, and vehicles can be rented by the hour or by the day.

6.10.4 Mobile Taxi Hailing Apps

In the last years, taxi hailing companies had a go at the business of normal taxi operators. In France for instance, 2014 street protest of taxi drivers who must pay a hefty license to operate but in exchange benefit from a protected market, have blocked and paralyzed for a few days the main French cities.

6.11 Change of Car Ownership Business Model

For many baby boomers and the following generations of men and women, owning a car was and still is perceived as a social status. Owning a fancy expensive car was a sign that the driver had succeeded in life. We will look at the main factors changing such perception (Fig. 6.7).

Fig. 6.7 The new generations prefer using public transport and owning tablets or smart phones rather than dreaming of owning a car. *Source* Author

6.11.1 Car's Negative Image

Young adults who would be in age to drive, or get their driver's licenses, and by extension buy their first car, are increasingly deciding that it's not a high priority. Reasons explaining such attitudes toward cars are:

- **Rise of mass transit**: It is today cool among young people to use mass transit. In the past only poor people would dare taking public transport to university or work, when they could afford not to;
- **The parent taxi**: Teens parental transportation is another reason why many young people hold off on getting their license or buying a car. Furthermore, from a young person's perspective, it makes good sense. Why go to the expense of driving your own car when your parents will give you a lift?
- **Alternative transport means**: Young people are increasingly combining mass transit with walking and cycling, instead of using cars to get around;
- **Peer-to-peer transport**: Young people can use their mobile to hail a ride from a friend, participate in car pooling program or just rent a car by the minute;
- **High cost of driving**: besides high acquisition and insuring cost, young people must also pay high gasoline prices, tolls and in big cities congestion charges;
- **Boredom of driving**: In the past, people could drive at high speed, drink and drive and race their friends. Though this attitude killed thousands of young people, it was still part of this image of excitement that cars carried on. Today, with all the radars, new laws and law enforcement officers ready to take out licenses, young people find more attractive other "extreme sports".
- **Connectivity factor**: While owning a car represented status and freedom for past generations, modern cell phones, internet and the instant connectivity they enable are the new freedom and success symbols of the "millennial" generation.

This disinterest in cars is generalized throughout developed countries. Even in countries like the USA or Japan that are among the biggest car market, this trend can be observed.

For instance a 2013 survey by the economic newspaper "Nikkei" indicated that young people in the city of Tokyo were less and less prone to own a car. Only 13 % of young Tokyo habitants own a car while they were 23 % in the early 2000. Worst for car manufacturers, only 25.3 % said they would want to buy a car against 48.2 % in a similar 2002 survey. This survey confirmed that young Japanese prefer to spend their money on flat screen television, digital cameras and other high tech electronic products.

6.11.2 Commoditization of Cars

Car manufacturers spend billions every year on marketing campaigns to position their products according to social trends and values that allow their brand to differentiate themselves from others.

The automotive industry is one of the industries that invest the most in advertising. For instance in the last few years, automotive spending on marketing costs represented between 8 and 9 % all spending of the $500 billion worldwide advertising market. This money is thought necessary to maintain attractive brands, but also to help selling higher value cars. As we all know, the industry members mainly make their profit on their high end products. Convincing a resident of a megacity that he needs a Ferrari to ride in gridlocked cities at the outstanding speed of 20 km per hours is all about selling a dream rather than the reality of jammed cities.

However, by 2030 most cars will be completely or partially driverless, depending on the rate of unmanned technology adoption. This means that by then, the special relationship that bonds fast car owners with their vehicle will have totally disappeared for a simple reason: car performance directly associated to acceleration and speed will have definitely vanished from a driverless environment and with it the values associated with fast or powerful cars. Furthermore, around 2020, speed will most likely be automatically monitored or controlled (at least for accidents), taking away some of the risks but also the feeling of freedom that drivers still enjoy. In other words, an entire spectrum of the imaginary space promoted by the car industry will have disappeared. This doesn't mean that no space will remain. Safety will remain a strong selling feature. Comfort and more generally the traveling experience will become increasingly important. No wonder, car manufacturers are insisting to maintain a close hand on the traveling experience, rather than agreeing to transfer such experience to the IT moguls. This is for them a matter of life and death in a five to ten year period. What is interesting to notice is that when we look more carefully to these key selling features, they look strangely like values that a bus or train operator are promoting. This is to say that more and more the difference between public and private transport will shrink.

6.11.3 Blurring Differences Between Private and Public Transport

An important question we should ask ourselves is what's the difference between public and private transport? The easy answer would be ownership and being driven by somebody else. In fact, it is a little bit more complicated. People associate public transport most of the time with electric vehicles and thus less pollution. Safety could also be a differentiator, as cars tend to be involved in many more deadly accidents, though in many countries taxis are also associated with public transport. Higher transportation capacity and speed are probably also features associated with trains. Now, as we've seen all along these pages, many of these features will change.

- *Electric cars*: As cars become electric, pollution will definitely cease to be an issue for restricting cars use within inner city boundaries.
- *Driverless cars*: As cars become controlled by software and not drivers, safety will increase to level comparable to public transport.
- *Capacity*: Capacity will continue to be a differentiator between public and private transport in megacities. However, with car pooling, platooning and lower interspacing distances between driverless vehicles, cars will become an attractive "public transport" solution for mid size cities.
- *Speed*: Speed will remain a crucial selling feature of trains for trips between 50 and 500 km. High Speed Trains will continue to increase their share of the regional and intercity travel market.
- *Ownership*: People and especially the younger generation are favorable to sharing their cars or renting by the hour (with insurance by the hour as well). Will these models become main stream or will they fade away?

6.11.4 Car Ownership Models

Why buy assets such as a car when you can rent it whenever you want one. In fact, this is probably the worst investment people do. Most cars sit idle 95 % of the time. Furthermore, brand new cars lose 25 % of their value when they exit the retailers' door.

New business models such as peer-to-peer car sharing or car rental sharing can bring huge benefits to the environment, the occasional car users and the car owners willing to share their assets. Some big car manufacturers have seen both the threat and opportunity that such new business models could offer and have launched their own competitive services. The most well known is probably Car2go.

Car2go is a subsidiary of Daimler AG providing services in European and North American cities. It offers exclusively its small electric car—Smart—for two people. Users get access via an App to wherever parked cars and are then charged by the minute. These cars have autonomy of around 120 km after full reload. Plugging the car into a charging station is only required at the end the trip, if the charge has fallen under a certain threshold.

Fig. 6.8 Example of a CAR2GO vehicle available at European airports. *Source* Author

Business model: Though rates may vary from city to city, the business model is similar in all markets. Car2go charges per minute rates (i.e.: $0.41 in a US city in 2014), with discounted fixed rates for hourly ($14.99) and daily usage ($84.99). This fee covers rental, gasoline, insurances and parking. Sometimes a low fixed annual fee is also charged. Users can park their car in designated parking places or standard parking areas with special city permits. Clients can choose refueling their car to gain free usage minutes (Fig. 6.8).
Business model attractiveness: Though in May 2014, Car2go pulled out of the UK for a lack of customers after less than 2 years in that market, it has been extremely successful elsewhere. For instance, in Milano and Roma, 110,000 Italians have decided to use their services. That same year, there was already 70,000 Car2go members in seven Germans cities. Worldwide more than 700,000 people had subscribed in 2014 to the car2go services.[5]

Change in car ownership business models won't just affect car sales it will also impact the conventional taxi market. By getting an easy and almost immediate access to cars, but also reducing the difficulty to find parking space, these peer-to-peer and rental by the minute car sharing models will shrink the taxi market. However, the biggest threat for conventional taxi services will come in the immediate future from peer-to-peer hailing companies.

[5]Automotive news Europe (29/5/14).

6.12 Change of "Public Service Car" Business Models

As we've just seen, cars are today considered private mainly because they are owned and driven by citizens for their personal interest. Taxis on the contrary are considered public first and foremost because they serve different customers. This section is dedicated to taxis and their inevitable changes of business models.

6.12.1 Conventional Taxi Business Model

Taxis riding along busy avenues or waiting in line to pickup travelers at airports or railway stations are part of the standard urban landscape. Everyone can picture the yellow cabs of New York or the black special London cars. As a matter of fact, these taxis' unique features are so well known that they have become themselves one of the main attractions of these cities.

Taxis' conventional business model has been the same for many years. Most drivers are independent contractors owning their car or renting it from an owner. The driver or the owner will pay a dispatch company a fixed monthly fee around 300 $. Taxi drivers wait for the dispatcher to tell them to pick up a customer but will also stand in queues at specific areas for potential clients to hop in, or drive along busy roads, hoping someone will hail them. In exchange for transporting people from their point of pickup to a specific address, they cover a minimum fee to which is added a mixture of cost per kilometer and time spent in traffic. In some specific destination such as airport, a flat fee is covered to both protect the drivers and the ingenuous travelers.

Depending on the country, the taxi market is regulated or free for everyone to operate in. In regulated market such as Paris, taxi drivers pay a hefty license that can cost as much as 300,000 euro. This license is purchased either from the municipal authorities or from previous taxi owners. In exchange for such expensive right to operate, the municipal authorities ensure that the quantity of licenses is maintained to a minimum to enable taxi drivers to repay the license's acquisition and ensure reasonable minimum revenue. Obviously, this maintains artificially high prices for those who could afford to buy the license in the first place. This also means that offer is always lower than demand.

In cities like Paris, owning the phone number of a taxi company was considered in the past an advantage that Directors of big corporation would get as a personal benefits. For commoners it was almost impossible to find a taxi after 22:00. In exchange for tightly regulated markets, drivers are usually expected to give a public service. This means in theory treating everyone with equal access and knowing the city sufficiently well to make sure that passengers would receive the fastest and most direct trip possible. For instance, in London drivers still are expected to know all city roads and the fastest way to get anywhere in the city. Many candidates postulating to become a taxi driver still need to pass a very competitive test, where knowing by heart most city roads is still a prerequisite.

E-mobility technologies are changing all this. GPS is obviously eliminating the need for taxi drivers to know all the city roads. Worst for taxis, new services such as real-time traffic update mean that taxi drivers themselves need to use GPS to know the most rapid way to destination. Commoditization of taxi services is now the rule in all developed countries. Furthermore, new taxi Apps are increasing this commoditization and changing durably the taxi industry's business model. However clever these Apps are, they wouldn't be so popular if it wasn't for the fact that potential clients feel frustrated about the poor service provided by conventional taxis. How many city dwellers around the world had to watch helplessly as several taxis passed by full of passengers, or found themselves leaving a bar at night without any way of getting home? How many innocent tourists have been ripped off by taxis? How many regular taxi users have suffered taxi driver strikes in regulated markets? Though many frequent travelers may say that this is more the exception than the rule, few taxi clients have the impression that this is really a public service. For most travelers, this is a normal business where taxis are small entrepreneurs who want to maximize their profits. As prices very often are decoupled from costs in these regulated markets, there is not much sympathy for taxi owners.

6.12.2 New Taxi Hailing Business Model

Peer-to-peer taxi Apps like Hailo, GetTaxi, MyTaxi, and Uber connect passengers directly to individual drivers using Apps. As both the numbers of mobiles and Apps have exploded in the last few years, it is now facing strong reaction from local conventional taxi services. General taxi strikes throughout Europe in June 2014 are a clear sign that these services are starting to change profoundly the market. In fact for those countries where they are already available, 5 % of all journeys are estimated to use these services that year. Considering that these services only started around 2011, it is easy to see that their growth has been exponential. Furthermore, these Apps are now starting to expand from their home market to the main cities of this world aiming at cities where inefficiencies are the highest: the most heavily regulated markets.

New business model: these services use smart phone tools to locate potential passengers, enter credit card details and provide a map to show the real-time location of the hailed taxi. Some Apps even send additional information such as the drivers name, the car registration plate and even the driver's phone number in case the driver or the passengers experience problems. In other words they give accountability in an industry renown for freewheeling.

The companies behind these hailing Apps need to recruit and sign taxi companies or independent taxi drivers, who accept to receive reservations. With every day that passes by and the growth in brand awareness of these Apps, this task is getting easier. Once a network is in place, the power of all taxi apps comes from their complex algorithms that track which cars currently in operation within this network can provide the quickest pick up. Furthermore, some newer algorithms are being developed to predict how many more rides will be needed on a rainy day or on any nights with special events. In big cities, Apps of areas where drivers are most likely to find their next fare are also produced to improve service and increase the likelihood to find nearby clients. In other words, these Apps are enhancing taxis' availability in areas of demand and improving the taxi offer.

Remuneration: Hailing companies take a cut on the trip fee of around 10–15 %. However, it requires a minimum fare per trip (i.e.: around 6 euro in cities like London or Paris) that all passenger will need to pay, to ensure that it is worthwhile for drivers. In some cities, services from companies like Uber's can cost 30 % less than conventional taxis. As competition intensifies, price is likely to drop even further. Competition will also ensure that such services enter less densely populated areas, benefitting other clients. In many big cities, when everything is added, solutions such as Uber are cheaper than owning a car.

Conventional companies aren't looking at the train passing by without employing themselves these technologies. Dozens of tech companies either promote taxi Apps that operate alongside existing infrastructure (i.e.: TaxiMagic or Cabulous) or make it affordable for taxi fleets to launch their own Apps for passengers to compete with the Uber and Hailo Apps (i.e.: Fleetbit, MTData, or GoFastCab). However, the point is that by doing this they are turning themselves also in internet hailing companies.

Impact of hailing service on society: Some experts predict that these services could renew interest in public transport as people usually take taxis in conjunction with other transportation means. Furthermore, in many cities and suburbs with those types of services, it is becoming increasingly easy to organize your life without a car. This doesn't mean that suddenly all families will stop owning their cars. This means that instead of having two or three cars, people will make better use of their unique car, and use these services for all unexpected events.

6.12.3 The End of Taxi Drivers

The shrinking of the conventional taxi market has started and might be perceived by taxi drivers as a calamity. However, "they haven't seen nothing yet." The reality is that even if drivers start improving their image and their service level, they just can't compete on price. This is already the case when competing against online hailing services, which can be 30 % cheaper. But this price difference is nothing compared to a driverless cab business operational cost. Three different studies have calculated that conventional taxi's operational cost in big North American or European cities would be about 8 times more expensive than driverless cab. How long can taxi drivers expect to still be in business? By 2030, a significant portion of new purchased cars will be driverless and most cabs will be unmanned. In fact the faith of taxi drivers could be played sooner. One of the specificities of taxi operation is that they usually don't run at high speed. When we look at it more closely, average taxi speed in big cities such as London, New York or Paris is around 15–20 km/h and in most area speed is limited to 50 km or less.

The Google cars introduced in June 2014 are designed to drive at a maximum speed of 40–50 km per hour. At that speed they would avoid the more rigorous regulatory requirements that come with higher speed. This is the perfect speed for unmanned cabs. This probably explains why Google Venture has poured in August 2013, $257 million in Uber, though relationship between both companies has soured since then. This strained relationship could be partly the result of Uber's 2015 announcing that it would invest in driverless car technologies. It comes as no surprise that the June 2014 taxi driver strikes were aimed mainly at Uber, even in some European cities where this company wasn't even present.

The real potential of Uber for Google is to get into the taxi multibillion dollar market avoiding the likely damage to its brand that would surely result from antagonizing taxi drivers. With this investment, Google could be in a position to propose unmanned taxis reserved via smartphone—that as we've seen are so cheap that they could make car ownership obsolete within inner cities. For a company that's investing in driverless technology, acquiring the technology to hail them makes a lot of sense. It also fits well with this IT's core strategy of developing tomorrow's algorithms, which will become crucial elements of day-to-day transportation.

Acknowledgments and Disclaimer The use of pictures or references made to studies or companies and their brand does not in any way suggest that the authors of such studies or the mentioned companies endorse in any way this book or its content. The author endeavors in respecting the copyright subsisting any of the graphics, and texts that he uses, to use graphics and texts he has himself created or to use graphics and texts not covered by copyright. All trademarks

Companies and Brands Stated in the Chapter

- Mckinsey & Company
- BNDES (Banco Nacional de Desenvolvimento Economico e Social)
- Paribas (Banque de Paris et des Pays-Bas S.A)
- Deutschbank
- UBS Group AG
- Blackstone
- Apollo Global Management, LLC
- KKR & Co L.P.
- Carlyle Group L.P.
- International Finance Corporation (IFC)
- World Bank
- Electricite de France S.A. (EDF)
- Toyota Motor Corporation
- Concessionária Porto Novo S/A
- OAS Engenharia S.A.
- Norberto Odebrecht
- Carioca Christiani-Nielsen Engenharia
- Google Inc.
- Apple Inc.
- Samsung
- Facebook, Inc.
- Google AdWords is a trademark of Google Inc.
- Moovit
- Tranquilien™ is the trademark of Snips
- Bluetooth (Low Energy)
- Waze is an App developed by Waze Mobile acquired by Google Inc.
- Microsoft Corporation
- Carpool World
- Zipcar is a subsidiary of Avis budget Group
- Ferrari S.p.A.
- Car2go is a subsidiary of Daimler AG
- Smart is a division of Daimler AG
- Hailo
- GetTaxi
- MyTaxi
- Uber Inc.

Chapter 7
E-Mobility Likely Winners and Losers

Acronyms

4G/LTE	4th Generation long term evolution
CAPEX	Capital expenditure
DSRC	Dedicated short-range communications
ERTMS	European rail traffic management system
EDA	Event driven architecture
M2M	Machine to machine
OPEX	Operational expenses
PRT	Personal rapid transit
SOA	Service-oriented architecture
TCP-IP	Transmission control protocol/Internet protocol
V2I	Vehicle to infrastructure
V2V	Vehicle to vehicle
VANET	Vehicular ad hoc network

7.1 Who Can Ignore the E-Mobility Revolution?

When we first started writing about the megatrends shaping the future of transport, we asked friends from very different backgrounds what were the main transportation issues they thought should be addressed. Two points of interests were invoked very often. When will car fly and will Elon Musk's Hyperloop initiative be implemented in the near future? Though both answers would probably require a dedicated chapter, we will try to give a simple answer on the probability that these two technologies will one day be popularized throughout the world. Though it might seem that answering such questions is far from our e-mobility main topic, we will use these two examples to illustrate how and why e-mobility implementation and fate will be different.

S. Van Themsche, *The Advent of Unmanned Electric Vehicles*,
DOI 10.1007/978-3-319-20666-0_7

7.1.1 Flying Cars

Why can't cars fly? After all, planes and helicopters have been around for more than 100 years with good safety record. Building a solid airframe is surely not the barrier to adoption of such technology, as the many flying leisure objects prove it. Obviously costs of owning, parking, and operating such equipment are much higher than a car but nothing crazy to impede their market introduction. After all a Cessna plane at around $150,000, is surely more expensive than a popular sedan, but cheaper than a Ferrari. Furthermore, many companies have claimed over the years that they were capable to design and build such technology.

The company PAL-V Europe NV is one of them. It recently designed a flying car with a short takeoff range of 165 m that could fly up to 1200 km at a speed of 180 km/h on roads and in the air. This two seat hybrid car and gyroplane would be equipped with three wheels, giving it agility on the road and would cost around $400,000. This is obviously not for the middle class but mass production could do wonders in reducing these costs.

If neither cost nor technology is the main barrier to adoption, could standard and personal requirements be? Not really. First, these cars could fly lower than commercial planes and thus wouldn't be controlled by visual flight rules traffic. Hence, drivers could takeoff without flight plans. Some might argue that piloting would be one of the main barriers to adoption. Though piloting in itself isn't cheap, anyone able to buy a 400,000 $ car could easily afford the pilot's $10,000 license costs. But earning a license is hard and requires years of time and study and could put off many potential consumers. True, but technology would most likely soon solve this problem, with the introduction of autopilot software. In fact, autopilot software for unmanned drones can already be bought for as little as $500. Moreover, flying outside of reach of pedestrians, dogs, or even trees is simpler than driving on the road. With the advent of new software, the piloting barrier should be overcome.

In reality, in the very near future only one barrier will be left but it will remain huge: safety. In a society adverse to risk and prone to litigation, any accidents that would most likely kill the passengers and bring fatalities also to the pedestrian or homeowners on whom the flying car would fall on, is hard to be adopted. Though unlikely to happen in a near future, if only the legal barriers could be overcome, flying cars might not become as common a sight as unmanned cars, but they could be more common than a Ferrari.

7.1.2 Trains and Hyperloops

Crazy transport ideas have been around for years. We remember meeting a few years ago serious scientists trying to find a technology provider wishing to develop underwater trains running in tubes along the American coast line. Though we might smile at such an idea, it is not crazier than imagining flying trains. Good science fiction is based after all on sciences. The fact is that true breakthrough technology does happen but takes time to mature. For instance, the original idea behind the German Transrapid technology started in 1922 and culminated with a 1934 patent for a levitation train. Though not a flying train, maglev technology actually allows trains to run contactless. However, it was only 80 years later that the Transrapid train system was opened, connecting Shanghai to Pudong airport, achieving 431 km/h operational speed.

Once breakthrough technology is proven, things usually get easier for their implementation but not always. It is true that levitated trains have found application in trains at lower speed (i.e., Korean technology running at 100 km/h). A more audacious system composed of pods attached to monorail tracks using magnets to reduce friction (SkyTran Company) could also potentially be implemented in Tel Aviv. Magnets used to reduce friction have even found new market applications. Thyssenkrupp, one of the Transrapid founding group, announced in 2014 a faster walkway based on levitated technology, which is used to solve the difficulty of stepping onto a rapidly moving surface without stumbling or slowing down the passenger flow. People might have smiled 10 years ago if any scientists would have mentioned flying carpets…

The reality is that it is extremely difficult to make money out of such breakthrough technologies. Germany, the birth place of Maglev trains, hasn't even been able to finance such technology on its own territory. On the other hand, no one in 2014 would doubt that the Japanese Government, 50 years after the opening anniversary of the world's first high speed train between Tokyo and Osaka, could build by 2027 a 500 km/h levitation technology between Tokyo and Nagoya (on test tracks it broke the 603 km/h barrier in April 2015). The point is that crazy ideas are just visions before an industrial or a country decides to back financially the project. Nowadays almost anything can be built. The problem isn't technology related. It comes from the difficulty to deliver such complex projects on schedule and within budget, with the legal authorization to operate.

Will the fate of the hyperloop concept be different from the maglev and be introduced quickly and successfully on the market? The hyperloop has some things going for it. First, the idea doesn't come from just anybody, but from Mr. Musk, one of the most successful entrepreneurs of these last 20 years. Second, the idea was already there. Mr. Musk proposes to revive an old science-fiction idea from the 1910s called the "vacuum train," with a few important tweaks. Instead of having a real vacuum tube with no air as originally envisioned, it is planned to remove sufficient air from the tubes to keep air drag to a minimum, dealing with it in other ways, such as pushing it to the sides to create a cushion, reducing friction between pod and tunnel walls. Third, and though the system would be completely new, it would rely on propulsion and other known technology (linear induction motor and booster motors). Finally, at a speed of 1200 km/h, each vehicle carrying 28 people would cover the route between San Francisco and Los Angeles in little over half an hour, making such journey attractive enough for passengers, and competitive in relation to airplanes.

With all this going for it, would we put our own money to invest in such project? Not really. The main issue as always is that the provisionary budgetary figure of around $6 billion will never be met and most likely end up costing twice or three time this amount. Thus, will this crazy technology ever be implemented? Once again the problem is that someone or some Government's entity needs to build a reasonable pilot project to give credibility to the technology. Mr. Musk has mentioned the possibility to do just that and build an 8 km prototype in Texas. However, and until an industrialist such as Mr. Musk or a Government commits resources, the hyperloop concept will stay just that way: a concept.

7.1.3 Can the E-Mobility Not Happen?

Unlike Maglev when it was first introduced or the hyperloop that is still an idea, e-mobility revolves around proven technologies that have leapfrogged from other industries (i.e., telecom and IT). There is no need for fundamental research breakthroughs for e-mobility's implementation. It is true that the improvement of two critical pieces of technology would allow earlier introduction of these e-mobility technologies.

- Batteries would benefit greatly from a breakthrough, but progress in performance has been steady and should be sufficient to make electrical cars attractive to the mass market in the very near future.
- Complex algorithms enabling the perception of the environment are still not capable of seeing and understanding a vehicle's surroundings as accurately as a human being can. Among the issues that IT providers must solve is the inferential thinking that drivers have, because they can combine stored memories and sensory input to interpret events and anticipate likely scenarios.

One big advantage of unmanned cars is that, unlike for public transport, there is no money issue involved in bringing all these e-mobility technologies to market. In fact, it is quite the opposite. The e-mobility revolution will bring tremendous revenue opportunity for those companies that will be able to anticipate the market trends and provide product and service offering that new technologies, regulations, and fulfillment of unattended needs will generate. With so much money involved (by 2020–2025 the market size for the unmanned technologies alone will be around $200 billion), there is no doubt in our mind that the automotive industry will go the same way as the railway industry: electric and unmanned. No big IT corporation, industrial group, car, or railway manufacturer, and infrastructure company can ignore such a fundamental shift in urban mobility. There will definitely be winners but also market losers that will see revenue, relevancy, or brand awareness shrink or rise as a result of these new offerings.

7.2 E-Mobility Technology Adoption

The railway, bus, and automotive markets are at different stages of their e-mobility development. We believe there will be many similarities between their evolutions. Cars and buses will become electric relying on more powerful batteries. At the same time, tramways will be more and more often seen in free-catenary operation, as the railway industry can benefit from the battery mass production generated by electrical cars. All transport means will at one point in time become unmanned.

7.2.1 Railway Industry

The railway industry has already started its e-mobility revolution. IP technologies have been introduced not only on the wayside but also on the onboard environment at the beginning of this century. Machine to machine (M2M) technology is now becoming pervasive throughout the railway maintenance sites and yards. Most railway operators and manufacturers have realized that by associating data mining capabilities with algorithms able to associate separate events meaningfully, failures can be detected before they happen. This new discipline called predictive maintenance will be reducing significantly the total cost of ownership.

Other signaling technologies based mainly on software are replacing the old electromechanical solutions. New technologies such as moving block, though not based on IP technologies, are enhancing the safety and enabling shorter headways. By reducing time required between two trains to around 60 s, mass transit capacity can be increased significantly. System compatibility is also being pushed at the European level through signaling software standardization, enabling compatibility between the various regional levels. ERTMS technology level 3 is indeed enabling the integration of intercity railway network.

7.2.2 Automotive and Bus Industries

In this book, we have rarely focused on the bus industry. The fact is that on the e-mobility revolution, automotive and bus industries are following the same path, almost at the same pace. It is true that the bus industry has for some time promoted noncarbon fuel technologies. As a matter of fact trolleybuses have been around for decades relying on electric energy like tramways. However, performance improvement to electrical buses will leapfrog from better and cheaper battery generated by mass production in the automotive industry.

Unmanned technology is also most likely to come from the automotive market. This automotive industry is only starting its e-mobility revolution and, in our view, will follow the same system approach as the railway industry, adopting similar safety concepts, as well as being based on a mix of TCP-IP protocols and non-IP technologies. Many barriers to e-mobility technology adoption exist but are being lifted or will be.

7.2.3 Lack of Industry Standard

For many years, car manufacturers have decided to go their own way, pushing their proprietary solutions as a way to differentiate their brands. However, a system approach requires adoption of standards throughout the industry to enable interoperability. Finally, after many years of fruitless discussions, the automotive industry has adopted its future product platform around the IP technologies. The first cars to hit the roads with onboard IP network should do so in 2015–2016. Such an open architecture should enable software developers to develop Apps and new service offerings across the industry.

7.2.4 Lack of Infrastructure

Unfortunately, not everything is perfect even when e-mobility technology is already available. For instance, though many manufacturers are investing massively in electric cars, many early adopters of such technology still complain about the lack of autonomy and even, more importantly, about the limited recharging infrastructure. This is a chicken and egg issue. Without recharging infrastructure, potential customers are discouraged from buying the technology. Without enough electric cars, potential investors don't want to equip roads or parking with recharging infrastructure. Governments and municipalities must at the early stage intervene and supply the minimum infrastructure to foster this new industry. Faster recharging technologies should also help, as cars will be able to spend less time at the recharging station. New charging plugs and high capacity DC recharging equipment are helping reduce charging time from a few hours to a few minutes. In fact, some manufacturers are working on charging technology which would reduce the necessary time to only five minutes. At 5 minutes per recharge, a time comparable to gasoline refueling, the market dynamics change and allow gasoline stations to see a benefit in providing electrical plug-in.

7.2.5 Unmanned Car Technology Adoption Barriers

Self-driving technologies is still in its infancy, but prototypes such as the Google car have proven that they can work. What are mainly still missing for truly unmanned vehicles are the sensor-based technologies at a reasonable price, connected-vehicle communications, and the fusion of these two technologies.

7.2.6 Sensor Technology

Advanced sensor technology like the ones that equipped Google cars—lidar, radar, and stereo cameras—are expensive and create a real barrier to adoption of unmanned vehicles. At a price estimated at around $150,000, the self-driving technology of a Google car is still far from reaching an acceptable price for its introduction in ordinary cars. It will require from car manufacturers and their OEMs hard work and benefits from economy of scale. Specialists forecast that mass production of these sensors should allow prices to fall to around 10,000 $ by 2025 and to less than $5000 by 2030. At that price, adoption on each and every car shall be systematic and by then probably legally mandatory. Time isn't playing against the auto part manufacturers. Onboard software, and car and wayside telecom infrastructure won't happen out of the blue and swamp the market overnight. On the contrary, car manufacturers will bring to market a continuum of new technologies, each one more autonomous as time goes by, till vehicles become completely driverless. Some of these solutions already exist, such as parking assistance, adaptive cruise control, lane—keeping and warning systems. Many others are in the manufacturers' pipeline but will take time to be developed, tested, and implemented.

7.2.7 V2V and V2I Connectivity Solution

Real-time wireless communication between vehicles (V2V) and between vehicles and infrastructure (V2I) will complement sensor-based technology. Such safety-critical communication must be reliable, immediate, secure, network and device agnostic, as well as operate in a dedicated and free or commercial spectrum. Similar to trains with radio technology, communicating cars will need to blend multiple networking technologies, including dedicated short-range communication (DSRC), 4G/LTE cellular wireless broadband (or by 2020 5G technology), mesh networking, precise mapping, and accurate geo-positioning. All these technologies already described in this book form a system that goes in the name of VANET. Some of these technologies already exist and can easily be implemented while others are still at the prototype stage. However, European and American

Governments have realized the potential of these technologies and are planning laws and incentivizing measures that should push by 2017 the spectrum of autonomous car technology we've described thoroughly in this book.

7.2.8 Putting All Pieces Together

To work, unmanned cars will require that many pieces of a large puzzle fit together. How quickly these pieces will be put together remains an open question because it depends on how car manufacturers, transport authorities, law makers, and consumers will work together to speed or inhibit progress. In our view, self-driving technology will follow a very different adoption process to Maglev and hyperloop's technology. In fact, such adoption process is more similar to the flying car technology. Many factors will influence the adoption pace but unlike for flying cars, they are much easier to solve. Let us now review how these factors that we've regrouped under three major categories—consumer acceptance, achieving critical mass, and legal framework—will be solved for self-driving technology to thrive.

Consumer acceptance One of the biggest barriers to adoption is the trust in technology. We shouldn't forget that experienced drivers will fear putting their life into the hands of onboard computers. Obviously, passengers won't have the same fear as if their car would be flying but they will still need to trust that received signals come from the right car, that the cars' position is reported accurately, and that no one is hacking the system. Trust will come with time, as the number of accidents and fatalities will fall drastically. Furthermore, two megatrends we've identified will facilitate this acceptance. The growing number of senior citizens will adopt driverless technology massively. The younger generation that doesn't cherish cars as much as the older generations, will also be early adopters. These highly computer literate generations are more than likely to see robot cars as cool.

Critical mass effect To work well, V2V technology requires a large fleet of vehicles equipped with interoperable communication systems, relying on V2I technology installed throughout important regional areas. The higher the density of cars using V2V technology the more reliable the network will become. As some level of cooperation will be required between cars, the overhaul of existing fleets will be a key issue to address early, to ease driverless technology implementation.

Legal and regulatory framework Lack of legislation will without doubt impact the speed of adoption. Some American states (i.e., Nevada, California) or countries like the UK have already passed restricted legislation allowing licensing and operation of autonomous vehicle. Governments will need to push for implementation of standardized V2V technology and safety concepts to allow interoperability between various car manufacturer systems. Incentives though less compelling than regulation would also have broad effects on the industry. In Chap. 3, we've discussed thoroughly the complex legal issues that come with self-driving cars and have identified the changes in liability responsibility. Litigation-related issues that come with widespread use of autonomous vehicles will be a challenge but unlike for flying car start-up, automotive manufacturers are well prepared to fight back such issues.

7.2.9 Growing or Shrinking Market?

There is no question in our mind that the global car market will keep growing in the next 10–15 years in volume, in great part thanks to millions of citizens from developing countries reaching middle class status and wealth. The semiautonomous technologies that will be brought to market should make also room for increase in value proposition with higher prices. In other words, at least for the next 10 years, the market for semiautonomous technology will grow exponentially and could reach these $200 billion we indicated earlier.

The issue about market growth or reduction will start being relevant after, when unmanned car technology will hit the road in significant numbers. The key question for car manufacturers is will consumers see benefits in buying their own car or will they feel comfortable with public cars? It is a difficult question to answer. Do people buy goods for their practicality or for their status? Today nobody needs a watch anymore, but still people love to wear them. Our opinion, which doesn't rely on any market research, is that people will still want to own a luxurious car and continue cruising at 20 km/h in the places to be seen. For leisure or for long trips, owning a car will probably continue to be the simplest, cheapest, and most comfortable form of travel. In the rural and remote areas, it will probably remain the only way to travel.

In fact, the issue will be restricted to commuting and inner city trips. However, as the number of city residents will represent more than 75 % of the worldwide population, this is no small issue for the automotive industry to address. In this urban environment, self-driving cars could have a significant negative effect on car sales in three different ways:

- **Family owned car reduction**: In many parts of the rich word, families own more than one car. An autonomous car would be able to drop one family member and return home to pick-up the other members.
- **Car ownership reduction**: In Chap. 6, we've seen that peer-to-peer and rental business models are becoming popular within the city limits. They ensure that cars don't stay idle 90 % of the time but instead be used by several clients during the day. Since these business models still require some kind of administrative work and have restrictions such as parking or refueling, the number of clients is still limited. Now let us imagine that these cars could be summoned whenever needed or programmed for daily pick-up with the help of some smartphone Apps. Doesn't it look like a taxi model promoted by companies such as Uber? The big difference, however, would be that the cost of a ride with an

autonomous car would be reduced by around 75 %! At that price it would make no economic sense to own a car for small trips.

- **Multi-points to multi-points traveling**: Train and buses are limited to point-to-point transport, following a defined journey with its stopping areas. Public Rapid Transit allows for point to multi-point transport, where pods can go directly to the selected bay area without having to stop at intermediary stations. Autonomous cars with the right Apps will allow for pick-up and drop-off of several passengers, based on optimization criteria linked to neighborhood or journey line. This would drop even further "public unmanned taxi" costs.

So how negatively will car sales be affected by these new consumer behaviors? Some specialists forecast a significant drop in sales volume, mainly based on economical reasons. With autonomous cars providing cheap transport services, it really doesn't make any sense to own two or three cars when a self-driving car can pick-up and drop-off several family members throughout the day.

We will now try to figure out what could be the impact by 2025. The US Census Bureau gives us the evolution of American families. From this census, we can estimate that there will be 137 million households by 2025. Using an average of 2 cars per household (2011 study by the Center for Automotive Research) we could project that by 2025, the total number of cars in the US would be 284 million.

Furthermore, using the 2007 American Household Survey, we know that 29 % of U.S. households were located in central cities. Using the 2010 census, we could estimate the risk to car sales volume in 2025 to roughly 100 million cars or around 36 % of the entire American fleet. This risk is linked to the second car owned by urban citizens by 2025. We don't believe that urban residents whose sole interests are located within the city limits, represent a risk for the industry, as they are already not buying cars.

We know that not all these families could live on only one car, but supposing half the families were to reduce their car ownership, this would still represent a 20 % reduction in car ownership, a percentage that could probably be applied to other rich countries. This doesn't mean that such reduction in volume would necessarily translate into a shrinking market value. Indeed with such savings, families would have more discretionary money and might decide to buy more luxurious cars, reducing overall revenue losses for car manufacturers.

Surprisingly, there is a large portion of transport means that will be impacted by autonomous cars. Our view is the day cars are electric, unmanned, and managed by a fleet owner there will be no differentiation between public and private transport. What we think is that taxi drivers will disappear but their services will thrive. By proposing journeys for one-fourth of the current taxi ride cost, fleets of autonomous cars will start competing with buses and light metro cars on price. As cars are more comfortable and offer more direct rides, public transport users might decide to change their transport behaviors. Furthermore and as mentioned, multi-points to multi-points transport services with mini vans or big cars will drastically reduce bus attractiveness, as fares could be shared among multi-point passengers. At one point in time, transport authorities will just conclude that rather than subsidizing big buses or tramways that run empty, it will make more sense to introduce such services, especially outside peak hour.

In other worlds, not only can't car manufacturers ignore the e-mobility revolution but bus and light train manufacturers are also forced to understand how their own business will be affected by such revolution in the car industry. After all, one of the principles we've seen is that all transportation means are competing against each other. The e-mobility revolution will disrupt the transport market and will bring opportunities and threats for many potential actors. Strategic consultants will probably claim that under such disruptive forces, car manufacturing might no longer be the core competency, as it could easily be subcontracted. Consumers would buy the new self-driving cars based on the proposed "signaling performance and connectivity solutions." In order to assess this possibility let us look at the recent history of two manufacturing industries: railway and mobile phone.

7.2.10 Railway Signaling Business Model

As we've seen, the e-mobility revolution already hit the railway industry a few years ago. However, if we look at the impact on the competitive landscape, we cannot identify new market entrants that have disrupted the industry. Nowhere is there to be seen any massive transfer of manufacturing to cheap Chinese manufacturers. Though Asian manufacturers have become global market players, they achieved this status by competing directly with the big European railway manufacturers, not as their subcontracting companies.

In fact, it is quite the opposite that has happened. We've indeed assisted to a consolidation of the industry, with mainly independent signaling companies being bought by the big railway manufacturers.

7.2.11 Smartphone Revolution

Who would have predicted at the turn of the millennium that the major phone manufacturer Nokia would sell its mobile phone division to Microsoft after getting ousted of the smartphone market? Who would have foreseen that Research in Motion would lose its grip on the smartphone market? Probably not many, as it was difficult to anticipate that IT companies such as Apple or Google would come up with such creative technology as the I-phone or the Google Android System.

To understand if some IT companies could do the same trick to the automotive industry, let's focus on these two companies' smartphone business model.

Apple designs the I-Phone hardware and develops the software based on a common platform for all its products. Clients can feel the software commonalities between the products and the interoperability between the technologies. The manufacturing is subcontracted to Asian equipment providers. Google on the contrary develops an operating system that all phone manufacturers can adapt and personalize to their hardware. They only sell their operating system licenses with the applications that can run on the phones.

7.2.12 Designer Model

Could we envision that big IT companies such as Apple, Google, IBM, or Microsoft decide to become a car designer, subcontracting the car building portion, for instance to a cheap Asian manufacturer? This might seem like a fantasy today, but let's not forget what happened in the smartphone business. First, the automotive market is much more fragmented than the mobile market was. Nokia held a little more than 50 % market share of the smartphone market in 2007, while the biggest automotive players Toyota or Volkswagen have less than 10 % currently. Second, there are loads of cheap Asian car manufacturers that would love to manufacture for one of these IT companies. Quality for some of the Chinese car manufacturers could still be an issue, but for most of them gaining brand awareness in and outside China is more problematical than providing the right level of quality. A Google or Apple car manufactured by them would solve their problem altogether. Third, many car manufacturers cannot seem to turn a decent profit. We shouldn't forget that during the subprime crisis, most car manufacturers had to be saved from the brinks by their national Governments. On the contrary these IT companies are huge cash machines that need to foster growth to maintain their stock valuation.

The competitive landscape is favorable but how likely is it that these IT companies decide to build unmanned cars? In our view, this is a remote but tangible possibility. In fact, Google has never ruled out that it wouldn't manufacture cars. Furthermore, it decided in 2014 to build its own prototype cars and could follow suit if they were successful. Apple has also announced that it wishes to enter the unmanned electric car market, without saying if it would build them or just sell the signaling technology. Obviously it would be much easier for Google and Apple to team up with one second Tier player from the rich or the developing world. If successful such an alliance could then sweep out through the world. Another interesting option that would save time and energy would be for these IT companies to buy one of the second-tier car manufacturers when the technology would be more mature. Any of these IT companies could certainly afford easily to acquire a car manufacturer.

Finally, we shouldn't forget that as time goes buy, it will become increasingly difficult for these big IT companies to find new markets attractive enough to make a difference in their annual turnover. With 2014 worldwide sales around $1.6 trillion (i.e., 82 million cars at $20,000 per car), the automotive market would definitely qualify as big enough to make a significant impact on these companies' financial results.

7.2.13 Licensing Model

Google has already pulled off such a strategy with its Android operating system. This model is also probably a less risky strategy for an IT company with no real manufacturing or mechanical competency. This safer market approach can still bring tremendous benefits for any IT company. Assuming a cost between $2000 and $5000 for the driverless operating system, this would still represent a market opportunity of $160–$400 billion, not including Apps and additional service offering directly related to cars. Furthermore, we shouldn't forget that IT companies can benefit enormously from a strong position in the car environment with potential huge additional sales due to search queries or download of infotainment content.

Major carmakers would most likely resist letting Google or Apple control the driving OS and signaling layer of their vehicles. Indeed many big car manufacturer companies have spent decades working on driverless cars. General Motors has a large driverless car testing facility outside of Detroit. Toyota opened its laboratory for self-driving car near Mount Fuji. Furthermore, the joint acquisition of Nokia's mapping solution by Audi, BMW and Daimler can clearly be associated with the perceived risk associated with these big IT companies. But, even if some carmakers could match Google's or Apple's technology proficiency, it is doubtful that every carmaker could. Furthermore, it is unlikely that those that couldn't would easily team up to develop such technology, as this a pretty divided industry. An OEM such as Bosch could maybe try and if successful would anyhow replicate such licensing model. So in fact it is quite possible that smaller players would turn to IT companies as a white knight in response to capabilities developed by these bigger automotive manufacturers.

7.2.14 Service Model

Unmanned vehicles could be dispatched by taxi and car service companies. In this model, companies that manage efficiently fleets of vehicles would be well positioned to capture a significant portion of the automotive market. Three types of companies with such capacities come immediately to mind:

- Rental companies that already do that for a living for more than 50 years, as well as the newer version that rents by the hour;
- Hailing service companies such as Uber, which announced its interest in the driverless technology; and
- Service providers of car fleet management for the big corporations.

What is less evident but also possible is that bus or railway operators could also decide to acquire such driverless fleet and bundle their services together. Why not imagine these operators providing self-driving services at bus stops or train stations for an extra fee to enable passengers to do the last kilometers quickly and comfortably. They after all are in close relation with many potential clients and have the most to fear about such service model. We shouldn't also lose sight that many of these companies are State owned. This would give them the political clout to be attributed the management of "public" transport taxi fleets.

7.3 E-Mobility Likely Losers

In the previous chapter, we've seen that the conventional taxi industry is the first likely fatality of the e-mobility revolution. It won't be the only one. In order to define who would be the other losers we would need a crystal ball to see which one of the three models would be dominant. Regardless of these future insights, there are at least a few industries that we can fairly assess will lose big time. For instance, in a world of ubiquitous cheap unmanned taxis or family own driverless cars, we surely won't need as many parking lots.

7.3.1 Parking Owners and Municipalities

Conventional cars can stand idle 90–95 % of the time. Making sure they could be used 24/7, transporting more than just car owners, would reduce significantly the need for parking space. Indeed, a fleet of cheap unmanned taxis could shuttle maybe 50–100 passengers a day. This would reduce significantly the need to park in inner city areas, where people most often need to park their car. Moreover, to avoid parking tickets, owned unmanned cars could adopt three strategies:

- Continue driving, which would make sense for short-duration visits;
- Park in neighboring roads, for longer period visits; and
- Go back home or park anywhere without parking obligations for long durations.

Furthermore, as cars proposed in programs such as Car2Go are usually electric and smaller, fewer parking spaces are even likely to be needed, reducing even further the need for parking area.

Many people might not know, but this is likely to impact a huge business. According to the International Parking Institute, it is estimated that in the USA alone, the million commercial parking spaces generate revenues in the \$25–\$30 billion range.

Parking companies aren't the only entities that offer parking space that will be affected. Municipalities that can cover a significant portion of their budget with surface parking will also lose. For example, London city earned in 2012,

600 million pounds ($1 billion) in parking revenues. And it's not the only money that will disappear for municipalities. Fine revenues will shrink, as self-driving cars will drive till finding the right available parking slot. On the other hand, parking space freed could be sold for other usage, compensating for some of this lost. Though each city is different, this could generate good money or improve quality of life, as surface parking can represent up to 25 and 10 % of big American and European cities respectively. Finally, ordinary homes, offices, and shopping centers wouldn't need the vast fields of parking that are required by law in many cities, opening the possibility to add more apartments or green areas as well.

7.3.2 Car Manufacturers

On one side, we are saying that the worldwide car market will double by 2025 and on the other hand, we are including car manufacturers in this section about losers. The reality is that history has not been kind to those who get stuck protecting the status quo in face of disruptive changes. Though they have decades of experience in designing and manufacturing vehicles, as well as in creating emotional bonds with consumers, they could lose easily their grip on the market to IT companies or as we've seen when unmanned technology becomes universal, see their sales volume drop as families stop buying their second car.

As in many industries, profits in the automotive industry have migrated downstream to the service providers. Car manufacturers haven't really been able to transform their business into a service industry like other industries facing such trends have been able to. Could they transform themselves into a fleet manager to recapture these profits having control over service, repairs, and every other new services that e-mobility will enable? Unfortunately, if the past is a sign of the future this isn't going to happen.

Among the car manufacturers there will also be winners and losers. The car manufacturers that could lose the most are the ones that are associated with images of powerful and fast cars. This doesn't mean that the Ferraris of this world won't be able to thrive in such a world, as rich people still need to find a way to splash money around. However, it will be difficult to sell the image of a fast car, when speed is limited to 50 km/h in urban environments and 120 km/h on highways and no one has the hands on the wheel. More generally, all the brands that sell the pleasure of driving will have to reposition themselves around new concepts. Brands targeting urban residents are also more likely to lose than those that sell to farmers or to small city residents. Brands associated with the second family car images are also likely to see their sales drop as their number will reduce significantly. On the contrary, brands associated with stylish design will thrive, as vehicle will be much less influenced by safety feature, such as visibility, window placement, strength and stiffness of materials, bumpers, etc. With much safer cars, design possibilities will explode (Figs. 7.1 and 7.2).

Fig. 7.1 Picture of the internal setting of the Mercedes Benz self-driving car presented in 2014. This unmanned car presented as "Luxury in Motion" F 015 shows some of the potential internal layouts. *Source* Mercedes Benz (Copyright of Mercedes Benz and kindly lent for this book)

Fig. 7.2 External view of Mercedes Benz' driverless concept car showed in January 2015, which gives a view of the unlimited opportunities for luxurious and stylish designs. *Source* Mercedes Benz (Copyright of Mercedes Benz and kindly lent for this book)

7.3.3 *Body Shops*

With much fewer accidents, the number of auto body shops will dramatically decrease, and the jobs inside these facilities will as well.

7.3.4 *Current PRT Manufacturers*

The few current manufacturers of PRT systems won't be able to face competition from all the big players that will want to create this market. This isn't to say that PRT will disappear. On the contrary, they will thrive under a different environment, based on fleets of unmanned cabs. It could be a clever move to acquire these companies for corporations wishing to position themselves early in this market.

7.3.5 *Steel Companies*

Steel represents more than 60 % of a car's materials, offering passengers protection in a collision. With massive decrease in accidents, cars will likely be redesigned out of lighter-weight material for greater fuel efficiency. This movement will be strengthened by the need to lower weight to increase electric car autonomy.

7.3.6 *Light Rail Vehicle*

In cities reaching 250,000 inhabitants, there is a need for public transport. The day passengers can hop in an unmanned cab for 25 % of the current taxi prices, meaning a price that is cheaper or at par with current public transport fares, people will start switching from public tramway or light rail vehicle to public unmanned cars. Municipal authorities might then decide to lower fares to make it competitive again but might also decide altogether that the service provided by these unmanned cabs is more than adequate.

7.3.7 *Conventional Bus Operation*

By 2025 buses won't just vanish from big cities, they will have become driverless, following the same trend as cars and trains. Per se, this isn't a threat for bus manufacturers or operators but probably an opportunity to sell added value. The issue though is that by then fleets of unmanned public taxis will compete directly with buses. Similarly as for tramway services, bus ticket will be more expensive or at par. However, unmanned taxis will be able to propose multi-point to multi-point transportation, adding the flexibility and comfort that buses cannot for less. Wherever road capacity won't be an issue, fleets of unmanned cars or vans will most likely substitute buses, especially in the evening or on weekends when demand is low. In fact, electric unmanned cabs could be the most serious threats that the bus industry has ever faced.

7.3.8 Conventional Car Rental Industry

Three main factors are disrupting the traditional car rental market. Smartphone Apps are used more and more to book cars, reducing significantly the number of rental over the counter, especially at airports and affecting the profitability of each transaction. New business models such as car-sharing and peer-to-peer sharing are reducing the need for short term car rental. Furthermore, renewed interest by car-makers (i.e., Daimler and Renault) in providing such services is likely to make competition worst. In the long run, if these companies cannot position themselves as a service provider of unmanned cars or taxis, they could also lose all their short term rental business.

7.3.9 Fossil Fuel Industry

Cars will be built from lighter-weight components and run on electric motors. Furthermore, carpooling and cars driving in convoys to reduce wind resistance are a few of the e-mobility solutions that will reduce demand for fuel per kilometer traveled. However, will petrol demand shrink in transportation is a completely different issue. At least in the short term, this isn't likely to happen, as the number of cars is likely to double in the next 20 years. Furthermore, as most car sales growth is to come from developing countries, it is more than unlikely that the infrastructure to allow for transport electrification will be available overnight. What is clear is that in the long run this electrification will happen and fossil fuel in cars or buses will be something of the past.

7.3.10 Health Sector

Throughout the world, emergency rooms will lose millions of patients a year resulting from less numerous and severe crashes. A consequential damage of the lower number of deaths of wealthy young men and women is that the number of transplants will collapse, as the number of potential donors will dwindle.

Additionally, hospitals and doctors will have much fewer people who will be hospitalized because of pollution-related diseases. Chiropractic will lose a big source of revenue as the number of handicapped or crippled people will be reduced significantly.

7.3.11 Personal-Injury Lawyers

Mainly present in the States, injury lawyers would see car-related cases all but disappear. In fact, the trend has already started as the spread of onboard cameras makes it much easier to identify the guilty party.

7.4 E-Mobility Likely Winners

In our view there will be more winners than losers in the e-mobility revolution and society as a whole will benefit greatly from this fundamental change in transportation.

7.4.1 Electric Car Manufacturers

The e-mobility revolution will be all about electrical cars. If the Service model is to be predominant, this trend will be quickened as the main present-day impediments to electric cars—range—will vanish faster. Indeed, most trips in urban areas are short and thus, building cars with battery big enough for long trips won't be necessary. Electrical vehicle manufacturers such as Tesla are likely to see their market share grow. As a matter of fact, the valuation of such a company is surely based on this assumption. Furthermore, it should capitalize also on its $5 billion investment in a new US plant to manufacture 500,000 batteries. Not only will it lower the current price of batteries helping Tesla's business model, it will provide additional profits from sales to traditional car manufacturers (or even grid providers) with new battery needs.

7.4.2 Environment

Transportation electrification will have a huge favorable impact on the environment. Not only are electrical cars much more efficient than fossil fuel vehicles and thus consume less energy, they also emit far less pollutant, as engine efficiency is highly correlated with pollution emission. This is obvious when considering wheel-to-wheel consumption but also taking well-to-wheel comparison. Furthermore, all electrical car pollutants are emitted outside of city limits where power plants are usually situated, reducing pollution's consequences on city residents' health. Furthermore, urban environments where cheap autonomous cabs and unmanned fleets are predominant will be much greener than today's cities. Without as much parking space required, they'd also be denser and more productive. Moreover, driverless cars and taxis could lower the cost of living in cities and reduce the environmental footprint of transportation, as car parks could be freed for green areas or new housing development.

But there could also be some downsides for the environment. As travel becomes cheaper, faster, and more comfortable, it will lower the cost of being transported ever further. People will consequently be able to live further from work, facilitating suburban sprawl. Additionally, those that can't drive because of physical incapacity, loss of driving licenses, or old age would likely travel more if autonomous cars were available. Finally, some passengers might decide to have their car travel until they are ready to hop back to their next activity.

7.4.3 Insurers

We've seen that by eliminating the human factor through driverless cars, most fatalities will be eliminated. It doesn't mean that accidents will disappear altogether, which would be obviously a calamity for such an industry. After all, there will still be potholes, small rocks hitting the windshield, falling branches, hail, snow storm, etc. It just means that there will be fewer crashes involving two vehicles and more importantly much less severe accidents. Unfortunately, for insurers this still means that when all cars will be driverless, premiums on car insurance should drop, as there is a direct relation between premium and covered risks. But reducing sales doesn't always mean reducing profit. In fact till the entire market is unmanned, somewhere around 2035, insurers are likely to see their profit rise. Indeed, with far fewer severe accidents, expensive medical reimbursements incurred at emergencies and during hospital treatments, as well as compensation for life disabilities and impairments, will be reduced drastically. Thus, during the introduction of self-driving technology, costs will be reduced much faster than revenue, increasing profit. Furthermore, the introduction of a no fault policy, which would make sense with unmanned cars, could reduce accident processing costs, reducing even further costs.

The risks for current insurers will come from two areas. New insurance models are likely to arise, with for instance insurance by the hour or even by the trip, eradicating the conventional brick and mortar insurance companies. Furthermore, if the Service model were to prevail, a few fleet providers would likely contract global insurance coverage reducing significantly the number of insurances sold and the negotiated premium value.

7.4.4 Government Authorities

We've just seen that municipalities that generate much of their budgets from traffic and parking tickets would have to look elsewhere for revenue. State or federal governments would lose fines, because cars would all obey traffic laws. At the same time, though, governments would save by not having to pay police officers to write tickets. The governments could spend less on prisons, because there would be no such thing as drunk or overspeeding drivers. However, the area where governments are likely to gain the most is in the social security sector or Medicare for the United States. With far fewer severe accidents, and lower pollution emission within cities, health cost will drastically be reduced.

7.4.5 Software Providers

System providers are software-driven companies, such as signaling company in the railway industry. In the automotive industry, the obvious players are the main IT companies that have financial resources and are well positioned within the mobile environment, such as Google, Apple, or Microsoft. IBM and the other IT service providers could also play a role in this market but probably more as an infrastructure provider.

Google is one of the obvious IT players with the most to win from this e-mobility revolution. Google's Pay-Per-Click business model is extremely competitive, as clients are basically using for free a service paid by advertisers. With the advent of 4G, Google should be in a position to follow its strategy to go after the potential billion information seekers of the mobile transportation market.

In the European public transport market, the big railway players have resisted the temptation to transfer their information system to the IT companies. By acquiring Waze, Google has gained cleverly access to crucial information, and thus is likely to be the preferred portal for mass transit users. This should in our view help this company strengthen even further its predominant position in the search market, providing geo-localized ads to passengers.

In the private transport markets, Google needs to seek alliances with car manufacturers to be selected as a service provider in the car dashboard. In 2014, it announced a deal with Audi, which will leverage Google's Android operating system giving passengers access to navigation tools, Apps and music. In this sector, it could face strong opposition from other IT companies such as Apple with its iOS which can push services enhancing the traveling experience, such as i-tune. Audi for instance also works with Apple on other in-car experiences. Blackberry with its QNX software has also a reasonable presence in the car industry. However, many car manufacturers are likely to feel unease at the possibility of losing the traveling experience control, especially knowing that when driverless cars will be predominant this is likely to be the main buying criterion. The high end car manufacturers may want to develop an interoperable system independently from Android, iOS or QNX.

Google is uniquely positioned today to sell Driverless Operational System licenses. However, repeating the success it achieved with cellular phones and impose its platform to the car industry will not be easy. Knowing what Google has done to the Nokia and Blackberry of the telecom world, transportation incumbent companies aren't just going to sit and wait for Google to commoditize them. Having said that, Google has strong cards to play. It possesses Google map™, a service that provides views of most streets in the world. This service relies on frequent picture update of houses and streets and physical positioning via GPS. Google has also invested in drone technology, which could be used for territories with bad internet and telecom coverage. Furthermore, this company has been pouring millions of dollars in developing its driverless technology. Unlike most of its competitors, its unmanned software platform has been tested successfully for years, as their accident free track record shows. It has invested in Uber, which doesn't only supply it with hailing Apps technology but also can give an access to the unmanned taxi industry. Moreover through collaboration with Uber, information resulting from datamining could enable new service offering or make them more attractive, (i.e.: certification of drivers, insurance coverage, risks assessment). As a matter of fact, for those who haven't yet noticed, Google now proposes online car insurances!

IT companies have in front of them a very favorable competitive landscape, with an automotive market extremely fragmented. However, their best opportunity relies probably on the fact that the automotive industry isn't software driven. Mechanical engineers and software engineers are just two different beasts. Mindsets and approaches to product development are so different that it is unlikely that such a software platform could be launched within the main engineering department of the automotive industry. Fiat, Peugeot, Ford, Hyundai, etc., will probably have problems getting the skills to develop and sustain in the long term their own software platform. Moreover, collaboration hasn't been a strong feature of this industry and will make it difficult to promote a common standard. As we've seen this will be crucial if, as we believe so, the system approach will prevail, with two-way communication with satellites, GPS, infrastructure signaling recognition (and thus picture updates) as well as real-time powerful algorithms that can analyze simultaneously thousands of changing environmental conditions.

7.4.6 System Integration Providers

Our strong belief is that the automotive driverless industry will be based on a system approach, like in the railway industry. This means that one player, usually called a system integrator, will need to make sure that all software from the onboard and wayside environments communicates instantaneously and seamlessly for unmanned systems to work. There are potentially in our view four types of companies that could play this role alone or in consortium. Though potential players in these categories are much more numerous than what we will indicate hereafter, the following companies are the type of corporations with a reasonable chance of success, if they were to decide to target such market:

- **IT service providers**, such as IBM that have the knowledge to manage big IT infrastructure project needed for unmanned systems, as well as propose SOA and EDA software solutions that are likely to be required;
- **Big security integrators** with defense knowledge, such as Northrop Grumann (USA), Thales Systems (France) that will have the political backing and the manpower to deal with system integration;
- **Telecom and network equipment providers**, such as Ericsson, Nokia Network (which is currently acquiring Alcatel-Lucent intending among other things to position itself on the 5G future market for driverless cars), Huawei, Cisco on which the V2V technology and other unmanned technology will depend; and
- **Railway signaling companies**, such as Siemens (Germany), Thales (France), and Alstom (France) that could adapt some of their existing railway technology to an IP platform and bundle this with wayside equipment. Siemens the world leader in signaling, which also possesses automotive products and still maintains strong traffic management activities, is uniquely positioned to play a key role.

7.4.7 Road Infrastructure Providers

Unmanned technology can only work if wayside technology is integrated. These technologies will need to include smart traffic management system, which will in our view still require traffic lights for pedestrians or cyclists, intelligent cameras to identify road blocking objects, and panel systems informing about changes in road conditions. Most likely, some form of intelligent lane indicator (i.e., cheap balise or special paint) will be needed to ensure that cars can manage the 2 m GPS imprecision level and stay within their lanes, at least in critical junctions or tunnels. Two types of players are likely to benefit from such position:

- Smart traffic light providers; and
- Road concessionaires that may decide to bring knowledge from their highway traffic management control center to normal routes.

7.4.8 Electrical Infrastructure Providers

All companies related to the power generation industry will be obviously clear winners in the electrification process of transportation. Besides the big electric utilities, companies building or supplying equipment for new electric power plants, be it nuclear or thermal, will see their market grow. The renewable energy market is also likely to be a huge winner of such process. Not only will its market grow significantly, but electric cars could solve one of this industry's biggest problem: with smart grid technology, cars could be used as a storage facility. By storing electricity when the wind blows or the sun shines and reinjecting it at night or when power demand is high, cars could help stabilize the electrical grid. We shouldn't also forget the electrical equipment providers for smart grids that will provide connecting and charging infrastructure as well as the software to manage such smart grids.

7.4.9 Society

We would like to conclude this book by transmitting a positive feeling about the evolution of transportation. For all of us confronted with physical standstill in our car or squashed on metro platforms, the e-mobility revolution will bring solutions against immobility. This revolution is already in movement and has started changing profoundly the car, bus, and train industries. The day isn't far when cities will start breathing again. Millions of deaths from pollution-related diseases will be avoided with a significant positive impact on governments' health budgets. Roads will still be noisy but the sound of motors running will be a thing of the past. The area dedicated to parking spaces will be reduced, allowing for more trees or wider walkways, creating a virtuous cycle for the environmental and a better quality of life of inner city residents. Electrical vehicles connected to smart grids will store energy generated by renewables, helping fight global warming.

The e-mobility revolution has shown that unmanned trains are much safer. Cars will follow the same path. Unmanned cars will save millions of passengers' life and reduce significantly the number and severity of injuries. It will also allow our graying society to live longer in the comfort of our home, as the older generation won't need any more to depend on family or friends to commute. New business models and innovative financing solutions, facilitated by e-mobility technology, will reduce transport infrastructure CAPEX and OPEX. With lower costs and less demand for newer infrastructure as existing road and rail lines will be optimized, e-mobility should help bridge the gap between the necessary infrastructure budget and available funding.

However, not everything will be rosy for everyone. There will be winners and losers and society will need to support people losing their jobs as the economy restructures. Some low skill jobs will be lost, such as taxi drivers, but thousands of new jobs will be created in new service offering that we aren't even aware of.

Companies and Brands Stated in the Chapter

- Cessna Aircraft Company
- Ferrari S.p.A.
- PAL-V Europe NV
- SkyTran Company
- Thyssenkrupp AG
- Google Inc.
- Uber Inc
- Nokia Corporation
- Microsoft Corporation
- Research in Motion
- Apple Inc
- Android is a trademark of Google Inc
- IBM/International Business Machines Corporation
- Toyota Motor Corporation
- Volkswagen Aktiengesellschaft
- General Motors Corporation
- Robert Bosch GmbH
- Car2Go is a subsidiary of Daimler AG
- Daimler AG
- Renault S.A.
- Tesla Motor Inc.
- Pay-Per-Click is a trademark of Google Inc.
- Waze is a company acquired by Google inc.
- Audi AG
- QNX is a brand of Blackberry
- Google Maps is a trademark of Google Inc
- Fiat/Fiat Chrysler Automobiles N.V.
- Peugeot S.A.
- Ford Motor Company
- Hyundai
- Northrop Grumman Corporation
- Thales S.A.
- Ericsson/Telefonaktiebolaget L. M. Ericsson

- Nokia Networks
- Alcatel-Lucent S.A.
- Huawei Technologies Co. Ltd.
- Cisco Systems Inc.
- Siemens AG
- Alstom S.A.

MIX
Papier aus verantwortungsvollen Quellen
Paper from responsible sources
FSC® C105338

If you have any concerns about our products,
you can contact us on
ProductSafety@springernature.com

In case Publisher is established outside the EU,
the EU authorized representative is:
Springer Nature Customer Service Center GmbH
Europaplatz 3, 69115 Heidelberg, Germany

Printed by Libri Plureos GmbH
in Hamburg, Germany